Perspectives on Ecological Integrity

Environmental Science and Technology Library

VOLUME 5

The titles published in this series are listed at the end of this volume.

Perspectives on Ecological Integrity

Edited by

Laura Westra
*Department of Philosophy,
University of Windsor,
Windsor, Ontario, Canada*

and

John Lemons
*Department of Life Sciences,
University of New England,
Biddeford, ME, U.S.A.*

KLUWER ACADEMIC PUBLISHERS
DORDRECHT / BOSTON / LONDON

Library of Congress Cataloging-in-Publication Data

```
Perspectives on ecological integrity / edited by Laura Westra and John
   Lemons..
       p.    cm. -- (Environmental science and technology library ; 5)
   Includes index.
   Includes index.
   ISBN 0-7923-3734-4 (hardbound : acid free paper)
   1. Ecosystem management.  2. Ecological integrity.
  3. Environmental policy.  4. Environmental ethics.  5. Conservation
  biology.  6. Sustainable development.   I. Westra, Laura.
  II. Lemons, John.  III. Series: Environmental science and technology
  library ; v. 5.
  QH75.P38  1995
  333.7'2--dc20                                                  95-35895
```

ISBN 0-7923-3734-4

Published by Kluwer Academic Publishers,
P.O. Box 17, 3300 AA Dordrecht, The Netherlands.

Kluwer Academic Publishers incorporates
the publishing programmes of
D. Reidel, Martinus Nijhoff, Dr W. Junk and MTP Press.

Sold and distributed in the U.S.A. and Canada
by Kluwer Academic Publishers,
101 Philip Drive, Norwell, MA 02061, U.S.A.

In all other countries, sold and distributed
by Kluwer Academic Publishers Group,
P.O. Box 322, 3300 AH Dordrecht, The Netherlands.

Printed on acid-free paper

All Rights Reserved
© 1995 Kluwer Academic Publishers
No part of the material protected by this copyright notice may be reproduced or
utilized in any form or by any means, electronic or mechanical,
including photocopying, recording or by any information storage and
retrieval system, without written permission from the copyright owner.

Printed in the Netherlands

To Peter Westra, thank you for your love and support. - L.W.

To the animals in my life. - J.L.

ACKNOWLEDGMENTS

The "Integrity Project" (SSHRC Grant No. 806-92-0019 and No. 806-93-0027) supported research on the "Ethical Implications of the Concept of Integrity" from 1992 to 1995. We are grateful for the support of the federal government of Canada and to all of our partners and collaborators for their individual contributions. We also would like to thank Cheryl Miller and Brenda Smith who provided valuable editorial and other assistance without which this book would not have been possible.

ABOUT THE EDITORS

Laura Westra received her Ph.D in philosophy from the University of Toronto in 1983. She is associate professor of philosophy at the University of Windsor, Windsor, Canada, where she teaches ancient philosophy, environmental philosophy, bioethics, and business ethics. She is the author of *An Environmental Proposal for Ethics: The Principle of Ethics*, *Freedom in Plotinus,* and many articles on environmental ethics.

Dr. Westra is secretary of the International Society for Environmental Ethics, and has served as a consultant to the Canadian Consulate General to prepare *Environment Services Resource Guide*. She also is recipient of a Social Sciences and Humanities Research Council of Canada Thematic Grant for 1993-1995 which has allowed her continued work on ecological integrity.

John Lemons is professor of biology and environmental science in the Department of Life Sciences at the University of New England, Biddeford, Maine, U.S.A. He served as Editor-in-Chief of *The Environmental Professional*, the official journal of the National Association of Environmental Professionals (NAEP), from 1990 through 1995. In 1994 he received a Distinguished Service Award from NAEP. Dr. Lemons has published numerous articles on nuclear waste disposal, national park management, climate change, conservation issues, and environmental ethics. He also is coeditor of *Sustainable Development: Science, Ethics, and Public Policy* which also was published by Kluwer Academic Publishers, and *Scientific Uncertainty and Environmental Problem-Solving.*

CONTENTS

	Acknowledgments	vii
Chapter 1	**Introduction** *John Lemons and Laura Westra*	1

CONCEPTUAL DIMENSIONS OF INTEGRITY

Chapter 2	**Ecosystem Integrity and Sustainability: The Foundational Value of The Wild** *Laura Westra*	12
Chapter 3	**Ecological Integrity: Reclaiming Lost Connections** *James R. Karr and Ellen W. Chu*	34
Chapter 4	**Embracing Complexity: The Challenge of the Ecosystem Approach** *James J. Kay and Eric Schneider*	49
Chapter 5	**Ecological Integrity and Sustainability: Buzzwords in Conflict?** *Reed F. Noss*	60
Chapter 6	**Ecosystem Integrity: A Causal Necessity** *Robert E. Ulanowicz*	77
Chapter 7	**Ecosystem Integrity in a Context of Ecostudies as Related to the Great Lakes Region** *Henry A. Regier*	88
Chapter 8	**Universal Environmental Sustainability and the Principle of Integrity** *Robert Goodland and Herman Daly*	102

INTEGRITY: SCIENCE, ETHICS, AND POLICY

Chapter 9	**Hard Ecology, Soft Ecology, and Ecosystem Integrity** *Kristin Shrader-Frechette*	125
Chapter 10	**Science for the Post Normal Age** *S. O. Funtowicz and Jerome R. Ravetz*	146

| Chapter 11 | **The Value of Integrity**
Mark Sagoff | 162 |

CASE STUDIES AND PRACTICAL CONSEQUENCES OF APPLYING INTEGRITY

Chapter 12	**Ecological Integrity and National Parks** *John Lemons*	177
Chapter 13	**The Importance of Landscape in Ecosystem Integrity:** **The Example of Everglades Restoration Efforts** *D. Martin Fleming, D.L. DeAngelis, and W.F. Wolff*	202
Chapter 14	**Integrity, Sustainability, Biodiversity and Forestry** *Peter Miller*	218
Chapter 15	**The Global Population, Food, and the Environment** *David Pimentel*	239
Chapter 16	**Sustainable Development and Economic Growth** *Joel E. Reichart and Patricia H. Werhane*	254
Chapter 17	**Ethical Obligations of Multinational Corporations to the Global Environment: The McDonald's Corporation and Conservation** *James D. Nations, Ray Cesca, J. Angus Martin, and Thomas E. Lacher, Jr.*	265

Index   276

Chapter 1
INTRODUCTION TO PERSPECTIVES ON ECOLOGICAL INTEGRITY

John Lemons[1]
Laura Westra[2]

1. Introduction

Recently, concepts of ecological integrity have been proposed to facilitate enhanced protection of biological and ecological resources against the threat of human activities. The promotion of ecological integrity as a basis for public policy and decisionmaking stems from scientists and others concerned about the threats of human activities to ecosystems and species, and from philosophers attempting to derive a more suitable ethic to the relationships between humans and the nonhuman environment.

Although ecological integrity has been proposed as a norm for public policy and decisionmaking, it is important to recognize that the concept is relatively new and therefore the scientific and philosophical rationales undergirding it have not been developed fully. Consequently, the goal of this book is not to present a definitive or comprehensive treatment of ecological integrity, but rather to offer a number of perspectives that might stimulate and inform future discussions. As work on ecological integrity progresses, our hope is that ultimately it will be possible to further assess the strengths and weaknesses of existing scientific, economic, and legal tools and ethical considerations in the context of ecological integrity problems and to determine whether new tools or approaches may be designed either to overcome limitations or take advantage of the strengths of current scientific, economic, legal, and ethical capabilities, much as has been done recently in the area of sustainable development (Lemons and Brown 1995).

The aim of various recent activities focusing on aspects of ecosystem integrity has been to relate the ecosystem approach and the goal to restore integrity to moral principles and norms in a way that they might serve as a basis for public policy. In general, the use of "integrity" language in various regulatory acts, treaties, and agreements implies a goal or preferred public policy choice. As Karr and Chu (Chapter 3) argue, ecosystem integrity and ecosystem health are not the same, and most of the contributors in this volume accept that the two concepts, while not defined precisely, can be separated and distinguished from one another. This also has been Westra's (1994) contention, and in this volume some of the implications of this distinction for public policy and morality are explained and discussed.

Whether explicitly and deliberately stated or implicitly understood, most of the papers in this volume present arguments that have clear consequences and implications for science, morality, and public policy, although some of the contributors chose not to address these

---

[1]Department of Life Sciences, University of New England, Biddeford, ME 04005, U.S.A.;
[2]Department of Philosophy, University of Windsor, Windsor, Ontario N9B 3P4, Canada.

topics directly. Our hope is that the contributions in this volume will help to clarify the role ecosystem integrity should play in these areas and, most important, why it should play such a role. Following sections of this chapter introduce readers to some of the scientific, philosophical, and practical dimensions of ecological integrity, and provide a brief summary or overview of subsequent chapters. Subsequent chapters are placed in one of three groups which focus on the following aspects of ecological integrity: (1)conceptual dimensions; (2)scientific, ethical, and public policy; or (3)case studies and practical consequences of application. However, as readers will find many of the chapters transcend their respective grouping.

2. Scientific Concepts of Ecological Integrity

Many concepts of ecological integrity have been derived from studies of ecosystems based upon complex systems theories. Such systems are described as nonlinear whose properties or behavior cannot be explained or predicted by knowledge of lower levels of hierarchical organization within them. In addition, complex systems have multiple organizational states and processes based upon nonequilibrium paradigms that include the following notions: (1)ecosystems are open, (2)processes rather than end points are emphasized, (3)a variety of temporal and spatial scales are emphasized, and (4)episodic disturbances are recognized. Most definitions of ecological integrity focus on the ability of ecosystems to cope with stress and maintain their self-organizational capacities (Kay and Schneider, Chapter 4); however, the concept is not defined precisely.

Westra (1994 and Chapter 2) has defined ecological integrity as including the following: (1)ecosystem health, which may apply to some nonpristine or degraded ecosystems provided that they function successfully; (2)ecosystems' abilities to regenerate themselves and withstand stress, especially nonanthropogenic stress; (3)ecosystems' optimum capacity for undiminished developmental options; and (4)ecosystems' abilities to continue their ongoing change and development unconstrained by human interruptions past or present.

Ulanowicz (Chapter 6) focuses his definition of ecological integrity on how a system might deal with an array of unforeseen circumstances in the future. In particular, he emphasizes the development and maintenance of positive feedback or autocatalysis as an inherent or fundamental attribute of ecological integrity. One example includes that of *Utricularia* (an aquatic vascular plant). Upon the leaves of the plant's bladderworts grows a film of bacteria, diatoms, and blue-green algae (periphyton). The *Utricularia* seems to secrete mucous polysaccharides to bind algae to the leaf surface and to attract bacteria. Bladderworts are never found in the wild without their accouterment of periphyton. Microscopic motile animals (zoophytes) feed upon the periphyton film and in the process of feeding, they sometimes bump into hairs attached to one end of the small bladders. When moved, trigger hairs open a hole in the end of the bladder and an animal is sucked into the bladder. Upon decomposition of the animal its nutrients are absorbed by the surrounding bladder walls.

It is important to understand that the individual components of these types of systems maintain some plasticity or indeterminacy insofar that the compositions of periphyton and zooplankton communities change as a result of competition and other interactions over time and with various habitats. In this sense, the maintenance of positive feedback in ecological systems has the potential for generating nonmechanical behaviors and for imparting cohesion or integrity to systems that contain mutualistic dynamics.

Cairns (1977) defines ecological integrity as "the maintenance of the community structure and function characteristic of a particular locale or deemed satisfactory to society." Karr and Dudley (1981) define ecological integrity as "the capability of supporting and maintaining a balanced, integrated, adaptive community of organisms having a species composition, diversity, and functional organization comparable to that of natural habitats of the region." Woodley (1993) defines the term as a state of ecosystem development that is optimized for its geographic location, including energy input, available water, nutrients and colonization history.

While recognizing that the characterization of ecological integrity in more precise terms in not easy, Noss (Chapter 5) nevertheless proposes that the concept can be made operational by selecting measurable and quantifiable indicators that correspond to the ecological qualities we associate with integrity. For Noss, these indicators include: (1)structural measures of patch characteristics; (2)structural measures of patch dispersion; (3) access, flow, and disturbance indicators; (4)structural measurers; (5)compositional measures; (6)functional measures; (7)composite indices; (8)measures of genetic integrity; and (9)measures of demographic integrity. There is, however, no a priori way to decide which of these attributes should be selected.

Kay and Schneider (Chapter 4) have proposed that ecological integrity encompasses three facets of ecosystems: (1)the ability to maintain optimum operations under normal conditions; (2)the ability to cope with changes in environmental conditions (i.e., stress); and (3)the ability to continue the process of self-organization on an ongoing basis, i.e., the ability to continue to evolve, develop, and proceed with the birth, death, and renewal cycle. By optimum operations, Kay and Schneider mean the situation where the external environmental fluctuations that tend to disorganize ecosystems, i.e., make them less effective at dissipating solar energy, and the organizing thermodynamic forces that make ecosystems more effective at dissipating solar energy are balanced. It is not clear what they mean by "normal environmental conditions" and, hence, to what extent ecological integrity can or should refer to human interventions in ecosystems. Westra (1994) proposes a definition slightly different from theirs, wherein she says that ecosystems can be said to have ecological integrity when they have the ability to maintain operations under conditions as free as possible from human intervention, the ability to withstand anthropocentric changes in environmental conditions, and the ability to continue the process of self-organization on an ongoing basis. She argues that concepts inherent in ecological integrity emerge from continuing scientific, legal, and ethical analysis and that while they correspond in her mind to more or less "pristine nature," they cannot be described or predicted precisely because ecosystems are constantly changing and evolving.

Regier (Chapter 7) provides an abstract definition stating that ecological integrity exists when an ecosystem is perceived to be in a state of well-being. In part, a more precise definition of ecological integrity is dependent on peoples' perspectives of what constitutes complete ecosystems. In addition to reflecting the concerns and values of scientists, definitions of ecological integrity also must reflect various social and ethical values relevant for public policy decisions regarding protection of ecosystems. One reason for inclusion of these various values is because there is no a priori scientific definition of ecological integrity, and therefore the concept encompasses perspectives or ways of viewing the world that inevitably reflect value-laden judgments. The ambiguity of ecological integrity is a recognition that its definition, like many ecological concepts, is determined, in part, on the basis of value-laden judgments and not solely on so-called value-free or precisely defined scientific criteria. Most of Regier's conclusions are based upon an analysis of the historical development of ecological and environmental views. Likewise, Karr and Chu (Chapter 3) examine various notions of progress that have contributed to a separation of humans from their ecological life-support systems and make recommendations to reestablish these connections.

To the extent that the definition of ecological integrity might be thought of as corresponding more or less to a natural or pristine ecosystem condition, the problem of ambiguity is exacerbated even further by questions regarding the meaning of "natural." The ability of the ecological sciences to specify the meaning of "natural" is highly questionable (see also Shrader-Frechette Chapter 9). "Natural" often is defined as a condition existing prior to human perturbation of ecosystems. Jorling (1976) argues that this definition ignores the fact that humans are a part of nature and therefore need be included in the definition of "natural." Given the historical and current global impacts of humans on ecosystems, there are probably few or no ecosystems in existence that have not been or are not affected by humans. Finally, it is difficult or impossible to know with reasonable certainty ecological conditions existing in ecosystems prior to human influence. Consequently, ecology cannot provide an unambiguous or noncontroversial definition of "natural." Ferré (1993) has argued that the interactions between humans and the environment result in ecosystems possessing different degrees of similarity to pristine ecosystems and that the task of those concerned with management policies based upon concepts of ecological integrity or "natural" need to ascertain what types of ecosystems count as being acceptable for management purposes. One of the practical difficulties of embedding concepts of "natural" in ecological integrity is that although we might be predisposed to consider certain ecosystems to be relatively pristine, many are partial cultural artifacts because of the consequences of historical and present uses and interventions. Consequently, not only have humans influenced present ecosystem structure and function but also their future evolutionary states.

According to Kay and Schneider (Chapter 4), ecosystems can respond to environmental changes in five qualitatively different ways: (1)after undergoing some initial structural/functional changes, they can operate in the same manner prior to the changes; (2)they can operate with an increase or decrease in the same structures they had prior to the changes; (3)they can operate with the emergence of new structures that replace or augment existing structures; (4)different ecosystems with significantly different structures can emerge; and (5)they can collapse with little or no regeneration. Although ecosystems can respond to environmental changes in one of these five ways, there is no inherent or predetermined state to which they will return. Further, none of the first four ways indicates a priori whether a loss of integrity has occurred. As Kay and Schneider point out, any change that permanently alters the normal operations of an ecosystem could be said to affect its integrity; accordingly, the last four types of ecosystem responses could constitute a loss of integrity. Consequently, some ecosystem changes might represent a loss of integrity, while others might not. While in theory science can inform decisionmakers about the responses of ecosystems to environmental change, it cannot provide a scientific or so-called objective basis for deciding whether one change is more desirable than another. In other words, the selection of criteria to use in such a decision must be based on human judgment regarding the acceptability of a particular change.

3. Ethical Dimensions of Ecological Integrity

Are there any explicit arguments that can be made to support the use of ecological integrity as ethically as well as practically preferable as a basis for public policy? Westra (Chapter 2) argues that integrity and its protection represents a moral choice for a host of human considerations, not only for the goals of conservation biology (see Noss, Chapter 5). The survival interests of all forms of life at the level of life-support systems are fostered by integrity, hence, at that level Westra believes it is not necessary to enter into an argument that polarizes anthropocentric and nonanthropocentric interests, namely, whether all human activities should be assessed in the light of the primacy of the integrity goal.

At least a derivative form of integrity should characterize areas of culture, thus even urban areas should be managed in the light of the requirements of ecological integrity and must be judged acceptable or not according to their compatibility with it. In essence, living in ecological integrity entails understanding that all human activities must be viewed as taking place within buffer areas or areas of ecosystem health, as no development or activities hazardous to humans or ecosystems should be permitted.

From the moral point of view, the normative aspects of integrity only can be supported if it can be said to represent a so-called objective good, that is, if it can be argued that areas possessing integrity have intrinsic value. If this is possible, then the argument can take the following form: (1)ecological integrity is required in order to maintain and protect life-support systems which in turn provide basic functions to both humans and nonhumans; (2)in contrast, the loss of ecological integrity and its component biodiversity is often seen to be hazardous to humans and nonhumans; (3)however, the threats to some species and nonhuman communities are far better researched, thus better supported scientifically at this time, than is the relation between unmanaged "wild nature" and human sustainability and health; and therefore (4)whether one acknowledges the existence of intrinsic value of ecological systems in a state of integrity or not, these systems are instrumentally valuable, although the weight of evidence to support this claim is not equal in all instances (see Shader-Frechette, Chapter 9).

As life support systems, ecosystems are analogous to the intrinsic and instrumental value of life at the level of individuals. Thus ecological integrity can be viewed as embodying intrinsic value even when the latter is analyzed in terms of what there is reason to desire or cherish or foster in virtue of the state or object concerned rather than for ulterior reasons (Attfield 1995). Although it is easier to make an argument or environmental concern based upon ulterior reasons, that is, individual or aggregate human self interest, this permits life and its support to be placed on the same level as all other human interests both economic and social. But to do so is to misunderstand the fundamental role that life and life-support systems play. As Goodland and Daly (Chapter 8) argue, ecological sustainability and integrity must be a first priority of human individual goals and social policies instead. The conjunction between life and integrity often is used to support normative policy claims, and if one considers all of the components of an ecosystem then it is easier to see where intrinsic value of the whole and instrumental value of and for all its component parts not only are compatible but even are difficult to separate.

In contrast, Sagoff (Chapter 11) points out another area where intrinsic and nonanthropocentric value coincides with anthropocentric instrumentality. For Sagoff, people only ascribe value collectively to something possessing a history or a tradition. Hence, scientific grounds are not by themselves sufficient to support value in nature; holding something to be valuable or sacred entails that something has a history that is meaningful to us. This might appear prima facie to be an anthropocentric position. However, biodiversity is an important component of ecological integrity because it represents information about the history of each specific species and ecosystem as evidenced by their various adaptations and changes through time. Consequently, present biodiversity within an ecosystem and its integrity represent far more than a mere collection of component parts; they represent a whole whose history includes an historical past and the capacity for future adaptation and evolutionary change which comprise an ecosystem's identity over time. This concept also is discussed by Ulanowicz (Chapter 6), who shows how past evolutionary paths taken by organisms and ecosystems define their history and future ascendant propensities. Ulanowicz considers ecological integrity to be a causal necessity reflecting the "necessary glue that imparts order to the observable world." He argues that integrity is an essential attribute of all living systems, in spite of the fact that some argue that it only is a metaphor or "construct" (see Shrader-Frechette, Chapter 9). Ulanowicz adds that integrity is such that, without it, ecosystems would not exist nor for that matter could life as we know it continue. Thus, whether we adopt Sagoff's position and recognize a system's historical traditions and sacredness, or accept as foundational Ulanowicz's explanation from complex systems' science, it is difficult to dispute that an objective value for ecological integrity exists.

Once that value is discovered and acknowledged, its normative aspects can be more readily outlined and emphasized. Scientists often are unwilling to draw normative conclusions, but many recognize that to embrace an ecosystem approach entails a commitment to ecological integrity that presents some difficult challenges. For example, we would have to recognize that older and more traditional scientific models and methodologies may no longer be appropriate as a basis for public policy decisions, and that our view of the world must be different from that of a reductionistic view which sees the world's workings in terms of mechanical interactions between the components of a system. This recognition in turn entails acknowledging both the pervasive uncertainty in ecosystem behavior (Lemons 1995) and the influence of thermodynamic considerations and chaos in influencing ecosystem changes over time.

The uncertainties and imprecision of scientific information and methodologies as well as the lack of clear scientific agreement on both the outcomes of human activities prompt Shrader-Frechette (Chapter 9) to argue that, scientifically speaking, concepts of ecological integrity may not be as useful for public policy and decisionmaking as we might believe or wish. Funtowicz and Ravetz (Chapter 10) also recognize that because of the uncertainties of science new approaches and uses of science should be developed for public policy purposes; they term such a new approach "post-normal" science.

Despite some of the problems posed by uncertainty, Kay and Schneider (Chapter 4) propose some prescriptions which emphasize some of the normative impacts of embracing the complexities of ecosystems and ecological integrity: (1) left alone living systems are self-organizing and our responsibility is not to interfere with this self-organizing process; (2) we must not destroy the genetic and biological information needed for life's regeneration processes which are continually ongoing; (3) we need to stop managing ecosystems for some fixed or predetermined state, whether it be an idealistic pristine forest or a corn farm; and (4) the notion of accepting ecological integrity as a normative principle for public policy means accepting all of this.

As to why humans should accept these prescriptions, Kay and Schneider argue that if they do not, then they will be selected out of the systems. Hence ours is a difficult and complex choice to accept the obligation to be stewards of integrity or to be the disrupters of integrity. The former will foster the continuation of life for present and future humans and nonhumans while the latter will curtail it. Although Kay and Schneider are not philosophers, they do make claims that clearly are normative and that can be supported on moral grounds.

For Kant life has infinite value, and moral goals such as happiness or justice for all humankind only make sense if life is present. As Westra (Chapter 2) has argued, therefore, the factual and scientific realities we can ascertain, even imprecisely, must provide guidelines about the limits of human activity that can be defined morally. And the more controversial and uncertain our scientific limits, the more extreme caution should be applied in our decisions and policies as they affect the environment and human health. Consequently, living in integrity entails understanding that many of our activities should take place in a buffer zone and that, additionally, areas as large as possible should be established to remain unmanaged and "wild," although it is clear that not all of the earth can be left in that state.

4. Case Study Approaches to Ecological Integrity

Several researchers have utilized a case study approach to exemplify the prospects and problems of using ecological integrity as a basis for public policy decisionmaking. Using a case study of national parks, Lemons (Chapter 12) has analyzed aspects of ecological integrity that depend upon concepts of "natural" or "pristine" ecosystems and their implications for management. The term "natural" often is used in a sense that implies freedom from human influence, especially that of modern humans. Used in this manner, the concept of "natural" would appear to be consistent with Westra's concept of ecological integrity, since the latter implies minimal human intervention in ecosystems (see Chapter 2). However, another meaning of "natural" may differ from Westra's insofar as some people argue that human use and influence should not necessarily be excluded from protected areas if such use is compatible and harmonious with the ecosystem (Lemons 1987). Ambiguity about the meaning of "natural" has permitted wide oscillations in park policy regarding the acceptability of phenomena such as nonhuman-caused fires, floods, and fluctuations of animal populations as well as of human intrusions in park ecosystems. Confusion also exists with respect to understanding of the word "natural" even when it is being used in a sense that implies freedom from human influence, especially that of modern humans. Consequently, the concept of "natural" can lead to differences of opinion regarding whether a management emphasis should be placed upon specific structural or functional attributes of ecosystems. Park policies designed to maintain ecosystems in particular structural states or that allow levels and types of human activities could conflict with concepts of ecological integrity that emphasize ecosystems' ability to continue the process of self-organization on an ongoing basis. Lastly, Lemons shows how national park legislation would have to be reinterpreted to allow for more strict policies of preservation if ecological integrity were to be a management goal.

Based upon their study of wading birds in the Everglades, Fleming et al. (Chapter 13) emphasize that ecological integrity consists of both structural and functional integrity and that both depend upon a functional landscape mosaic of appropriate spatial extent, heterogeneity, spatial arrangement, and connectivity of landscape or terrain patterns representative of the ecosystem upon which the organisms depend.

Miller (Chapter 14) defends a social and ethical imperative that gives priority to a norm of both sustainability and the preservation of ecological integrity. According to Miller, this combined norm is stronger than the standard of sustainability alone because it makes explicit a biocentric norm which is important in its own right and as a condition for the long-range sustainability of human activities. However, he finds that the combined norm of sustainability and ecological integrity is not sufficient or adequate as an informed, enriched, or enlightened imperative. Humanistically, Miller believes that the combined norm is troubling because it does not offer a sufficient account of human needs and interests. Biocentrically it is troubling because the concept of ecological integrity is sufficiently abstract that it might be compatible with numerous and significant loss of species. Miller does not advocate abandoning the norm of ecological integrity, but rather suggests that we acknowledge the plasticity of the concept and the political struggles to which they are prone and proceed to define them and build upon them in ways that can strengthen public commitments to the human, natural, and ecosystemic values we seek to protect. In his chapter, Miller uses examples from public policy debates and management concerning Canadian forests to demonstrate his points.

Pimentel (Chapter 15) discusses the importance of ecological integrity as the ultimate support for biological resources required for agricultural systems upon which the world population depends. He maintains that because no one knows how much ecosystem integrity can be reduced and the systems continue to function there is a need to take a conservative approach and plan for a reduction of the human population to about two billion. Goodland and Daly (Chapter 8) also discuss the prospects and problems of reducing human population growth, but include in their discussion an analysis of the need for public policy to address three types of sustainability: (1)environmental, (2)social, and (3)economic. While there is some overlap between these three forms of sustainability, Goodland and Daly argue that their meanings are clearest when kept separate, that is, ecologists should be left to produce their own meaning of environmental sustainability while social scientists and economists should be left to produce theirs, respectively. However, Goodland and Daly also maintain that ecological integrity must precede and support both economic and social sustainability, neither of which can exist without it. In order to achieve intergenerational and intragenerational sustainability and protect ecological integrity, Goodland and Daly demonstrate the need to assess and understand the interactions between population, affluence, and technology and their impacts upon ecological integrity.

Reichart and Werhane (Chapter 16) also discuss the role of business in promoting ecological integrity. To begin with, they point out that Westra's (see Chapter 2) definition of ecological integrity appears to equate human survival and well-being with that of all species and that restoring ecological integrity is a primary goal that perhaps overrides certain questions of human rights and justice. They also argue that achieving ecological integrity would require a limitation and control of human preferences and human freedoms. To overcome these perceived problems, Reichart and Werhane suggest that problems of ecological integrity and sustainability need to be reformulated in a way that environmental issues are taken into account at the same time as are economic and political ones. To demonstrate some ways in which this might be accomplished, Reichart and Werhane discuss specific examples of applied ecoefficient technologies that can help restore ecosystems while at the same time providing for sustainable economic growth.

Nations et al. (Chapter 17) utilize a case study to demonstrate the emerging role of multinational corporations in terms of their impacts upon the environment and their ethical obligations to protect the environment better. Specifically, they focus upon the Amistad Conservation and Development Initiative in Central America which attempts to: (1)improve agricultural yields of farmers by conserving soil and water through improved agronomic techniques, (2)reforest degraded land through the establishment of community tree nurseries and tree planting campaigns, and (3)help community members who organize into farmers' cooperative groups to increase income by cutting out high-cost middle business people who charge usurious fees to transport crops to market.

5. Some Practical Consequences of Ambiguity Surrounding Ecological Integrity

Despite whatever values-based or ethical justifications might exist for attempting to base decisionmaking on principles of ecological integrity, there are two problems that are understudied and bear directly on its utility as a basis for policy and management decisions. One, the burden for demonstrating environmental impact or risk from existing or proposed activities stems from two powerful prescriptions. The first prescription is based upon standard scientific norms governing acceptability of scientific evidence, namely, to accept conclusions at a 95 percent confidence level and minimize type-I error which leads to the acceptance of false positive results. In natural resource problems, a type-I error would be to conclude that harm to resources will result from existing or proposed human activities when, in fact, no harm will result. The second prescription is based upon rules of evidence and procedure in legal proceedings and agency administrative decisionmaking that generally place the burden of proof upon the moving party (i.e., natural resource agencies) and thereby force it to support its requirements with scientific certainty. While these two types of prescriptions might reduce speculative conclusions, they are inconsistent with a precautionary approach regarding the protection of natural resources. A precautionary approach would be to minimize type-II error which errs on the side of a desire not to impose unnecessary costs of environmental protection or restrictions on resource use. Two, pervasive uncertainty confounds scientists' understanding of ecological integrity. This uncertainty makes it extremely difficult to fulfill the above scientific and legal prescriptions.

Most statutory environmental law in the United States is proscriptive and therefore influences the consideration of the legislation in legal forums. In the context of public policy, the fact that most environmental legislation functions proscriptively in practice has implications for the role of science in the use of ecological integrity in decisionmaking. Given the ambiguous interpretations of ecological integrity and their generalities, how may we say that science informs decisionmakers? Because in practice environmental statutes function in a proscriptive manner, scientific uncertainty makes it extremely difficult for decisionmakers to demonstrate harm to environmental resources and therefore to identify activities that should be prohibited or restricted in order to protect ecological integrity. The problem of scientific uncertainty in public policy decisionmaking is discussed by Shrader-Frechette (Chapter 9) who maintains that the methods and tools of ecology are too general and limited in their precision and predictive capabilities. Therefore, she argues that the value of ecology primarily is heuristic and capable of providing useful information for decisionmaking if studies are based upon practical case studies and not upon generalizable theories.

In the United States, the rules of the use of scientific evidence in legal proceedings differ depending upon the type of proceeding (Brown 1995). The U.S. Supreme Court in *Daubert v. Merrell Dow Pharmaceuticals, Inc.* (1993) announced the following four pronged analysis to assist courts in determining whether evidence is relevant and reliable: (1)scientific methods used by experts to derive an opinion must be capable of being tested and capable of being shown to be false, (2)publication and peer-review of scientific methods used by experts to derive an opinion strengthens the admissibility of evidence but nonpublication does not impart inadmissibility, (3)admissibility of evidence is strengthened by the use of methods that have a known and low error rate, (4)admissibility and acceptance of expert opinion will be enhanced if the opinion is based upon methods that have been accepted generally within the scientific community (the court did make clear that this analysis was not a definitive checklist, however). Under *Daubert*, evidence that establishes a reasonable basis for concern about harm but does not conclusively establish causation is not admissible.

In addition to the problems of admissibility of uncertain scientific evidence described above, other proscriptive manifestations of science also make it difficult to fulfill burden-of-proof arguments to enhance protection of natural resources based upon concepts of ecological integrity.

In order to obtain more accurate and credible results, scientists often will define problems in a narrow manner by isolating and studying selected variables under controlled conditions. This approach leads to formulating environmental problems in particular ways, but it presents what is almost an intractable problem due to the fact that it attempts to understand complex systems by isolating a few variables so that they can be studied under narrowly controlled and simplified conditions. This approach becomes problematic because when dealing with more complex systems (such as ecosystems), the understanding of the interactions among variables that determine the way in which individual variables express themselves is not able to be discerned. Consequently, because of the pervasive scientific uncertainty inherent in complex environmental problems, establishing cause-and-effect relationships between the activities and impacts at the 95 percent confidence level is precluded in many instances. Further, by simplifying and reducing parts of complex environmental problems to a more manageable scale, scientists often end up studying a scientific problem that may be very different from the more complex environmental problem from which it stems.

The act of a scientist positing a hypothesis and deducing and testing principles therefrom or favoring minimization of type-I rather than type-II error is analogous to the setting of the burden of proof in law. So long as logically deduced principles do not contradict each other, a scientist's hypothesis is considered valid. The inherent and implicit power or right that a scientist has to posit a hypothesis is not a scientific process, but it is analogous to placing the burden of proof in law because it sets the stage for identifying and describing certain facts about nature but not others. Those who may wish to contest the validity of principles logically deduced from a tested hypothesis generally are assumed to have the burden to prove a competing hypothesis. Further, when scientists tend to favor minimization of type-I error as being the most conservative course of action under conditions of scientific uncertainty, they are establishing normative rules of conduct; namely, that an experiment's results cannot be assumed to be valid if the probability of their being due to chance is, say, greater than five percent. With respect to determining whether a harm exists to an ecosystem's integrity due to an activity, the effect of such a rule is to act as if it is better to minimize type-I error (rejecting a false null hypothesis that there is no harm to integrity) than it is to increase the chances of concluding a finding of harm when there is none (type-II error). Such a rule increases the chance that a scientist will make credible scientific conclusions, but it also increases the chance of a conclusion that there is no harm to integrity when, in fact, there is. The standard of proof that should be required of regulatory decisions is a public policy and ethical question, not a scientific one.

Shrader-Frechette (1994) analyzes the reasons why there should be an ethical preference for minimizing type-II error in natural resource issues (thereby increasing the risk of type-I error) and why the burden of proof for demonstration of no adverse environmental harm from development or human activities should be placed upon those calling for such development or activities. She bases her conclusion on: (1) minimizing the chance of not rejecting false null hypotheses with important public policy consequences is reasonable on the grounds of protecting the present and future public; (2) the proponents of development or activities that potentially threaten environmental resources typically receive more benefits from the development or activities than do members of the public and, consequently, minimizing type-II error would result in a more equal distribution of benefits and risks; (3) natural resources typically need more risk protection than do promoters of development or human activities because the advocates for protection usually have fewer financial and scientific resources than developers or promoters of activities that potentially can harm the resources; (4) the public ought to have rights to protection against decisions that could impose incompensable damages to natural resources; (5) public sovereignty justifies letting the public decide the fate of development and human activities that potentially threaten natural resources; and (6) minimizing type-II error would allow enhanced protection of nonhuman species that typically receive inadequate consideration in decisionmaking based upon cost-benefit methods.

Given the prevalent legal burden of proof and the existence of pervasive ambiguity and scientific uncertainty surrounding the concept of ecological integrity, if such a concept is going to be used as a basis for decisionmaking then an alternative role for the use of science in decisionmaking might have to be found. Such an alternative role might include a recognition that: (1) problem-solving should not reflect an emphasis on data acquisition per se, (2) more data will not necessarily solve problems, and (3) the scientific method is not always objective or value-neutral. An alternative role for science might also emphasize: (1) adequate formulation of problems so that data will contribute to public policy goals; (2) that most results from scientific studies will not yield reasonably certain predictions about future consequences of human activities and that many problems of protecting ecological integrity therefore should be considered to be "trans-science" problems requiring research directed toward useful indicators of change rather than precise predictions; and (3) the need to evaluate and interpret the logical assumptions underlying the empirical beliefs of scientists with a view toward ascertaining more fully the validity of scientific claims and their implications. This type of alternative role for science is not easy to characterize- it seeks a broad and integrated view of problems and places more emphasis on professional judgment and intuition and is less bound by analytically derived empirical facts. Its claims might be more amenable for practical public policy purposes than the claims of more predictive science approaches (Miller 1993, Bella et al. 1994). One result of basing decisionmaking upon such an alternative role for science is that type-II error would be minimized and a more conservative or protective approach to protecting ecological integrity would therefore be achieved. The approach likely will require a more precautionary legislative approach that allows protection of ecological integrity when there is a reasonable basis for concern but where scientific conclusions are uncertain about the consequences of human activities.

Following this line of reasoning, if environmental decisionmaking is to be based upon a norm such as ecological integrity, Funtowitz and Ravetz (Chapter 10) argue for a "post-normal" approach to science. They state that in environmental matters, where facts are uncertain, values in dispute, stakes high, and decisions urgent, scientists often need to follow methods that might not be appropriate in other scientific endeavors. In deciding such matters, decisionmakers will have to apply values to fact-findings. That is, the norms that scientists should follow in serious public policy matters are different than those that should be followed in scientific research matters where important public policy matters are not at stake. Consequently, the use of ecological integrity as a basis for decisionmaking would imply that a substitution of the use of traditional science by post-normal science be made.

6. References

Attfield, R. 1995. Post-Modernism, Value and Objectivity. Paper presented at the International Conference on Environmental Ethics and Metaphysics, Department of Philosophy, University of the Atlantic, Boca Raton, FL., May 28.
Bella, D.A., R. Jacobs, and L. Hiram. 1994. Ecological Indicators of Global Climate Change: A Research Framework. *Environmental Management* 18: 489-500.
Brown, D.A. 1995. The Role of Law in Sustainable Development and Environmental Protection Decisionmaking. In *Sustainable Development: Science, Ethics, and Public Policy*, J. Lemons and D.A. Brown, eds. Kluwer Academic Publishers, Dordrecht, The Netherlands, pp. 64-76.
Cairns, J., Jr. 1977. Quantification of Biological Integrity. In *The Integrity of Water*, R.K. Ballentine and L.J. Guarraia, eds. U.S. Environmental Protection Agency, Office of Water and Hazardous Materials, Washington, D.C., pp. 171-187.
Daubert v. Merrell Dow Pharmaceuticals, Inc., 113 S. Ct. 2786 (1993).
Ferré, F. 1993. Persons in Nature: Toward an Applicable and Unified Environmental Ethic. *Zygon* 28: 441-453.
Jorling, T.C. 1976. Incorporating Ecological Principles into Public Policy. *Environmental Policy and Law* 2: 140-146.
Karr, J.R. and D.R. Dudley. 1981. Ecological Perspective on Water Quality Goals. *Environmental Management* 5: 55-68.
Lemons, J. 1987. United States' National Park Management: Values, Policy, and Possible Hints for Others. *Environmental Conservation* 14: 329-340, 328.
Lemons, J. ed. 1995. *Scientific Uncertainty and Environmental Problem-Solving*. Blackwell Science, Inc., Cambridge, MA.
Lemons, J. and D.A. Brown. eds. 1995. *Sustainable Development: Science, Ethics, and Public Policy*. Kluwer Academic Publishers, Dordrecht, The Netherlands.
Miller, A. 1993. The role of Analytical Science in Natural Resource Decision Making. *Environmental Management* 17: 563-574.
Shrader-Frechette, K. 1994. *Ethics of Scientific Research*. Rowman and Littlefield, Lanham, MD..
Westra, L. 1994. *An Environmental Proposal for Ethics: The Principle of Integrity*. Rowman and Littlefield, Lanham, MD.
Woodley, S. 1993. Monitoring and Measuring Ecosystem Integrity in Canadian National Parks. In *Ecological Integrity and the Management of Ecosystems*, S. Woodley, J. Francis, and J. Kay, eds. St. Lucie Press, Delray Beach, FL, pp. 155-176.

Chapter 2
ECOSYSTEM INTEGRITY AND SUSTAINABILITY: THE FOUNDATIONAL VALUE OF THE WILD

Laura Westra[1]

Today, we enthusiastically participate in what is in essence a massive and unprecedented experiment with the natural systems of the global environment, with little regard for the moral consequences.
Al Gore, *Earth in the Balance*

1. Introduction

The concept of ecosystem integrity figures prominently in a large number of regulatory and legislative documents. From the time of the 1972 Clean Water Act to the recent vision and mission statements following the Earth Summit in Rio de Janeiro, the term and concepts of "integrity" have appeared in the Great Lakes Water Quality Agreement (1978), the Canada Park Service Regulations (1988), the Great Lakes Science Advisory Board Report (1991), Agenda 21, Ascend 21, the draft Montana Environmental Protection Act (1992), Environment Canada's Mission Statement (1992), UNCED documents (1992), World Bank reports (particularly in the bank's 1992 discussion of biodiversity), and many other documents, including the Constitution of Brazil (Chapter 6, Meio Ambiente).

I believe that the challenge that faces us now is three-fold. First, it is vitally important to define integrity in general terms-that is, in terms applicable, at least prima facie to any area within the biosphere, aside from landscape-specific requirements. Second, it is equally important to decide not only what integrity is but also what size of land area should be set aside to preserve ecological integrity worldwide. Third, it is vital to relate integrity to sustainability as such and also as required in what I term I_b, or buffer areas (that is, healthy ecosystems available for respectful, careful manipulation) (Westra 1994a). In essence, I have argued that an ecosystem can be said to possess integrity (I_a) when it is wild, that is, free as much as possible today from human intervention, when it is an "unmanaged" ecosystem, although not a necessarily pristine one. This aspect of integrity is the most significant one; it is the aspect that differentiates I_a from ecosystem health (I_b), which is compatible with support/manipulation instead. (See Appendix A for definition of integrity.)

I have discussed the meaning of health in this context elsewhere (Westra 1994a). Some have spoken of health as the capacity to resist adverse environmental impacts at the present time and as "the imputed capacity to perform tasks and roles adequately" (Parsons 1964). But the capacity to perform certain roles need not be dependent on specific or complete ecosystem structures. A carefully managed monoculture (such as a plantation forest), for instance, may

---

[1]Department of Philosophy, University of Windsor, Windsor, Ontario N9B 3P4, Canada.

fit the health model quite well, yet it may have few parts or little structure other than those for which it is specifically managed.

Another point of divergence between health and integrity is the time frame within which they are viewed. The health paradigm is concerned with the present time and perhaps the immediate future. The integrity perspective poses no time limits and envisions birth, maturity, and death cycles that may also produce different paths and trajectories, according to the largely natural, evolutionary development of the system. As we will see below, Karr (1994) acknowledges that even unmanaged ecosystems, in today's world, will not be pristine; further, the precise point where an unmanaged system can no longer be said to possess even diminished integrity and has-at best-only health instead is open for debate.

But whether or not integrity (as a perfect paradigm case) exists, any meaningful debate must start with separate definitions for the two states of health and integrity (I return to this point below). In other words, I may not be at my "perfect weight" at this time (for my height, bone structure, age, gender, and so on)-in fact, no one might be in that enviable position-yet I must have in mind the best possible definition or understanding of what that weight should be before I may debate whether it is attainable or how close to it I might be at this particular time.

I have also argued that ecosystem integrity (I_a) is defined through the aspects and characteristics it exhibits of health, the capacity to withstand anthropogenic stress, and primarily a system's undiminished optimum capacity for the greatest possible ongoing developmental options within its time/location. This capacity is fostered by the optimum possible biodiversity (dependent on contextual natural constraints), in its dual role as basis for genetic potential and as locus of relational information and communication, both actual and potential; thus a system in a state of integrity will retain its ability to continue its ongoing change and development and will therefore retain its excellence (ergon/function in the Aristotelian sense) or capacity for an optimum number of options (Westra 1994a). Practically, the requirements of integrity will: (1) affect the protection of sustainability as such, in large, wild areas, thus also ensuring the protection of habitats and the goals of conservation biology; (2) address the need for food production to eliminate world hunger; and (3) specify the limits of all other nonessential human activities in culture. I call the latter I-c, to indicate the fact that these activities must at least reflect I_b in a derivative way. No completely natural evolutionary processes may persist in an urban area (although reproduction of some species continues), but successful work has been done to show that such areas can be made compatible with both I_a and I_b (Regier 1993; Kay and Schneider 1994). Although an abundant literature exists on the topic of sustainability, from the standpoint of the principle of integrity (PI), all three requirements of integrity can be viewed differently, and the role of integrity in all three will clarify my point (Kay and Schneider 1994; Westra 1994a).

Ecosystem integrity (I_a) represents the necessary condition for the support of areas of ecosystem health (I_b) and for areas in less natural settings, that is, areas of human culture (I-c). It represents both goal and benchmark of restored/restorable ecosystems, and as such, it is required for areas covering at least one third of the global landscape (Naess 1991; Noss 1992; Noss and Cooperrider 1994) contrary, for instance, to the World Wildlife Fund's (Canada) figure of 12 percent required for Canadian landscapes, which has been adopted by the Liberal government policy book. This is by no means unproblematic, because even wild areas such as nature preserves or national parks, as Karr (1994) indicates, at best can be said to indicate "a condition at sites with little or no influence from human actions." Thus it is clear that neither totally pristine nor totally untouched areas exist in today's world. But once the meaning of integrity as benchmark and goal for certain areas is understood, it is possible to see it partially instantiated. For instance, some unsafe precipitation or leaching may be present, and yet it may still be the case that "the resident biota is the product of evolutionary and biogeographic processes at a site" (Karr 1994).

However, several problems remain. First, the conceptual definition of integrity does not guarantee its instant recognition on the part of ecologists when faced with a specific landscape, any more than understanding what happiness or justice are would ensure their automatic recognition in cases, persons, or events. This indicates that the concept of integrity, like many others in science, is not a concept that can be defined in a purely mechanistic manner. Complex related problems still exist about the practical applications of the concept's definition. (1) Can we practically recognize a landscape that has the capacities we believe are necessary for integrity to be present? (2) If integrity is not pristineness but may apply to a somewhat affected ecosystem-that is, to one with little if not no influence from human actions-then at what point of influence does it cease to have integrity? (3) Given the difficulties inherent in (1) and (2), how can we successfully use integrity in practice and in public policy?

These are all serious difficulties, and it is because of them that areas as large as possible should be protected and that, in general, we should be governed by the precautionary principle in both our moral and practical understanding. It is also why both philosophers and policymakers should persist in an open and collaborative dialogue with science.

In essence, there is a basic difference between landscapes utilized (however carefully) for the implementation of some human goals, hence managed rather than protected, and viewed as instruments and valued as such, and those that are not. The latter, in turn, can be said to be valued as life is, both in itself and as the basis of all other "goods" that can be experienced. The existence of such landscapes is valuable, as it supports not only the life of all biota within the landcapes, but also the life of everything else in various ways.

Some of these ways are discussed in what follows. They include: (1) support for all biota, even that which lives beyond the limits of wild areas; (2) support for human health, through general assistance in human maintenance, including agricultural sustainability; and (3) support for some carefully and respectfully chosen human activities that are technological and economic and for which, minimally, a healthy environment is necessary. The latter, of course requires areas of integrity for its own sustenance. For this reason I have argued that ecosystem integrity is foundational to all three areas mentioned above, hence, that it is primary and must be supported and protected.

Therefore, I argue that all other aspects of managed human activity within ecosystems gain their life support, hence their identity, from I_a (or core, wild areas), according to varying scales appropriate to different biomes, and that the elimination of all such areas would have severe negative impacts on all human life.

2. Sustainability and the Wild

Much of today's concern among both the scientific community and policymakers addresses the question of ecosystem health, with a comparative lack of focus on ecosystem integrity. In a sense, that is appropriate, because managed and supported ecosystems are the only ones we should consider altering for sustainable use such as alternative agriculture or sustainable forestry.

But in another sense, this is a dangerous development, because by its exclusive focus on instrumentally valuable, or "resource-use," ecosystems, it draws attention away from the relation between the wild (I_a, including both structure and function) and its foundational role for sustainability. In a discussion of ecosystems and sustainability in fisheries, Hammer et al. (1993) state:

> Whereas species diversity is a property at the population level, the functional diversity, what the organisms do and the variety of responses to environmental changes, especially the diverse space and

time scales to which organisms react to each other and the environment, is a *property of the ecosystem.* (my emphasis)

To limit ourselves to dealing with the areas where our interest lies (e.g., areas of ecosystem health, viewed and treated as instrumentally valuable, or sustainable in a referential sense) is to ignore the larger picture and the life-support and benchmark functions of the wild in landscapes of appropriate geographical size (biomes). Hence, the primary concern must focus on the wild (core areas, or I_a), even when sustainability is the question at issue. To put it plainly, sustainable agriculture, sustainable forestry, and sustainable fisheries make little sense unless sustainability of wild ecosystems is addressed first, at least in the long term anticipated and in fact required by most North American and global regulations and treaties, all of which include future generations in their reach.

When sustainability is discussed in our context, it is intended purely as ecological sustainability (ES). This does not mean, however, that either economic sustainability (ECS) or social sustainability (SS) are ignored or deemed to be unimportant. The thrust of my argument is that, although some may view ecological sustainability as potentially inimical to ECS and SS, in the long term it is not. For the most part, it is the short term economic advantage that is both sought and often contrasted with ecological imperatives. I will return to this problem in closing, as well as in the next section.

The distinctions among the different senses of sustainability are necessary, because "sustainability is presently viewed as a catch-all for everyone's environmental and social wish list" (Goodland 1994). I have argued that integrity was also a concept kept deliberately fuzzy in order to avoid forced behavior changes which are instead implicit in the correct understanding of the concept (Westra 1994a). Goodland argues that there are three types of sustainability and that they are "clearest when kept separate."

Goodland defines social sustainability (SS) as using resources in a way that "increases equity and social justice, while reducing social disruptions." SS also entails promoting and sustaining empowerment, self-control of people, and other moral and cultural ideals through strong "community participation." Economic sustainability refers to the maintenance of capital in four senses: (1)human-made, (2)natural, (3)social, and (4)human. Goodland remarks that environmental capital was externalized and not properly considered, because it was and is hard to value in money terms and because only recently are environmental costs starting to be internalized. Finally, ES focuses on preventing physical harm to humans, by "protecting the sources of raw materials...and ensuring that the sinks for human wastes are not exceeded." Goodland concludes, "Humanity must learn to live within the limitations of the physical environment." This is precisely the message my emphasis on integrity is intended to convey.

Some may object to the use of the term "integrity" in this way, because: (1)no area is truly pristine, and we are not really sure what the conditions were in any area before anthropogenic stress intervened to alter the system; and (2)even if we could reconstruct what was there through paleoecology, the climatic conditions would be too different to enable us to say we have restored integrity without reference to a specific, arbitrarily chosen point. Unfortunately, the same lack of specific reference applies to the health paradigm. For instance, the health of an organism may be quite different when we consider an organism's age, and the point of reference also must therefore be chosen with a certain degree of arbitrariness (healthy at age 15? or 50?). Further, in a recent paper outlining the history of the ecosystem concept, Bocking (1994) says, "Ecosystem health, a concept frequently invoked, varies in several respects from the concept of ecosystem integrity; ecosystem health emphasizes a stable state of well-being, not a process of change, response to stress, and self-organization." Karr (1994) also separates the two concepts; speaking of the recent calls for the protection of ecological integrity, he says:

> In these calls, *integrity* and *health* are often used synonymously, but I propose to apply the terms in different contexts. Integrity most appropriately refers to the condition at sites with little or no influence from human actions; that is, the resident biota is the product of evolutionary and biogeographic processes at a site. The natural integrity of national parks, nature preserves, and most water bodies should be protected. In contrast, health describes the preferred state of sites modified by human activity (e.g., cultivated areas, plantation forests, and cities). Such sites do not have integrity in an evolutionary sense, but they may be considered "healthy" when their management neither degrades the site for future use nor results in degradation beyond their borders.

Although legislation emphasizes integrity (and some examples were listed in the first paragraph of this paper), it is clear that respecting untrammeled birth-to-death cycles in ecosystems (Kay and Schneider 1994) is only possible in the wild, and not, for instance, in areas where agriculture must be practiced. Hence my emphasis is not on the sole role and relevance of integrity, but rather on its foundational role for other environmentally legitimate enterprises. I attempt to go beyond Karr, by proposing three interdependent but distinct versions of integrity: I_a, I_b, and I-c, the last two of which can be areas of ecosystem health and hence not incompatible with appropriate management. The two last versions are capable of supporting some chosen human goals without adverse impact on the core areas (I_a).

Without denigrating the goals of the proponents of ecosytem health, I believe that the lack of differentiation within the concept represents a major difficulty when one attempts to apply it in policymaking. Is a naturally preserved area healthy as well as an organically based farm or a carefully monitored city? What are the differences among these areas, if one concept alone is used to describe them? What is their common link?

Using the integrity paradigm instead, large wilderness is necessary-that is, it is not simply an aesthetic option or a preference to be weighed against competing developmental, economic, or social claims. Rather, all competing economic, social, and developmental claims should be understood in the context of the primary necessity of large wilderness preservation. I am convinced that Ehrlich (1980) is not overstating his case when he says of our present impact on the biosphere:

> First, unless these trends can be reversed, the most ingenious tactics on the part of the conservation movement will, at best, slightly delay an unhappy end to the biotic Armageddon now underway. Second, without dramatic changes in the socio-political and, especially, the economic systems that dominate society today, these trends will not be reversed. And third, a nontrivial consequence of the failure to reverse these trends will be the disappearance of civilization as we know it.

The experiential aspect to the value of the wild remains valid. But the existence of the wild goes beyond the aesthetic or the specific habitat or population supports it provides. If all aspects of managed human activity within ecosystems gain their life support and their sustainability, hence their very identity, from areas of integrity (I_a or the wild), then we must learn to manage and limit human activities in harmony with this basic reality.

The experience of the beautiful, of the awesome, is part of the patrimony or the "capital" of all humankind, and hence it has a deep value, as Sagoff (1989) has argued, and we are all entitled to its protection. Further, recognizing our connection to a specific place, is basic to the way we interact with it (Sagoff 1992). An example taken once again from fisheries illustrates this point (without, however, the aesthetic component Sagoff injects in his argument). In their discussion of sustainable fishing practices, Hammer et al. (1993) say:

> Small scale fishers have sometimes been referred to as *ecosystems people*, who live within a single ecosystem (or at most within about 2 or 3 adjacent or closely related ecosystems), are aware of the constraints of the natural local resource base....In contrast, offshore fishers would fit in the category of "biosphere people," tied in with global markets and using more sophisticated and effective fishing technologies.

The cardinal difference between the two is that the latter "more easily subscribe to the myth of superabundance, the feeling that there are always other ecosystems to exploit."

The lesson here is a simple one: Activity that "has been detached from ecological understanding and knowledge" in the end plunges everyone "into a social trap of short-term solutions, counteracting long-term sustainability"-that is, it plunges us into a world of virtual, not actual, reality (Hammer et al. 1993; cp."the Jari case," in Westra 1994a; Ch. 6). In essence, then, our first concern should not be the preservation of this or that species or this or that human activity but the conservation of the resilience of the ecosystems on which human activity depends and the ability of these systems to continue to provide valued ecological services to all biota.

In order to achieve this, not only the quality of ecosystems in the wild needs protection but also the quantity of these ecosystems (in the sense of the size of the areas required), yet this is one of the least discussed questions in public policy, perhaps as too radical to gain public support. Noss (1991) points out:

> The re-establishment of huge, wild, functional ecosystems replete with large carnivores and their prey is the pinnacle of restoration ecology and human harmonization with nature.

Brown (1981) also supports this position. In a section entitled "Preserving the Web of Life," he recognizes that "bans or quotas on the killing of forms of life prized by hunters, tailors or furriers" may address some of the symptoms but leave the fundamental issue untouched. Nor is a collection of different, piecemeal regulations addressing some specific threat sufficient. We must, therefore, seek a holistic approach, so that the separate channeling of legislation affecting pollution, agricultural practices, fisheries, or forestry can be viewed instead through a perspective that starts from the necessary centrality of integrity and only then moves on to particularities in this or that area, all the while never losing sight of both qualitative and quantitative (or area) limits.

3. Sustainability Beyond the Wild: Living in Integrity

This section considers the consequences of supporting the establishment of large protected areas for one biotic community, humankind. Is it possible for us as a species today to live in integrity? What changes are indicated not only in our approach to natural landscapes but also in our lifestyles in order to comply with the requirements of integrity? I address first the most basic human need, that of food, in the context of integrity, tracing the connection between wilderness protection and sustainability in agriculture. I then turn to our cultural and social needs in the same context, in order to discuss which can be termed needs and what can be classed as wants. The latter may have to be redirected, put on hold temporarily, or eliminated from consideration altogether.

4. Wilderness Protection and Sustainability in Agriculture

Much has been said about the value of wild habitats for conservation (Soulé and Wilcox 1980; Wilson 1989; Noss 1992). However, only Noss has strongly emphasized the size of the required wilderness areas, although this emphasis is necessary. Not only are large native carnivores indicative of a relatively healthy ecosystem, where both functions and processes are not severely affected, but also all tiny, noncharismatic organisms thrive there, if the areas are undisturbed and roadless in order to foster integrity (I_a), as defined in the first section of this chapter. Noss (1991) adds:

> Probably no single feature of human-dominated landscapes is more threatening to biodiversity (aquatic and terrestrial) than roads.

If we settle for smaller reserves, two problems will emerge: (1) no wholeness or integrity (I_a) can be found in areas that only protect some elements of biodiversity; and (2) small reserves usually require a considerable amount of manipulative management in order to maintain what diversity they have (Pyle 1980; Sennen 1980; Noss 1991). In fact, the earlier definition of integrity, in contrast with (and in addition to) ecosystem health, points to human support and manipulation, no matter how well-intended, as the two major factors providing the dividing line between I_a and I_b, or that between core areas of integrity and areas of ecosystem health. In contrast, I_a (or big wilderness) "is essential. It has intrinsic value" (Noss 1991). We cannot even begin to understand what we should support and manipulate in organic food production and other compatible healthy or buffer areas and activities without understanding as best we can what true, pristine integrity (I_a) is and can be. The main model for the zoning requirements I have in mind comes from the Wildlands Project (Noss 1992), which many take to be the statement of an extreme position. It is interesting to note that the same mainstream journal, *Science*, which ridiculed the project as economically and ideologically impossible in June 1993, less than a year later reports the existence of a complementary project intended to render the Wildlands Project operational. The Lifelands Project, which calls for "establishing a protected network of connected public and private lands to preserve ecosystems," proposes supporting the project through economic policies to compensate affected landowners. Wilson (1993) says, "Lifelands is a potentially revolutionary idea," and Schaeffer (1994) adds, "If the science is right, then we need a network of protected areas." But even if the Wildlands Project's position can be supported for its own goals, the implication of its acceptance for humankind in general needs to be addressed. Thus the next question is: Even if life-support systems and habitats need certain landscape percentages and zoning regulations to survive and to fulfill their function, is integrity in this sense not in conflict with, or at least irrelevant to, sustainability in food production and other human activities that should be preserved? What is the relation between I_a and I_b or ecosystem health and appropriate management?

The call to integrity is foundational for sustainability in food production in at least two senses: (1) the loss of ecosystems' capacity (that is, disintegrity) that follows upon intensive petrochemically based agricultural practices; and (2) intrusive and manipulative activities such as the introduction into ecosystems not only of exotics, but also of aliens (such as transgenics), both of which affect integrity both at the macro level (ecosystems) and at the micro level (single organisms).

Wild areas support sustainable agriculture directly and indirectly. Directly, the biodiversity they foster remains a supply depot in which alternatives may be found for lost species; indirectly, wild areas provide the only exemplar or benchmark for comparable areas, for what is appropriate and necessary within an ecosystem to support its health and foster its function. Hence, if we continue to exploit certain areas without reference to integrity and wild areas, sustainability will be lost through the scientific incapacity to understand and predict the effects of technical interference and alterations.

Agricultural examples abound. Studies concerning chemically based agriculture show the dangers of reliance on short-term solutions. A wealth of literature points out the human risks and dangers of chemical products from cradle to grave, so to speak-that is, in their manufacture, distribution, use, and eventual disposal, as well as through possible ingestion by humans (e.g., fruits and vegetables), plus other grave concerns (Draper 1991; Pimentel et al. 1991; Shrader-Frechette 1991; Westra et al. 1991; Pimentel et al. 1992). Pesticides protect economic interests and may seem to ease hunger, in the short term. But as they eliminate unwanted species, they also eliminate countless other species that are necessary to preserve ecosystem health and sustainability in agriculture. In essence, we have no idea just how vast is the damage we inflict when we interfere with the natural functioning of an ecosystem for the purpose of exploitation. We do know that soil erosion, desertification, decreasing productivity, lack of nutrients in the agricultural products, and the need for higher and higher pesticide applications to counteract more and more resistant species of pests are part and parcel of these intensive, petrochemically based agricultural practices (Pimentel et al, 1991; Westra et al. 1991; Meadows et al. 1992; Pimentel et al. 1992; Goodland and Daly 1993; Pimentel et al. 1995).

Further, I have also argued elsewhere (Westra 1995b) for the necessity of a total reevaluation of Western dietary habits in the interest of both sustainability and equity. The importance of diet changes is also emphasized in a recent paper by Kendall and Pimentel (1994):

> About 38% of the world's grain production is now fed to livestock. In the United States, for example, this amounts (to) about 135 million tons per yr. of grain, of a total production of 312 million tons per yr...., sufficient to feed a population of 400 million on a vegetarian diet.

Kendall and Pimental also recommend maintaining biodiversity, as required by the principle of integrity, and the approach I have suggested (Westra 1994a).

5. Cultural and Social Needs

Humankind needs more than food. We are a species that appears to require more than biological survival and reproduction: We need to learn, to congregate and interact, and to support and explore our common cultural heritage. Naess (1991) suggests a goal to be "poor in means but rich in ends." So the question we must ask ourselves is not whether we should have a culture, but how should we live our cultural and social life and whether our ends be they happiness, contentment, or the fulfillment of our natural capacities-could be served by ecologically sound means.

Is it necessary, in order to have a good life, to live a highly materialistic lifestyle? The previous section does not address in detail an important issue, which I have discussed elsewhere: the question of diet (Westra 1995b). Yet that is at least one area where the ecological requirements of integrity and our specific cultural human requirements mesh: Our environmentally wasteful diet based primarily on animal products is as hazardous to our health as it is to the environment. The elimination of our present excessive dietary requirements would serve our environmental and biological needs equally well.

The same, however, cannot be said for a host of other human activities and practices, not all of which should be simply eliminated. Hospitals, office buildings, universities, and malls are all environmentally unfriendly to varying extents, in the use of products of all kinds in building, decorating, cleaning and maintenance, as well as in the specific products and technologies they employ. The difficulties of adjudicating among all competing factors and demands, are daunting, and they cannot be resolved a priori, from a purely philosophical point of view, unless the circumstances and details are taken in consideration.

But if the concept of integrity and the principles derived from it are intended to be used practically, not only in moral argument but also in public policy, they must be capable of generating second order principles and to offer some guidance in the resolution of these problems. The problem can be framed both in the context of developed and developing countries: We will start with the latter, because, although this adds further difficulties, in some sense it permits us to better conduct an intellectual experiment. In Plato's *Republic*, in order to define the meaning of justice in the city (in the context of the character of the citizens and leaders alike), Plato "constructs" a city, from an initial nucleus consisting of a small community of citizens with minimal survival needs, and proceeds to trace its growth through the addition of various consumer and lifestyle preferences. We have acknowledged that the establishment of large wild core areas, surrounded by appropriate buffers is viewed by some as a goal in contrast with the legitimate aspirations of citizens in less developed countries. This is neither the *intended* result of the use of integrity's guidelines, nor an acceptable *by-product* of it.

Like Plato, let us ask ourselves how we should add to the basic community, in a constructive way, although in our case, it is not the virtue or the character of the citizens and their institutions that is our main goal, but ecological integrity and sustainability. As Plato saw, the addition of more goods beyond basic needs and requirements would necessitate balancing measures, such as doctors to offset the preferred richer diets, and soldiers to ensure additional territories from which to extract more resources and greater trading markets. These measures were based on the experience of the altered living conditions. Similarly, the economic and social development in presently less developed countries should, ideally, be governed by the limits indicated by the environmental problems in which developed countries are presently mired.

I suggest that the guidelines of ecosystem integrity also require a holistic approach for the common good, understood beyond Plato's model, as both a communitarian and a universal but nonanthropocentric goal. Further, integrity would not block or preclude development where sorely needed; it would simply suggest the shape such development should have in order to be sustainable in the long term and nonhazardous in the immediate present. For instance, this approach would still require the protection of wild areas and the establishment of buffer zones, but this would not preclude suitable human activities, in the sense that humans-in-the-biosphere projects, for instance, also support such activities. Of course, all activities in buffers are limited by the requirements of the core or wild areas. This may not be sufficient to meet the expectations of some. If a country's citizens would prefer activities that are closer to the Northwestern ideal of a strongly consumerist lifestyle, in our decisions to aid and support, we should be guided by a decreased lifestyle goal that we ourselves are willing to adopt.

Integrity must be viewed as the control measure: Nothing can be either moral or appropriate to public policy that contravenes the requirements of noninterference and the protection of wild areas (I_a) and appropriately sized buffers (I_b) (Westra 1994a; cp. Lemons and Morgan 1995). That is the reason why I have referred to human economic development and culture as taking place in areas where I-c should prevail; it is because even economic development and social/cultural activities should be such that, even though no natural, evolutionary processes obtain there, still no activity is permitted that might adversely affect the processes in I_a areas.

As in the Platonic vision in the *Republic*, it is easier to build such communities where human errors and wrongs have not yet happened, so that we can ensure that economic and social activities, as they unfold, are judged first from the standpoint of their possible impact on areas of integrity. Is this approach limiting? Of course it is, at least if one shares, as

Shepard (1995) has it, the "modern hallucination of no limits," whereby the latter are neither acknowledged nor accepted. But in that case, our life takes place within a world of virtual, not actual, reality.

It is worth noting that the limits I propose are for the common good of those in less developed countries, but they are also for the common good, universally, provided that we in the Northwest are prepared to move sincerely in the opposite direction from the one in which the Southeast is moving-that is, provided that we are prepared to withhold our approval from all further material growth and actively seek alternatives to it.

At the risk of mixing authors and metaphors, perhaps we should borrow not only from the global perspective of Plato's *Republic* (modified from his culture-bound social holism) but also from the Aristotelian quest for a golden mean. This should not be viewed as a compromise between ecologists and developers/economists, as it is supported by the present status quo. The mean I suggest is the optimum or most appropriate point of sustainable consumption, between Western style preference satisfaction and the underdeveloped and deprived lifestyle in the Southeast. This is by no means a mechanically computed midpoint; on the contrary, it is intended as an ecologically informed point, where the best possible material level of aggregate and individual choice, combines with the respectful sustainability of I-c and I_b areas, that is, where our consumption and lifestyle are compatible with the requirements of both buffer and core areas. In principle, then: (1)ecological sustainability should be the benchmark of its economic counterpart; and (2)the (sustainable) development of less developed countries should proceed apace with Northwestern sustainable, deliberate underdevelopment, or cut-back development.

As we have used and abused the world's natural resources with careless abandon for too long, and we are now forced to recognize the just claims of Third World people, it seems only fair that we acknowledge our mutual interdependence so that for every step forward they can make, perhaps even with our support, we are willing to take one step back. This should enable others to move forward without further devastation to our common world. In fact, given that our initial steps would not lead us anywhere near the abject poverty which is the lot of others in less developed countries, perhaps our steps should be accelerated so that the Southeast's movement toward development could increase while, at the same time, the overall stress to the earth would reduce.

The difficulty that persists even if this approach is accepted is the nonpredictability of science, so that the precise form these steps might take, their limits and results, are all to be understood in terms of the precautionary principle, and the appropriate policies guided by it (Lemons and Morgan 1995). The existence of transboundary problems is an additional complication. For instance, climate changes are the result of our previous mistakes. Even after the Montreal Protocol, a lot of time will have to elapse before CFCs no longer affect the ozone layer, and thus also ecological and biological processes, and during that time, we will all have to act to protect ourselves and the earth and-if possible-to counter to some extent the damage we have done. Jamieson (1990) says:

> The role of the physical science is to produce information regarding the physical effects of increasing concentrations of greenhouse gases. Physical effects include climatological effects, hydrological effects, and biological and ecological effects.

But, as he acknowledges, uncertainties about all those effects abound, and it is a mistake to assume that science can, in all cases, give an answer, let alone a precise one (see also Lemons 1995). That is why some have argued that it is inappropriate to permit public policy to be decided simply by scientists, that a dialogue with all stakeholders should take place instead, at least for all-important issues involving public safety or public health (Shrader-Frechette 1991; Lemons and Brown 1995).

Jamieson (1990) says, "The first thing we should do is to give up the idea that all problems can be managed." Kay adds that we must abandon the idea that all ecosystems can be managed as well (Kay and Schneider 1994). It is at this point that the role of integrity and that of the zoning requirements it engenders can be clearly seen. Had the precautionary principle been used, in conjunction with the understanding that wild areas (I_a) are central to all our public policy decisions and that we all must face the fact that halting and reversing ecological damage entails accepting life as appropriate to buffers (I_b and even I-c), then the most serious ecological damage may not have happened. Jamieson (1990) also suggests policies that "focus on incremental changes," and all those he proposes are indeed the sort of step-back that I recommend for the Northwest, to work in conjunction with the step-forward policies that justice would demand for the Third World.

A recent case, which may or may not be related to the global warming situation we used as an example of the complex problems with which the ethics of integrity must deal, is that of the flooding in several European countries, including the Netherlands. In my recent work and in this paper as well, I have used material from the Dutch government's environmental plans and regulations as an example of what can be accomplished in a Western democracy which, because of its size, location, and history, is now particularly vulnerable from the environmental point of view. Because all Dutch systems have been managed for so long, wild nature is almost entirely absent from the Netherlands. The hazards of intensive petrochemical agriculture and of human management in all areas (including extensive roads, dikes, and other human controls) have resulted in problems too numerous to mention, from poor food production to increased hazards to human health from increasing quantities of manure and excessive chemicals to unknown threats from the fragmentation of nature. As we will see below, the Dutch term the latter a form of pollution, and they have consistently tried to withdraw from their vigorous management and control policies, to permit nature restoration through legal veto powers on the part of environmental government bodies, in regard to social and economic programs which have not appeared to be ecologically sound.

During the last week of January 1995, the dikes in the south of Holland were threatening to burst. *Time* (1995) magazine said that, "250,000 people were evacuated, three died, and the damages may reach $2 billion." How does this disaster relate to the Netherlands' strong environmental goals? The Canadian newspaper the *Globe and Mail* (1995), says:

> The Dutch environmental movement, which has opposed dike reconstruction and maintenance projects, has been left a wreck in the wake of outrage over the government's dilatory efforts to protect the countryside from water.

But until the flood, the Dutch government was responding appropriately to the environmental crisis, with strong planning and legislation (RIVM 1989; 1992), in order to restore natural processes and reduce management. But this approach required the preservation of the less efficient earthen dikes, on line with the conservationists' successful campaigns to preserve the Netherlands' scenic countryside. To switch to the more efficient concrete dikes and to turn "meandering rivers...into straight-line water expressways to speed the flow [of rivers] to the North Sea" (*Globe and Mail* 1995) might be socially and economically preferable, but it is ecologically disastrous.

Lemons and Morgan (1995) argue that my proposed approach could "conflict with the goals of social and economic sustainability," not only in less developed countries but also in developed nations, where "support of large wild land areas is controversial." What is left out of Lemons's argument is the time frame within which we view our actions. In the short term, no doubt both Lemons and the Dutch movement against environmentalism are quite correct. But in the long term, only the integrity approach is sustainable. Further, at the public policy level, the integrity approach presupposes that: (1) governance in environmental matters should transcend fragmented national choices and (2)individual and collective preferences should be considered but not permitted to prevail when grave issues connected with public health and even human survival are at stake (as they were, for example, overridden in the case of CFCs) (Jamieson 1990; Westra 1994a).

Moving from the ethics of integrity to a Rawlsian approach would support this summary setting-aside of democratic processes in defense of ecological integrity: The practices I propose should be mandated as necessary, to protect the least advantaged, both presently and in the long term. The poor and the underprivileged in both Northwest and the Southeast are not insulated from the worst effects of environmental degradation in the way that more affluent persons and countries are. Therefore, the ethics of integrity does not conflict with justice for the poor, globally, at least in principle, although the complexity of the practical applications need study and discussion.

Finally, a great deal of evidence exists about the harm regularly inflicted on people in technologically advanced Western countries, because the operation of "risky business" (Westra 1994b) is not strictly regulated from the standpoint of core and buffer zones. It is commonplace to find homes and schools in the United States operating within half a mile of hazardous manufacturing or service operations, and equally close to waste facilities. In many cases a further component is present, that of "environmental racism," or perpetuating "brownfields" in areas affecting individuals and communities of color in a disproportionate way. I have discussed this problem elsewhere (Westra and Wenz 1995a), from the standpoint of "environmental racism," and the rest has been amply documented by many as well. For my purpose, the point now is that the imperative to impose zoning requirements, which start from the necessity for undisturbed core areas, and judges the rest of human culture with relation to that imperative and limits it accordingly is judged, is precisely what I term the imperative of integrity.

6. Conclusion

What does the ecosystem perspective recommend? One can find at least three recommendations in Kay and Schneider's (1994) recent work: (1)that we look at ecosystems "from a hierarchical perspective, with careful attention to scale and extent"; (2)that we must recognize the "spatial, temporal, thermodynamic and information (dynamics) aspects of these systems"; and (3)that we need "to stop managing ecosystems for some fixed state, whether it be an idealistic pristine climax forest or a corn farm." The authors conclude that "the notion of serving ecological integrity means accepting all this."

I have argued that: (1)biodiverse, naturally evolving habitats are valuable in themselves, that is, for all their known and unknown components and capacities; (2)we are also among the supported component forms of life, as our most basic survival necessities-nontoxic food, water, and air-must be understood as integrated and dependent in many ways upon those habitats, so that these necessities might be obtained in a sustainable way; and (3)all other nonbasic human needs in culture must be pursued with reference to wild habitats (I_a) primarily but also with concern for equity and respect for all humans.

Some may object that different forms of land use are already in place in most countries' regulations; yet present regulations are insufficient even from the standpoint of human health alone (Westra 1994a; cp. Bullard 1994). The connection between human health and the zoning and regulation required by the principle of integrity are clear and implicit in a recent publication of RIVM (1989) (National Institute of Public Health and Environmental Protection of the Netherlands). It shows how "acidification, eutrophication and dispersion of toxics" gravely endanger "functions that can no longer be fulfilled properly." The approach recommended for new regulations is "area oriented" and starts by dividing the Netherlands into "ecodistricts." It also speaks of "the functions, nature, agriculture, forestry, and...water" as examples of the most vulnerable areas of concern, for which function-oriented environmental requirements are necessary. Hence, what I term "wild," RIVM terms "nature," and it is described by RIVM as a function distinct from human goals yet necessary to support them. Various maps show the country's vulnerability to toxics, acidification, agricultural/pesticide impacts on groundwater, and so on. Two RIVM maps are particularly relevant for our purpose (Figures 1 and 2). Figure 1 shows areas where nature is a main function, while Figure 2 shows the vulnerability of nature and landscape to fragmentation.

Several points about these RIVM maps are worth noting: (1)"nature" is described as a "function" distinct from other goals or purposes (compare the "capacity" of I_a areas); (2)there are many so-called "nature areas," but they are scattered and separated, rather than connected; and (3)Figure 2 cites fragmentation, together with noise and chemicals, as a form of pollution. Both fragmentation and noise severely affect wild animals, and hence they destroy nature's capacity, or function. Figure 3, taken from the Wildlands Project (Noss 1992), shows a network in which fragmentation is no longer a problem, with core areas, buffers, and corridors, all designed to maintain and protect the wild and, with it, the function or capacity it possesses.

I use an example from the Netherlands because it is a small country which has been totally managed and manipulated for a long time, so that the problems we encounter are magnified and even more urgent and severe in that context than they are in many other places. We thus see in the Netherlands the public policy and concerns that we should adopt now, before we, like Holland, are forced to them, with a far less unified political structure to support urgent changes.

Stringent regulations that include clear zoning requirements are also supported by the admitted lack of precision in standards and other assessments connected to human health. Another chapter of the RIVM publication describes the hazards of chemical agents, through oral exposure, inhalatory exposure, and various physical agents. In all cases the emphasis remains on the incompleteness and vagueness of standards proposed. For example, exposure to cadmium can cause renal damage, and the scientific standard to determine allowable exposure to the hazard is based on intake per kg of body weight. When this standard is communicated and used in public policy, it is not nuanced appropriately for different populations. For instance, it does not indicate that not only some optional activities, like smoking, but also eating crops from an area that has not been zoned for safety can make cadmium intake two to three times higher, as can even so-called safe intake for persons suffering from iron deficiency. The authors of the RIVM document add, "This makes it difficult to establish clear exposure-response relationships."

If standards are not precise enough to be effective in the present regulatory context, and if it is not likely that science will become more exact and specific (Funtowicz and Ravetz 1992, 1993; Shrader-Frechette and McCoy 1993), then we should provide for a greater margin of error in several ways. One way is to do much more to protect the supporting function of the wild by limiting our disturbances and our encroachment through strict zoning. Another way is to appreciate that our activities in landscapes that are not zoned as wild or core areas also must conform to the mandated noninterference for the wild. Thus, in a sense, all that we can do in managed and urban areas must be considered as taking place in buffers.

Hence, the consideration I propose for the foundational role of the wild (I_a) is supported by the well-known imprecision of risk assessment (Shrader-Frechette 1991), and the hazards of insufficient zoning requirements for wild habitats, their populations, and human health. The fragmentation of nature, as we saw, is also in itself a hazardous practice, with strong negative impact, both direct and indirect, on all life and health.

The aim of this paper has been to show the foundational value of integrity in its primary sense, as the wild, and to argue for its corresponding primacy in the same sense, in decisionmaking and public policy and ethics, in spite of difficulties about pinpointing and

Figure 1. Areas Where Nature is a Main Function in the Netherlands (Source: RIVM 1992).

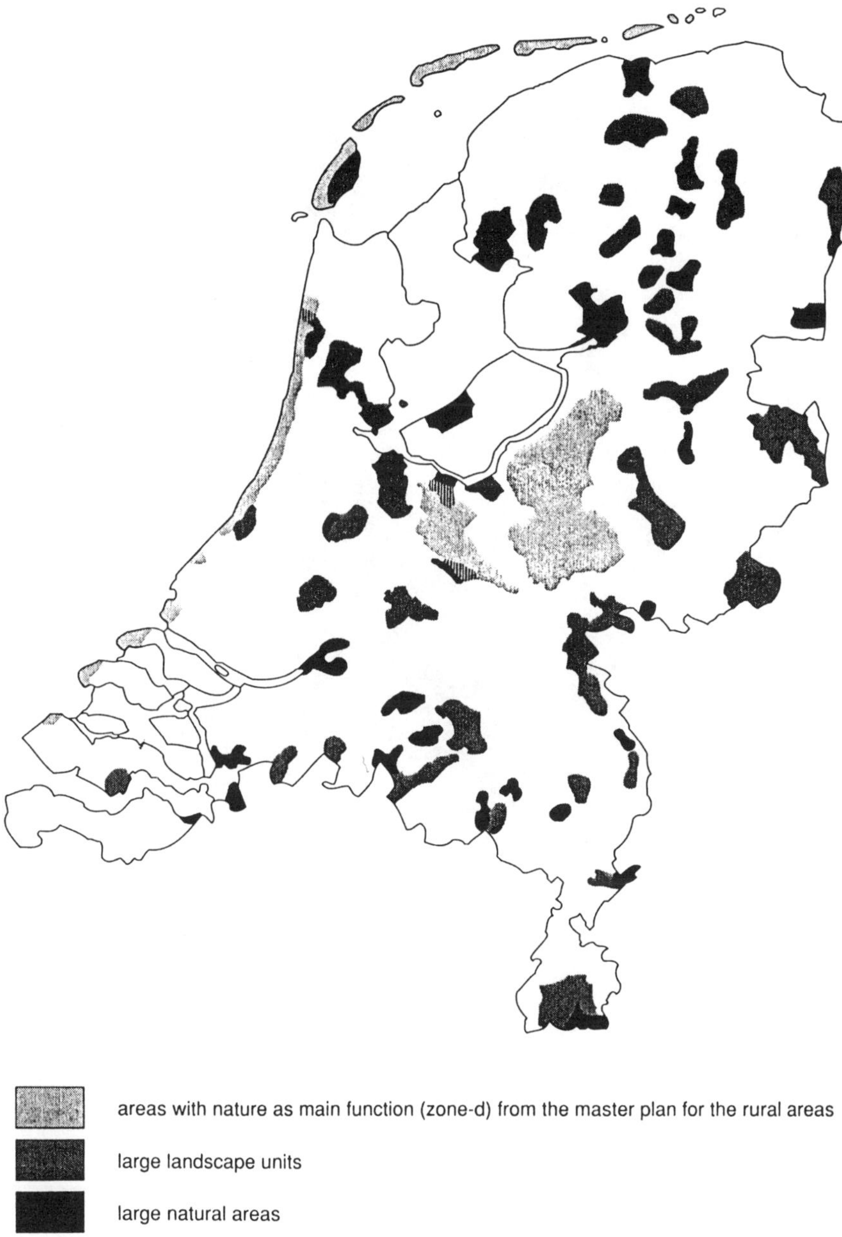

areas with nature as main function (zone-d) from the master plan for the rural areas

large landscape units

large natural areas

Figure 2. Fragmentation Vulnerability in the Netherlands (Source: RIVM 1992).

defending the value of integrity and of the ecosystem approach. Perhaps the newest and strongest defense of the necessity to preserve large wild areas comes from Noss's work in "The Wildlands Project"" (Noss, 1992) and his latest, *Saving Nature's Legacy* (Noss and Cooperrider 1994). The latter presents a strong plea to "protect and restore biodiversity," with a primary focus on the wild and the creatures that inhabit it; hence, it proposes "area requirements" consonant with the needs of large carnivores (Noss and Cooperrider 1994). My position is complementary to theirs because, for instance, both respectful, nontoxic buffers and equally carefully monitored urban and culture areas will be necessary for their goal to succeed, and also because I have indicated that there is a symbiotic relation between the wild and us. Living in integrity will respect the wild, and the wild will continue to provide life support not only for wild creatures but also for those who put their caves and burrows beyond it.

Noss and Cooperrider (1994) acknowledge that many will judge their requirements for over one third of the earth to be left wild (in various landscape-specific percentages) to be utopian, visionary, and antihuman. Shrader-Frechette and McCoy (1993) also remark that pleas for ecosytems that exclude humans are somehow not appropriate, if I understand them correctly. Some clear zoning is not antihuman as such. To say that humans do not belong in wild systems is obviously true in one sense, but it needs qualifications: the exclusion of humans does not set one species aside against the others, nor does it separate humans from the rest of nature. This simply recognizes that one species only refuses to enter such systems in its own natural state. In fact, neither humans nor panthers nor frogs have any place in wild systems, if they come riding Jeeps and carrying computers and electrical generators and insist on using and then dumping alien/toxic matters within the wild. It is a fallacy to assume that "human" equals "technological human," and it is only the latter who is not welcome in the wild. People may walk in the forest; a busload of tourists requiring paved roads, restaurants, souvenirs, and flush toilets may not.

We would do well to heed Ehrlich's (1980) final caveat for conservationists intent on aggressive conservation strategies when confronted by politicians, economists, engineers,

Figure 3. Regional Wilderness Recovery Network (Source: Noss 1992).

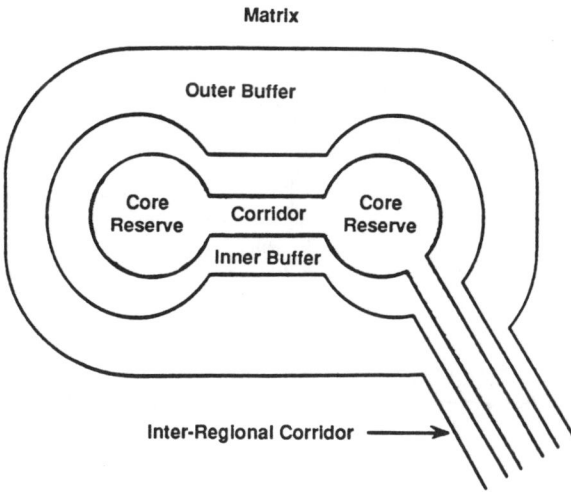

Figure 4. The Global Dimensions of Integrity: Sustainability, Biodiversity, and Environmental Justice.

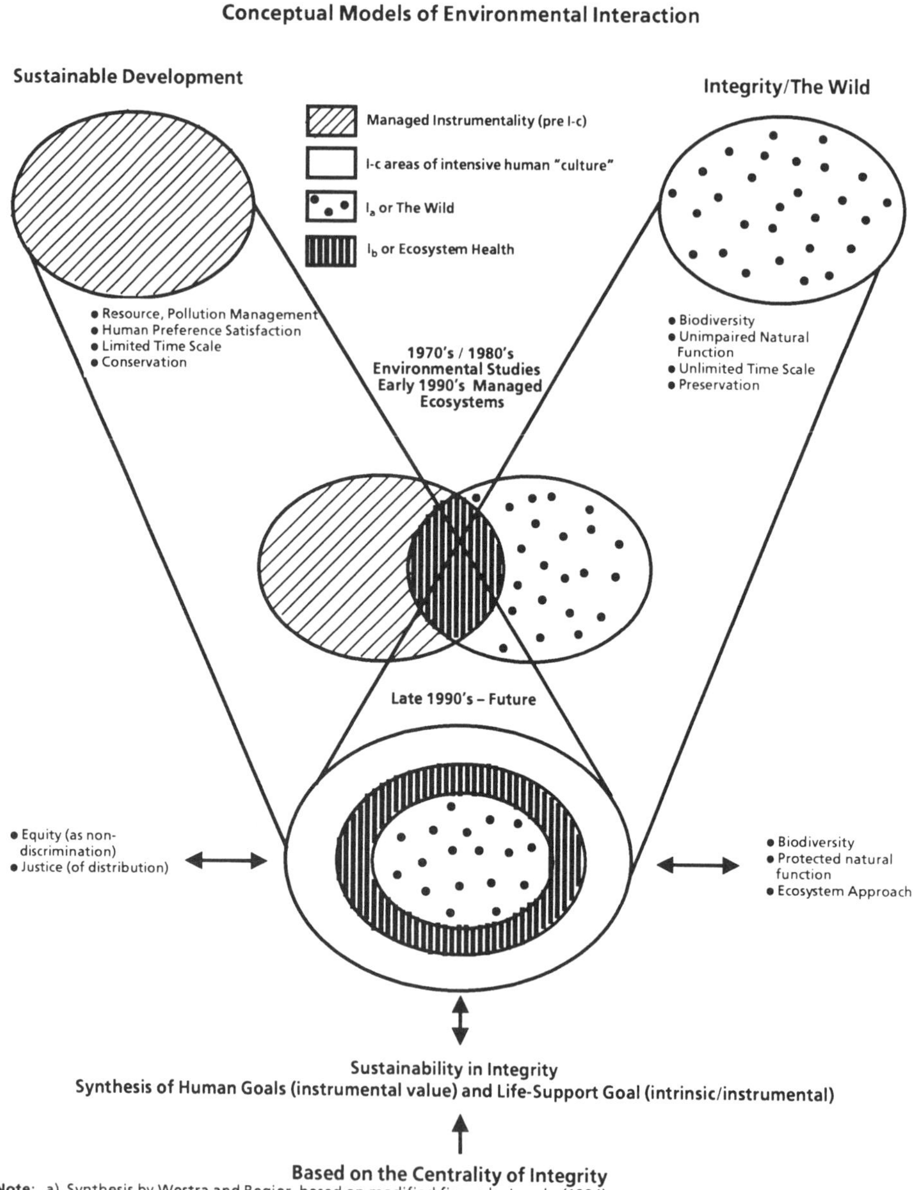

Note: a) Synthesis by Westra and Regier, based on modified figure by Loucks (1994).
b) I_a area is modeled on Noss's core/corridors concept (Noss 1992).
c) I_b area represents buffers; and I-c, areas of intensive human habitation and use, which remain compatible with the nontoxicity and safety required by I_a and I_b.

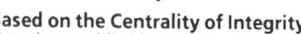

developers, and so on, asking one to be reasonable, responsible, and to make compromises. These people, he says, are often intelligent, attractive, well-meaning people who only want to keep on behaving in a manner that, after all, has been acceptable for a couple of hundreds of years. They are, says Ehrlich, "the enemy."

These are strong words, but no stronger than the current movement to propose amendments aimed at criminalizing environmental abuse. My vision of an appropriate future goal for humankind is to make the wild (I_a) central to both ethics and public policy (Figure 4). Nor is this a totally "ideal" or purely philosophical suggestion. In Canada, for instance, the B.C. Commission on Resources and the Environment proposed a "sustainability act," as it "seeks strict control on human settlement" (*Globe and Mail*, 1995). The reporter quotes the head of the Commission, Stephen Owen:

> We simply cannot harvest more than we grow, just as we cannot spend more than we can earn. This act is really a challenge to government to constrain itself and to ensure that society lives within its ecological means.

It is worth noting that, hard as it may be to achieve consensus on this sort of legislation, as it addresses present and future choices, such legislation is still much easier to implement than are decisions affecting people retroactively, that is, to reverse previous public policy decisions (hence the use of Plato's *Republic* as an example above).

But we cannot hold our policies to the standards defined by old knowledge and old science. In the past we did not understand the implications of postmodern science; we did not fully understand science's true capacities. After chaos theory, we now know why we need to use the precautionary principle in all decisions involving public policy. Schneider and Kay (1994) say:

> Chaos theory...notes that any real dynamic system, even one described by a set of deterministic equations, is ultimately not predictable because of the accumulation of individually small interactions between its components. This applies to balls on a billiard table and the planets in heavens. This means that our ability to forecast and predict will always be limited regardless of how sophisticated our computers are or how much information we have.

I propose the integrity approach and the ethics of integrity to ground public policy, and the implementation of core and buffer zones, as well as control in non-evolutionary systems (that is, the application of I_a, I_b, and I-c in various sizes and proportions) as a first step and partial remedy to the reality of the gap between our expectations and what science, even if unmanipulated by economic interests, can in fact deliver. I also suggest a strong drive to equity based on steps progressively receding from our present untenable and unsustainable lifestyle in the Northwest toward a golden mean that is ecologically sustainable as well as economically fair.

7. References

Bocking, S. 1994. Visions of Nature and Society, A History of the Ecosystem Approach. *Alternatives* 20(3): 12-18.

Brown, L. 1981. *Building a Sustainable Society*. A Worldwatch Institute book, W.W. Norton & Co., New York.

Bullard, R. 1994. *Dumping in Dixie* (Race, Class and Environmental Quality), Westview Press, Boulder, CO.

Draper, E. 1991. *Risky Business.* Cambridge University Press, Cambridge, MA.
Ehrlich, P.R. 1980. The Strategy of Conservation, 1980-2000. In *Conservation Biology*, M.E. Soulé and B. Wilcox, eds. Sinauer Associates, Sunderland, MA, pp. 329-344.
Funtowicz, S.O., and J.R. Ravetz. 1992. The Good, the True and the Post-Modern. In *Futures* December: 963-976.
Funtowicz, S.O., and J.R. Ravetz. 1993. Science for the Post-Normal Age. In *Futures* September: 739-755.
Goodland, R., and H. Daly. 1993. Why Northern Income Growth Is Not the Solution to Southern Poverty. The World Bank Environment Department Divisional Working Paper, No. 1993-43, May. World Bank, Washington, DC.
Goodland, R. 1994. Environmental Sustainability and the Power Sector-Part I: The Concept of Sustainability. *Impact Assessment* 12(3): 275-304.
Hammer, M., A.M. Jansson, and B.O. Jansson. 1993. Diversity, Change and Sustainability: Implications for Fisheries. *Ambio* 22(2/3): 97-105.
Holling, C.S. 1986. The Resilience of Terrestrial Ecosystems: Local Surprise and Global Change. In *Sustainable Development in the Biosphere*, W.N. Clark and R.E. Munn, eds. Cambridge University Press, Cambridge, MA, pp. 292-320.
Holling, C.S. 1992. Cross-Scale Morphology, Geometry and Dynamics of Ecosystems. *Ecological Monographs* 62(4): 447-502.
Jamieson, D. 1990. Managing The Future: Public Policy, Scientific Uncertainty and Global Warming. In *Upstream/Downstream*. Temple University Press, Philadelphia, PA, pp. 67-89.
Karr, J.R. 1994. Landscapes and Management for Ecological Integrity. In *Biodiversity and Landscape: A Paradox of Humanity*, ed. Cambridge University Press, New York, pp. 227-249.
Karr, J.R., and P.L. Angermeier. 1994. Protecting Biotic Resources: Biological Integrity Versus Biological Diversity as Policy Directives. *BioScience*, 44(11): 690-697.
Karr, J.R. 1995. Forthcoming. Ecological Integrity and Ecological Health Are Not The Same. In *Engineering Within Ecological Constraints*, National Academy of Engineering, National Academy Press, Washington, D.C.
Kay, J. 1992. A Non-equilibrium Thermodynamis Framework for Discussing Ecosystem Integrity. *Environmental Management* 15(4): 483-495.
Kay, J.J., and E. Schneider. 1994. The Challenge of the Ecosystem Approach. *Alternatives* 20(30): 1-6.
Kendall, H.W. and D. Pimentel. 1994. Constraints on the Global Food Supply. *Ambio* 23(3): 198-205.
Lal, R. 1976. *Soil Erosion Problems on an Alisol in Western Nigeria and Their Control.* International Institute of Tropical Agricture, Ibadan, Nigeria.
Latowsky, G. 1993. The Woburn, Massachusetts Case. Presentation for the American Association for the Advancement of Science, Boston, MA. February 14.
Laudan, L. 1984. *Science and Values.* University of California Press, Berkeley, CA.
Lemons, J., ed. 1995. *Scientific Uncertainty and Environmental Problem-Solving.* Blackwell Science, Cambridge, MA.
Lemons, J. and D.A. Brown, eds. 1995. *Sustainable Development: Science, Ethics and Public Policy.* Kluwer Academic Publishers, Dordrecht, The Netherlands.
Lemons, J., and P. Morgan. 1995. Conservation of Biodiversity and Sustainable Development. In *Sustainable Development, Science, Ethics and Public Policy*, J. Lemons and D. Brown, eds. Kluwer Academic Publishers, Dordrecht, The Netherlands, pp. 77-109.
Meadows, D.H., D.L. Meadows, and J. Randers. 1992. *Beyond the Limits.* Chelsea Green Publishing, Post Mills, VT.

Naess, A. 1991. *Ecology, Community and Lifestyle*, ed. D. Rothenberg. Cambridge University Press, New York, NY.
Noss, R.F. 1991. Wilderness Recovery: Thinking Big in Restoration Ecology. *The Environmental Professional* 13: 225-234.
Noss, R.F. 1992. The Wildlands Project: Land Conservation Strategy. *Wild Earth*, Special Issue, pp. 10-25.
Noss, R.F., and A.Y. Cooperrider. 1994. *Saving Nature's Legacy*. Island Press, Washington, DC.
O'Neill, R., D. De Angelis, J.B. Waide, and T.F. Allen. 1986. *A Hierarchical Concept of Ecosystems*. Princeton University Press, Princeton, NJ.
Oreskes, N., K. Shrader-Frechette, and K. Belitz. 1994. Verification, Validation and Confirmation of Numerical Models in the Earth Sciences. *Science* 263: 641-646.
Parson, T. 1964. Definitions of Health and Illness in the Light of American Values and Social Structure. In *Social Structure and Personality*, E.G. Jaco, ed. Free Press of Glencoe, New York, NY.
Pimentel, D., L. MacLaughlin, A. Zepp, L. Benyamin, T. Kraus, P. Kleinman, F. Vancini, J.W. Roach, E. Graap, W.S. Keeton, and G. Selig. 1991. Environmental and Economic Effects of Reducing Pesticide Use. *Bioscience* 41(6): 402-409.
Pimentel, D., et. al. 1992. Conserving Biological Diversity in Agricultural/Forestry Systems. *BioScience* 42(5):354-362.
Pyle, R.M. 1980. Management of Natural Reserves. In *Conservation Biology*, M.E. Soulé and B. Wilcox, eds. Sinauer Associates, Sunderland, MA, pp. 319-328.
Regier, H.A. 1993. The Notion of Natural and Cultural Integrity. In *Ecological Integrity and the Management of Ecosystems*, S. Woodley, J. Francis, and J. Kay, eds., St. Lucie Press, Delray Beach, FL, pp. 3-18.
RIVM-National Institute of Public Health and Environmental Protection. 1989. *Concern for Tomorrow, A National Environmental Survey, 1985-2010*. I.F. Langeweg, ed. Bilthoven, The Netherlands.
RIVM-National Institute of Public Health and Environmental Protection. 1989. *National Environmental Outlook 1990-2010* . O. Hunter and Partners. Leiden University, Department of Environmental Biology, Groningen University, IVEM, Bilthoven, The Netherlands.
Sagoff, M. 1989. *The Economy of the Earth*. Cambridge University Press, Cambridge, MA.
Sagoff, M. 1992. Settling America, or The Concept of Place in Environmental Ethics. *Journal of Ecology, Natural Resources and Environmental Law* 12(2): 349-418.
Schaeffer, M. 1994. The Lifelands Project. *Science* 264: 1078.
Schneider, E.D., and J.J. Kay. 1994. Complexity and Thermodynamics, Towards a New Ecology. *Futures* 26(6): 626-647.
Senner, J.W. 1980. Inbreeding Depression and the Survival of Zoo Populations. In *Conservation Biology*, M.E. Soulé and B. Wilcox, eds. Sinauer Associates, Sunderland, MA, pp. 209-224.
Shepard, P. 1995. *The Company of Others* (Essays in Honor of Paul Shepard), Max Ochlschlaeger, ed. Kivaki Press, Durango, CO.
Shrader-Frechette, K. 1991. *Risk and Rationality*. University of California Press, Berkeley, CA.
Shrader-Frechette, K., and E.D. McCoy. 1993. *Method in Ecology*. Cambridge University Press, New York.
Soulé, M.E. 1980. Thresholds for Survival: Maintaining Fitness and Evolutionary Potential. In *Conservation Biology*, M.E. Soulé and B. Wilcox, eds. Sinauer Associates, Sunderland, MA, pp. 151-170.

Soulé, M.E. and B. Wilcox. 1980. Conservation Biology: Its Scope and Its Challenge. In *Conservation Biology*, M.E. Soulé and B. Wilcox, eds. Sinauer Associates, Sunderland, MA, pp. 1-8.

Troeh, F.R., J.A. Hobbes, and R.L. Donahue. 1991. *Soil & Water Conservation*, 2nd ed. Prentice Hall, New York.

Westra, L. 1994a. *An Environmental Proposal for Ethics: The Principle of Integrity*. Rowman, Littlefield, Lanham, MD.

Westra, L. 1994b. Risky Business: Corporate Responsibility and Hazardous Products. *Business Ethics Quarterly* 4(1): 97-110.

Westra, L. 1994c. Biodiversity and Food Production: The Perspective of Ecosystem Integrity. Read at the American Association for the Advancement of Science meeting, Feb. 11, 1994.

Westra, L. 199a. The Faces of Environmental Racism: Titusville, AL., and BFI. In *Faces of Environmental Racism: Issues in Global Equity*, L. Westra and P. Wenz, eds. Rowman Littlefield, Lanham, MD, in press.

Westra, L. 1995b. Paul Shepard's "The Wild" and Biocentric Holism: A Convergence of Ends. In *The Company of Others.*, Max Oehlschlager, ed. Kivaki Press, Durango, CO.

Westra, L., K. Bowen, and B. Behe. 1991. Agricultural Practices, Ecology and Ethics in The Third World. *The Journal of Agricultural Ethics* 4(10): 60-77.

Wilson, E.O. 1993. Threats to Biodiversity. In *World Scientists' Warning Briefing Book*. Union of Concerned Scientists, Cambridge, MA, pp. 155-160.

8. Appendex A*

For all its common usage and its twenty-year history, "ecological integrity" has not been exhaustively defined or conceptually analyzed in the field of public policy or in that of science, ethics, or the philosophy of science, although scientists have attempted to design various indices of integrity for both land and water. James Kay and I have developed a provisional definition of "integrity," which was further enriched by the input of Henry Regier, Robert Ulanowicz, and Don DeAngelis. Our intent was to manifest its meaning and value, hence the reasons for its use in legislation. We have provisionally defined integrity as follows:

"Ecosystem integrity" is a concept that cannot be defined precisely but that includes in various proportions the following:

1. Ecosystem health and its present well-being (defined below). This condition may apply to even nonpristine or somewhat degraded ecosystems, provided they function successfully as they presently are. Some examples might be (a) an organically cultivated farm, or a low-input operation; but also (b) a lake which, having lost its larger species because of anthropogenic stress, now functions with a larger number of smaller, different species. Hence, ecosystems that are merely healthy may encompass both desirable and undesirable possibilities, and may be more or less limited in the capacities they possess (or have become artificially or accidentally constrained by humans). It is for this reason that health alone is not sufficient.
2. The ecosystem must retain the ability to deal with outside interference, and, if necessary, regenerate itself following upon it. This clause refers to the capacity to withstand stress. But nonanthropogenic stress is part of billions of years of systemic development. Anthropogenic stress, on the other hand, may be severely disruptive to the system in that it may contain realities that are radically new to the natural

components of the system, or it may operate at intensities that are unprecedented in evolutionary history, hence the system may not have ways to compensate or adapt to such stress.
3. Integrity is achieved when the system's optimum capacity for the greatest possible ongoing developmental options within its time/location remains undiminished. The greatest possible potentiality for options is also fostered by the greatest possible biodiversity. Biodiversity contributes to integrity in at least two ways:
(a) through genetic potential, based on the size and diversity of populations and their respective gene pools (hence supporting and enhancing both structure and function of these populations);
(b) through biodiversity's dimensions as purveyor and locus of both relational information and communication, of which existing populations and ecosystems manifest and embody only a small proportion. We can only theorize about the immense capacities for diverse qualitative interactions among individuals and species, which are not presently existing or knowable.
4. The system will possess integrity if it retains the ability to continue its ongoing change and development unconstrained by human interruptions past or present.

*Source: Westra (1994).

Chapter 3
ECOLOGICAL INTEGRITY: RECLAIMING LOST CONNECTIONS

James R. Karr[1]
Ellen W. Chu[1]

1. Introduction

Human history, like evolution itself, has been marked by relative stasis punctuated by periods of rapid change. Harnessing fire, making and using tools and weapons, and inventing the wheel were early mileposts signaling, we are told, an unbounded human ingenuity. These and other innovations allowed humans to tap natural capital and spread virtually throughout the world, living year-round from sea level to mountain tops, from equatorial heat to polar cold. The success of humans in these diverse natural settings resulted directly from the ability to adapt to diverse regional conditions and to develop and modify culture and religion.

Throughout evolutionary time, the success of living things has depended on the accumulation of information passed from generation to generation in the genetic blueprints of DNA. Humans, though, perfected another connection: the legacy of knowledge and culture passed from parents to their children and their children's children across hundreds, even thousands, of generations. During early human evolution, important knowledge was primarily biological—how to find food and shelter, escape from predators, avoid disease. Humans, like all other organisms, had to know their regional environments and how to support their communities within these environments.

But with the agricultural revolution, these connections began to fray. By the nineteenth century, scientific and societal specialization combined with rapid, massive industrialization and free-market economics and seemed to promise escape from dependence on, or even connections with, other living systems. Now the "information age" gives us "virtual reality," completing our isolation from the rest of the living world and, some claim, clinching an end to human need for the biological knowledge so important to our ancestors.

Alas, we have not escaped our dependence on nature; we have succeeded only in dominating nature over the short term. We have created a hybrid world—one neither entirely natural nor entirely mechanical. Consider, for example, the Columbia River in the Pacific Northwest. Environmental historian Richard White (1995) aptly characterizes the Columbia as an "organic machine"—a living system, mechanized by humans and maintaining a few natural, "unmade" qualities. But whereas once the Columbia supported some of the world's largest salmon runs and the human cultures linked to these fish, now it supports nonnative species like American shad and carp, and machines mediate the human relation to the river. The mechanization of the Columbia severed the connections within living systems, and

---

[1]Institute for Environmental Studies, Box 352200, University of Washington, Seattle, WA 98195-2200, U.S.A.

Columbia River salmon runs are down by 96 to 98 percent after only a century (Nehlsen et al. 1991). Many more unique salmon stocks will be lost over the next two decades. As humans further mechanize the river, they spend more and more money to maintain less and less of the living system. And, technological optimists all, we see the mechanisms as "workable" and "practicable" when in fact we have failed to comprehend the vital importance of the severed connections.

Touting the uncanny ability of humans to be innovative and thus improve their lot, optimists—including economists, technologists, and futurists—see progressive improvement in the human condition as an inevitable outcome of human ingenuity. But the world we have created may not be the ideal world we intended to create. Nature continues to challenge us through the very by-products of our own ingenuity. We have assumed that the connection through adaptation, ingenuity, and success was axiomatic, but it is in fact more tenuous: human ingenuity may not inevitably produce an improved quality of life.

On the contrary, human ingenuity has had serious unintended consequences, which can no longer be ignored. The core of the problem is biotic impoverishment, the systematic reduction in Earth's ability to support living systems (Woodwell, 1990; Karr, 1995a). Concern over the implications of this trend for the quality of human and nonhuman life is now widespread, and the concepts of ecological health or ecological integrity are, in some circles at least, being invoked as guiding principles for policymaking. The mutifaceted concept of integrity requires the integration of disciplines from science to philosophy and adds "a totally new note in the discourse of environmental concern" (Westra, 1994).

2. Unintended Consequences

A major consequence of human "progress" is the homogenization of global society; human language, technology, and culture are becoming more homogeneous as we become more independent of the idiosyncrasies of local natural systems. The rich diversity of human cultures is disappearing even more rapidly than the natural systems that nurtured that diversity. English is becoming a global language, and linguists are predicting that at least half of the world's languages will go extinct in the next century (Miller, 1995). These disappearing languages "are beyond endangerment," says Michael Krauss; "[t]hey are the living dead" (Haney, 1995). Knowledge of indigenous medicines and other indigenous cultural adaptations, too, are fast being lost while cellular telephones, gasoline-powered engines, and computers spread to every corner of the globe.

Another consequence of human "progress" is the impoverishment of Earth's life-support systems (Woodwell, 1990; Karr, 1993, 1995a). Biological diversity declines as natural systems are degraded and destroyed, and ubiquitous pests, diseases, and weeds homogenize the biological surroundings of humans and their industrial society. The loss of diversity reveals humans' flawed planetary stewardship, but, more important, it represents a loss of the unique life-support systems, including human culture itself, that the human species needs for survival. Simply put, the integrity of the entire biosphere is threatened.

The combined homogenization of human culture and global biotic impoverishment means that we are losing the adaptive complexes that once tied each human culture to the geographic region where it evolved. Our ingenuity and hubris let us forget the importance of these connections. It chained us instead to clever ways of extracting resources from environments that are too often depleted by our actions. And so the legacy we inherit and the one we pass on continue generations of toxic effluents, destroyed and fragmented landscapes, depleted forests and fisheries, and collapsing cultures throughout the world. The failure to maintain human bonds with place, biology, and culture—our connections to living systems—is likely the single most important challenge that future human generations will face.

3. We Must Learn from History

We first saw planet Earth from space nearly thirty years ago and suddenly realized just how isolated we were and how dependent on a small, yet unique, piece of space debris. Until then, we had only seen ourselves up close. From the distance of space, we see ourselves and our planet exposed in an unexpected fragility and vulnerability.

We would do well to remember that, like Earth alone in space, Easter Island is an isolated place, separated from other land by more than 1800 kilometers of ocean. When first settled in about 400 A.D. by some two dozen Polynesian explorers, Easter Island was densely forested, with ample natural resources (Ponting, 1991; Catton, 1993). By the seventeenth century, the population of Easter Island had burgeoned to about 10,000 people (Catton, 1993), but less than a century later, the island's human society had collapsed. When Western Europeans arrived in 1722, they found a treeless island and a small population living in primitive conditions. These Easter Islanders had no cultural memory of the society—their own ancestors—that only a few generations before had carved and placed the island's famous massive stone monoliths.

We should see Easter Island as a metaphor for human society on the globe today. In contrast to the optimists' view of an inevitable and continuous advance to human society, many others believe that humans have overshot Earth's carrying capacity—as a thriving Easter Island society overshot that of the island (Catton, 1993)—and that nothing short of substantial change in human behavior will reverse this trend. Such concerns are not merely extremist hand wringing. Although notable progress has been made in a few areas of environmental protection (e.g., agreement to limit release of chloroflourocarbons; Litfin, 1994), current public policy and legislative initiatives will not protect either natural or human environments.

The complex reasons for this inadequacy lie in the unrelenting hubris of a society that behaves as if it could repeal the laws of nature. Plans generated by economists, technologists, engineers, and ecologists have too often assumed that lost or damaged components of ecological systems are inconsequential or can be repaired or replaced. Yet we see the consequences of this attitude everywhere: In the Pacific Northwest, hatcheries are expected to sustain salmon stocks while little is done to restore degraded habitat, curtail harvests, or protect seasonal river flow (Karr, 1995d). Throughout the world, expensive fertilizers are expected to replace nutrients in depleted soils. Groundwater is depleted to supply unsustainable amounts of water to thirsty crops, livestock, and people. These consequences and many others point to the folly of maintaining the status quo.

Multidisciplinary initiatives seeking to improve environmental policy are cropping up in many contexts, driven by goals such as environmental justice (Bullard, 1994), protection of biodiversity (Wilson, 1992), and pollution control (Colborn and Clement, 1992). They are grounded in concepts such as ecological economics (Costanza, 1991; Jansson et al., 1994), conservation biology (Meffe and Carroll, 1994), and industrial ecology (Allenby and Richards, 1994; Richards and Fullerton, 1994).

Fruitful use of these concepts requires the human species to recognize its fundamental dependence on living systems and to develop a core societal vision capable of integration—a vision that should be similar to the Socratic vision of medicine to protect the health of individual humans. This larger goal has been variously expressed as the protection of biological integrity (Karr, 1991), ecological integrity (Karr and Dudley, 1981; Karr, 1993; Nash, 1991; Westra, 1994), or ecological health (Costanza et al., 1992; Rapport et al., 1995). Although the terminology varies, all of these visions focus on the reality that healthy biological systems are critical to the success, and survival, of the human species.

4. Growing Concern

Concern about the integrity of life-support systems has evolved over nearly two centuries. For most of the twentieth century, the most visible demonstration of "environmental awareness" was the conservation movement in the developed world. But voices now coming from all corners of society draw attention to the severity of present ecological crises. A Health of the Planet Survey by the Gallup Organization (Dunlap et al., 1993) asked more than 28,000 individuals in 24 countries (developing as well as industrialized nations) about their environmental attitudes. The results showed "strong public concern for environmental protection throughout the world, including regions where it was assumed to be absent."

Scholars too are calling for shifts in human behavior. A worldwide collection of 1575 scientists, including 99 Nobel Prize winners, noted that "human beings and the natural world are on a collision course.... A great change in our stewardship of the Earth, and life on it, is required if vast human misery is to be avoided and our global home on this planet is not to be irretrievably mutilated" (Union of Concerned Scientists, 1992). In the same year, the National Academy of Sciences and the Royal Society of London (1992) issued a joint statement recognizing the need for industrial countries to modify their behavior radically to avoid irreversible damage to the Earth's capacity to sustain life. A 1993 Population Summit held in New Delhi explored issues of population growth, resource consumption, socioeconomic development, and environmental protection; 58 of the world's national academies of sciences (Science Summit, 1993) called for action to turn 1994 into "the year when the people of the world decided to act together for the benefit of future generations." The Ecological Society of America (Lubchenco et al., 1991) and the International Association of Ecology (Huntley et al., 1991) called for research initiatives to move society toward sustainable use of ecological resources; so have Sigma Xi (1992) and the Carnegie Commission (1992a,b,c; 1993).

Universities and governments have also joined the chorus. In the 1990 Talloires Declaration (Cortese, 1993), the leaders of hundreds of universities from throughout the world expressed their deep concern "about the unprecedented scale and speed of environmental pollution and degradation, and the depletion of natural resources." The 1992 Earth Summit in Rio de Janeiro was the largest meeting ever of world leaders, an unprecedented gathering of representatives from 170 nations. World leaders and grassroots organizations explored the international dimensions of environmental issues to define steps necessary to run our economies and secure our future (Keating, 1993).

Business and labor also recognize the need for change. Forty-eight international industrialists and business leaders from more than 25 countries recently called for renewed efforts by business and government to make ecological imperatives part of the market forces governing production, investment, and trade (Schmidheiny, 1992). In 1990, the *Sunday Times* of London reported, "Sir James Goldsmith—corporate predator extraordinaire, scourge of board rooms, one of the most feared men on Wall Street—[is] retiring from business. From now on, he said, he would devote his energies and much of his fortune of more than $1 billion to ecological and environmental causes." Great wealth, the 57-year-old billionaire argued, was of no value in a crumbling world (Fallon, 1990).

The United Steelworkers of America (1990) overwhelmingly endorsed a report that says, "We cannot protect steelworker jobs by ignoring environmental problems." Further, the "greatest threat to our children's future may lie in the destruction of their environment," and "the environment outside the workplace is only an extension of the environment inside." At the August 1993 Parliament of World's Religions (Briggs, 1993), the leaders of Christianity, Buddhism, Islam, Judaism, Hinduism, and other faiths developed a "global ethic." Among other things, that ethic condemns environmental abuses. In an age of unparalleled technological progress, poverty, hunger, the death of children, "and the destruction of nature have not diminished but rather have increased."

A recent report by the Commission on Life Sciences of the National Research Council (1993) concludes that society possesses many of the "tools to address environmental problems of enormous consequence to our social and economic well-being. But we are not using those tools most effectively." In fact, while societal concern is growing, knowledge is not. A survey by Chicago's National Opinion Research Center shows that U.S. citizens "only have a middling knowledge of the environment and what affects it" (Holden, 1995). Citizens recognize extinction as a problem but attribute it to cosmic events, not human influence.

5. Biotic Impoverishment Goes beyond Extinction

These organizations and the constituencies they represent recognize that all is not well on planet Earth, but they do not explicitly define the main problem: biotic impoverishment (Woodwell, 1990). Biotic impoverishment is visible today in three major forms: indirect depletion of living systems through degradation of the chemical and physical environment; direct depletion of nonhuman living systems; and direct depletion of human systems (Table 1; Karr, 1995c).

Table 1. Types of Biotic Impoverishment, with Examples.

A. Indirect Depletion of Living Systems

1. Soil depletion and degradation (erosion, degradation of soil structure, salinization, desertification, destruction of soil biota)
2. Degradation of water (pollution, surface water and groundwater depletion, extinction, homogenization of aquatic biota)
3. Alteration of global biogeochemical cycles
4. Chemical contamination (bioaccumulation, cancer, immunological deficiencies, developmental anomalies, intergenerational effects)
5. Global climate change and ozone depletion (global warming, alteration of rainfall distribution and amount, ozone depletion)

B. Direct Depletion of Nonhuman Living Systems

1. Renewable-resource depletion (overfishing, excessive timber harvest)
2. Extinction of species
3. Habitat destruction and fragmentation (homogenization of biota, destruction of landscape mosaics and connectivity)
4. Red tides, pest outbreaks, and homogenization of crops
5. Introduction of exotics

C. Direct Depletion of Human Systems

1. Epidemics and emerging and reemerging diseases
2. Reduced human cultural diversity (genocide, loss of knowledge, loss of languages)
3. Reduced quality of life; economic deprivation (malnourishment, failure to thrive)
4. Environmental injustice (environmental racism, economic exploitation, intragenerational inequity, intergenerational inequity)

5.1 INDIRECT DEPLETION OF LIVING SYSTEMS

The primary physical systems that humans depend on are air, soil, and water. The productive potential of soils is degraded by erosion, salinization, desertification, and compaction. But soil is much more than its physical constituents: depletion of the organic activity in soil is also serious. Degradation of water resources—including chemical pollution, surface and groundwater depletion, and flooding—is pervasive. Myers (1993) rightly notes, "Our future will be deeply compromised unless we learn to manage water as a critical ingredient of our lives."

Chemical contamination of air, soil, and water has for many years been the primary focus of government and the public; the primary concern has been the threat to human health from a diversity of chemical pollutants. Special attention has been directed to the narrow problem of contaminants that induce cancers in animals and humans. Other contaminant red flags include bioaccumulation, immunological and developmental deficiencies, and a growing number of reproductive and intergenerational effects.

Historically, the consequences of human activity were limited in space and time, but the increase in toxic chemicals and radioactive materials during the twentieth century has created global problems and left legacies that will be present for thousands of years. Some of these affect people directly; many will have long-term indirect impacts on biological systems by altering biogeochemical cycles, global climate, and ozone concentrations.

5.2. DIRECT DEPLETION OF NONHUMAN LIVING SYSTEMS

Humans directly deplete renewable natural resources by harvesting fish, timber, and other products. Habitat destruction and fragmentation associated with harvest, urbanization, and other activities have perhaps the farthest-reaching effects on biological systems. Yet relatively little attention has been paid to habitat loss except when it threatens species with extinction. Human activities may even be responsible for increased frequency of red tides in coastal environments and insect and disease outbreaks in forests. Especially devastating to regional living systems is the homogenization of plant and animal communities through extinction and the spread of nonnative species, particularly commensals of human society.

5.3. DIRECT DEPLETION OF HUMAN SYSTEMS

The advances of modern medicine over the past three decades have lulled us into a false sense of security about human health and the environment. To be sure, antibiotics to control pathogens, and pesticides to control pests, have helped check many diseases. Yet, with a few exceptions like smallpox, diseases have not been conquered. Virulent forms of *E. coli*, tuberculosis, influenza, yellow fever, and malaria are becoming more difficult to control. In addition, emerging "new" diseases caused by bacteria (legionnaires' and lyme diseases), viruses (ebola, hantavirus, HIV/AIDS), and parasites (*Cryptosporidium*) are cropping up. Human population growth and behavior, global travel patterns, resistance to antibiotics, reductions in natural immunity in human populations stressed by other environmental degradation (e.g., global warming), and destruction of natural habitats all contribute to this trend. The impoverishment of human systems is also manifest as reduced cultural diversity (genocide and loss of knowledge), reduced quality of life and economic deprivation (failure to thrive in infants, malnourishment in 20 percent of people), and environmental injustice (racism, economic exploitation, lack of intra- and intergenerational equity).

Collectively, this broad sweep of issues illustrates the magnitude of the environmental challenge facing all members of the human community. It also reminds us of the close association and common underpinning of environmental and social concerns, and it provides an opportunity to exercise the ingenuity that has brought us this far. The loss of species; the destruction of agricultural lands; and the differential exposure to environmental hazards of economically disadvantaged people, often people of color, degrade the quality of human life. As human influence expands, the limits of technology, especially unintended consequences of technology, become more obvious. Depletion of water supplies cannot be "fixed" by engineers making water to refill aquifers; lost salmon spawning grounds cannot be "fixed" by adding gravel or "large woody debris."

Citizens and political leaders, humanists and scientists must work together to develop creative solutions. Failure to do so will relegate the world to continued biotic impoverishment and threaten the sustainability of human society. Ecologists are especially critical in these partnerships. In the same sense that medical doctors must be trained to recognize and understand the attributes of a healthy human, ecologists and environmental scientists must understand the attributes of healthy biological systems—systems that must be sustained over the long term, in the service of humans and for their own sakes.

6. Ecological Integrity and Ecological Health

If biotic impoverishment is the problem, then protecting the integrity of biological systems must be the goal. But how do we define *biological integrity* in a world that is increasingly altered by the actions of humans? How do we reconcile the inevitable changes required to accommodate a growing human population and the proliferation of modern technology while guarding the planet from irrevocable biotic impoverishment? Answering these questions in clear and explicit terms is especially important as we seek to bring scholars from diverse disciplines together to focus on common problems.

What do *health* and *integrity* mean? How do we integrate concepts of integrity in their philosophical, legal, biological, cultural, and ethical senses (Westra, 1994)? What kind of health or integrity do we seek? Are we seeking "environmental health," or is that phrase too narrowly associated with the effects of toxic substances on human health? As a societal goal, biological integrity suggests a meaning beyond human health. And the sum of physical, chemical, and biological integrity is ecological integrity (Karr and Dudley, 1981).

Aldo Leopold (1949) was the first to invoke the concept of integrity in an ecological sense: "A thing is right when it tends to preserve the integrity, stability, and beauty of the biotic community. It is wrong when it tends otherwise." Twenty-three years ago, as the United States Congress drafted the Water Pollution Control Act Amendments of 1972, it sought a broad statement reflecting a vision absent from earlier water resource legislation. "Can we afford clean water? Can we afford rivers and lakes and streams and oceans which continue to make possible life on this planet?" asked the late Senator Edmund Muskie (Congressional Research Service, 1972). "These questions answer themselves." Congress explicitly included "integrity"—"to restore and maintain the physical, chemical, and biological integrity of the nation's waters"—as the underlying goal of its legislation. Two major aims are clearly incorporated in this congressional language (Adler et al., 1993): active protection of remaining high-quality aquatic systems and a return of the nation's waters to a state of health. Other uses of the integrity concept include the binational Great Lakes Water Quality Agreement of 1978 and the amendment to Canada's National Park Act passed by Parliament in 1988, which makes" [m]aintenance of ecological integrity" its first priority.

Integrity implies an unimpaired condition or the quality or state of being complete or undivided; it implies correspondence with some original condition. The term most appropriately refers to the condition at sites with little or no influence from human actions; the organisms living there are products of the evolutionary and biogeographic processes influencing that site. Biological integrity (Frey, 1975; Karr and Dudley, 1981; Angermeier and Karr, 1994) refers to the capacity to support and maintain a balanced, integrated, adaptive biological system having the full range of elements (genes, species, assemblages) and processes (mutation, demography, biotic interactions, nutrient and energy dynamics, and metapopulation processes) expected in the natural habitat of a region. Although somewhat long-winded, this definition carries the message that: (1)living systems act over a variety of scales from individuals to landscapes, (2)a fully functioning living system includes items one can count (the elements of biodiversity) plus the processes that generate and maintain them, and (3)living systems are embedded in dynamic evolutionary and biogeographic contexts that influence and are influenced by their physical and chemical environments.

An evolutionary foundation ties the concept of integrity to a benchmark against which society can evaluate sites altered by human actions. The complex biological systems that evolved at a site have already proved their ability to persist in, and even modify, the region's physical and chemical environment. Their very presence means that they are resilient to normal variation in that environment. Species abundance, for example, changes as a function of changing physical environment and changing interactions among species in a local assemblage. But the bounds over which systems change as a result of most natural events are limited when compared with the changes imposed by human activities like row-crop agriculture, urbanization, or dam building.

Human society sets aside extensive areas in parks and reserves to protect their natural state, to protect their integrity. These areas deserve protection because of the diverse values they provide to society. Water bodies, both on the surface and underground, deserve special protection as well, because they provide water to drink and support recreational and other values. Most important, rivers are the lifelines of a continent, reflecting the condition of surrounding landscapes, linking landscapes across great distances.

Because of the demands of feeding, clothing, and housing more than 5.7 billion people, few places on Earth maintain a biota with evolutionary and biogeographic integrity. The growth of human populations in the last few centuries has made our species the principal driver of global change. Humans appropriate the equivalent of 40 percent of Earth's annual terrestrial production (Vitousek et al., 1986). Providing for human needs has required massive alteration of the planet in ways that preclude a return to the pristine environments of the preindustrial era. Thus, biological integrity is lost on a large share of the planet and is unlikely to be regained. Yet loss of ecological integrity for all lands and waters in all regions of the world is unacceptable on scientific, economic, aesthetic, and ethical grounds.

Health implies a flourishing condition, well-being, vitality, or prosperity. An organism is healthy when it performs all its vital functions normally and properly; a healthy organism is resilient, able to recover from many stresses; a healthy organism requires minimal outside care. The concept of health applies to individual organisms as well as to national or regional economies, industries, and natural resources such as fisheries.

Ecological health describes the preferred state of sites modified by human activity— areas cultivated for crops, managed for tree harvest, stocked for fish, urbanized, or otherwise intensively used. At these sites, integrity in an evolutionary sense cannot be the goal. Healthy land use, with or without active management, should not degrade a site for future human use or degrade areas beyond that site. Soils, for example, should not be eroded or otherwise transformed in ways that reduce future productivity (Pimentel, 1993). Groundwater should not be depleted. Land use should not have deleterious effects beyond a site; atmospheric contamination should not result in downwind effects, such as tree death or ozone depletion. Healthy sites should not release contaminants or eroded soils that degrade sites elsewhere.

According to these two criteria—no degradation of a site for future use and no degradation of areas beyond that site—most modern agricultural and urban land use, for example, are not healthy. Recent initiatives for sustainable, or healthy, communities and agriculture recognize this reality. Like losing ecological integrity everywhere, failing to protect the ecological health of intensively used lands is also unacceptable on scientific, economic, aesthetic, and ethical grounds. We must choose to develop a conceptual framework to define "acceptable" and "sustainable" uses.

These concepts are easily applied to small parcels of land; scaling up to large landscapes is another matter. What proportion of a landscape should be protected under a biological integrity goal? The World Commission on Environment and Development (1987) recommended 12 percent, but that percentage seems inadequate. Why 12 percent? Should that proportion vary with regional ecosystem type (desert, forest, grassland)? Which sites deserve the highest priority for protection? Water bodies, fragile sites such as steep slopes, or sites with the highest biodiversity? How shall we decide?

7. Measuring Health and Integrity

Ecologists have not been especially adept at defining or measuring either ecological health or ecological integrity. The track record of freshwater management provides an instructive example. Human society depends on fresh water and on the resources associated with freshwater and marine systems. Yet the United Nations estimates that 2 billion people will lack access to safe drinking water, and more than 2 billion will lack access to sanitation services by the year 2000 (Gleick, 1993). Improved water conservation, and treatment and recycling programs, can delay crisis, but human ingenuity will remain on a treadmill trying to keep up with expanding demand created by an expanding population. Even where supplies of fresh water remain adequate, resource degradation continues because society chronically undervalues the products and services, besides water, provided by aquatic ecosystems.

Although rivers are in many ways the lifeblood of society, prevalent attitudes toward rivers reflect disdain for their value and arrogance about our ability to replace or repair them. Despite the mandate in the Clean Water Act for protecting the integrity of the nation's waters, for example, it took nearly two decades to begin to incorporate that concept into water resource protection, largely because appropriate benchmarks were not defined for evaluating success in attaining these goals. That failure led to six realities about the condition of water resources (Karr, 1995b): (1) water resources, especially their biological components, are in steep decline; (2) degradation stems from more than chemical contamination, the primary focus of conventional water quality programs; (3) long-term success in protecting water resources requires careful thought about goals and benchmarks, including development and uses of criteria for protecting ecological integrity; (4) the legal and regulatory framework in place today does not respond soon enough to continued degradation; (5) the quantitative expectations that constitute biological integrity vary geographically; and (6) because ecological and biological systems are complex, multiple components of these systems should be protected.

These realities are a consequence of using the wrong indicators. In water resources, for example, the major indicators have been pollution permits issued, fines levied, and counts of chemical contaminants released into waterways. Managers made little or no effort to evaluate the biological integrity of water resources (Karr, 1991). The use and development of biocriteria are altering approaches to water resource planning and decision making throughout the world (Davis and Simon, 1995; Harris, 1995). Pioneering efforts are changing the way we measure environmental trends as well. The index of sustainable economic welfare (ISEW), for example, adjusts gross national product (GNP) to account for pollution effects, environmental services, depletion of nonrenewable natural resources, and long-term environmental damage (Daly and Cobb, 1989).

Others too are seeking imaginative ways to express national (Hammond et al., 1995; Alperovitz et al., 1995), regional (Washington State Department of Ecology, 1995), and local (Sustainable Seattle, 1993) environmental trends. The index of environmental trends (Alperovitz et al., 1995)—a comprehensive assessment of twenty-one key environmental indicators in nine advanced industrialized countries over the past two decades—is an effort to provide a composite picture of environmental quality. The image is clear but not very appealing. For all nine countries studied, the composite long-term index of ecological change is negative (Table 2). The quality of the environment has undeniably declined over the past twenty years of "environmental awareness."

Clearly, we need to stop paying heed to the wrong indicators or to stop ignoring the indicators we have. Complex systems are regulated by negative feedbacks, the way a thermostat regulates a furnace. Society pays attention to economic feedback signals, as it does when it uses monetary policy to reduce interest rates or control inflation; to a lesser extent, it also pays attention to social signals. But under the assumption that we can repair ecological damage or replace ecological systems, we fail to respond to signals that should change our behavior toward those systems. Failure to take biotic impoverishment as a warning to change may thus destroy what we depend on. We have pushed ecological as well as social support systems beyond their self-organizing capacity, and biological and social dysfunction is the result.

8. Reclaiming Lost Connections

Traditionally, human actions have not been judged in terms of their influence on ecological health or integrity. When human populations were small, resources were abundant, technological skills were less advanced, and environmental degradation was less extreme. As a result, society did not notice or understand ecological integrity or health. Degradation was local and usually short-lived; an undeveloped frontier was always available. But the frontier is gone now—as it was for Easter Islanders. Supplies of many renewable resources have been depleted; chemical pollution is pervasive; and the global atmosphere is changing under human onslaught.

Table 2. The Index of Environmental Trends for Nine Industrialized Countries. This composite index incorporates ratings of air, land, and water quality; chemical and waste generation; and energy use since 1970. Source: Alperovitz et al. (1995).

Country	Index of Environmental Change
Denmark	-10.6%
Netherlands	-11.4%
United Kingdom	-14.3%
Sweden	-15.5%
(West) Germany	-16.5%
Japan	-19.4%
United States	-22.1%
Canada	-38.1%
France	-41.2%

If we are to stem biotic impoverishment and reverse environmental degradation, we must: (1)set societal goals based on broad concepts of ecological integrity and ecological health; (2)forge partnerships among scientists, engineers, policymakers, resource managers, and citizens to develop approaches for attaining those goals; (3)revise the legal framework guiding environmental policy to ensure that both ecological risks and threats to human health are minimized; (4)protect existing resources; (5)restore resources that are degraded; and (6)reduce resource consumption.

In essence, we must rediscover and reclaim the connections to living systems that nineteenth-century science and twentieth-century technology let us pretend we could forget. We must go back in our thinking to the legacy of Baron Alexander von Humboldt, late eighteenth-century explorer, geographer, writer, naturalist, "and the man most responsible for bringing the practice of science into mainstream Western culture" (Sachs, 1995). In Humboldt's day, "science" meant "natural science," and it took interdependent relationships in nature, including the place for humans, for granted. Aware, from his explorations, of the diversity in nature and among humans, Humboldt nevertheless believed that "a knowledge of that interdependence was the 'noblest and most important result' of all scientific inquiry" (Sachs, 1995). He clearly understood the links between natural resource use, human welfare, and social justice.

For most of the past century, human energies have been directed toward economic growth, to the production and distribution of goods and services, to the production of wealth. Our consumption-centered political economy reflects those goals. Just as advancing technology was viewed as progress, economic growth and the generation of wealth have become not only products but a sign of our species' ability to exercise its self-bestowed birthright to monopolize an expanding fraction of Earth's resources. Political economy reduces everything to social, especially economic, constructions, "blatantly disregarding all that is not human" (Greenberg and Park, 1994) and ignoring the dependence of humans and their economies on the larger natural economy. It treats economic systems as self-contained, disconnected from their surroundings, rather than as subsystems embedded within a finite ecological system (Daly, 1991).

To protect that ecological system—the ultimate source of all human wealth—a "political ecology" needs to take shape as a guiding principle for public policy. Political ecology would incorporate ecological concepts into human culture and politics, and consider culture and politics within ecological realities. Political ecology would recognize the folly of behaving as if price mattered and cost did not. We pay a price at the checkout counter, but we often pay a higher cost when we ignore things of value that cannot be measured in numbers, when we ignore the value of things that can be measured, or when we lose something that we did not know was important until it was gone (Orr, 1994). Recognizing the dependence of society on natural resources—explicitly adding political ecology to our thinking—will require us to ask not only if we *can*, but if we *should*. Remembering the connections will at least make us consider more carefully whether price reflects both economic and ecological costs.

We cannot avoid the use of technology, but we can no longer adopt technology without careful evaluation of its ecological effects. We can no longer adopt a technology because of its short-term advantages; we must first evaluate its long-term consequences. We must ensure that protecting ecological health and integrity plays a central role in decisions about consumer goods and development of technologies, including when, where, and how to apply them.

For most of the past century, politics, economics, and engineering have been the driving forces in societal decision making. If the price seemed right to build a dam, for example, and the technology was there to do it, society built the dam in the interest of "progress" even though some questioned the wisdom of that action. In effect, society based its decisions on narrow aims of those who forgot the fundamental lessons Humboldt understood 200 years ago. They did not understand that, like frosting on a layer cake, human economies take their shape from society and the Earth's natural resource base. Society—its formal and informal rules, laws and regulations, and ethics and morality—forms the top layer of the cake; the natural resource base forms the bottom layer. Without that natural resource base—without other living organisms—neither human society nor human economy will flourish.

9. Acknowledgments

Portions of this paper were first presented at a National Academy of Engineering Workshop titled "Engineering within Ecological Constraints" held in Washington, DC, April 19-21, 1995, and published by the National Academy Press (Karr, 1995c).

10. References

Adler, R. W., J. C. Landman, and D. M. Cameron. 1993. *The Clean Water Act 20 Years Later.* Island Press, Washington, DC.
Allenby, B. R., and D. J. Richards, eds. 1994. *The Greening of Industrial Ecosystems.* National Academy Press, Washington, DC.
Alperovitz, G., T. Howard, A. Scharf, and T. Williamson. 1995. Index of Environmental Trends: An Assessment of Twenty-one Key Environmental Indicators in Nine Industrialized Countries over the Past Two Decades. National Center for Economic Alternatives, Washington, DC.
Angermeier, P. L., and J. R. Karr. 1994. Biological Integrity versus Biological Diversity as Policy Directives: Protecting Biotic Resources. *BioScience.* 44: 690-697.
Briggs, D. 1993. World's Clerics Draft Global Ethic: Violence, Sexism, Environmental Abuse Are All Targeted. *Seattle Times*, September 1.
Bullard, R. D., ed. 1994. *Unequal Protection: Environmental Justice and Communities of Color.* Sierra Club Books, San Francisco, CA.
Carnegie Commission. 1992a. International Environmental Research and Assessment: Proposals for Better Organization and Decision Making. Carnegie Commission on Science, Technology, and Government, New York, NY.
Carnegie Commission. 1992b. Environmental Research and Development: Strengthening the Federal Infrastructure. Carnegie Commission on Science, Technology, and Government, New York, NY.
Carnegie Commission. 1992c. Partnerships for Global Development: The Clearing Horizon. Carnegie Commission on Science, Technology, and Government, New York, NY.
Carnegie Commission. 1993. Science, Technology, and Government for a Changing World. Carnegie Commission on Science, Technology, and Government, New York, NY.
Catton, W. R., Jr. 1993. Carrying Capacities and the Death of a Culture: A Tale of Two Autopsies. *Sociological Inquiry* 63: 202-223.
Colborn, T., and C. Clement, eds. 1992. Chemically Induced Alterations in Sexual and Functional Development: The Wildlife/Human Connection. *Advances in Modern Environmental Toxicology* 21: 1-403.
Congressional Research Service. 1972. *History of the Water Pollution Control Act Amendments of 1972*, ser. 1, 93rd Cong., 1st sess.
Cortese, A. D. 1993. Building the Intellectual Capacity for a Sustainable Future: Talloires and Beyond. In *Environmental Literacy and Beyond,* B. Wallace, J. Cairns, Jr., and P. A. Distler, eds. President's Symposium, vol. V. Virginia Polytechnic Institute and State University, Blacksburg, VA, pp. 1-9.

Costanza, R., ed. 1991. *Ecological Economics: The Science and Management of Sustainability*. Columbia University Press, New York, NY.

Costanza, R., B. G. Norton, and B. D. Haskell, eds. 1992. *Ecosystem Health: New Goals for Environmental Management*. Island Press, Washington, DC.

Daly, H. E. 1991. Elements of Environmental Macroeconomics. In *Ecological Economics: The Science and Management of Sustainability*, R. Costanza, ed. Columbia University Press, New York, NY, pp. 32-46.

Daly, H. E., and J. B. Cobb, Jr. 1989. *For the Common Good: Redirecting the Economy toward Community, the Environment, and a Sustainable Future*. Beacon Press, Boston, MA.

Davis, W. S., and T. P. Simon, eds. 1995. *Biological Assessment and Criteria: Tools for Water Resource Planning and Decision Making*. Lewis Publishers, Boca Raton, FL.

Dunlap, R. E., G. H. Gallup, Jr., and A. M. Gallup. 1993. Of Global Concern: Results of the Health of the Planet Survey. *Environment* 35(9): 7-15, 33-39.

Fallon, I. 1990. The Jolly Green Giant. *The Sunday Times (London)*, October 21.

Frey, D. 1975. Biological Integrity of Water: An Historical Perspective. In R. K. Ballentine and L. J. Guarraia, eds. *The Integrity of Water*. Environmental Protection Agency, Washington, DC, pp. 127-139.

Gleick, P. H. 1993. An Introduction to Global Fresh Water Issues. In *Water in Crisis: A Guide to the World's Fresh Water Resources*, P. H. Gleick, ed. Oxford University Press, New York, NY, pp. 3-12.

Greenburg, J. B., and T. R. Park. 1994. Political Ecology. *Journal of Political Ecology* 1: 1-12.

Hammond, A., A. Adriaanse, E. Rodenburg, D. Bryant, and R. Woodward. 1995. Environmental Indicators: A Systematic Approach to Measuring and Reporting on Environmental Policy Performance in the Context of Sustainable Development. World Resources Institute, Washington, DC.

Haney, D. Q. 1995. Experts Say World May Lose Half Its Languages. *Seattle Times*, February 19.

Harris, J. H. 1995. The Use of Fish in Ecological Assessments. *Australian Journal of Ecology* 20: 65-80.

Holden, C. 1995. Environmental Knowledge Gap. *Science* 268: 647.

Huntley, B. J., and eighteen coauthors. 1991. A Sustainable Biosphere: The Global Imperative. *Ecology International* 1991: 20.

Jansson, A., M. Hammer, C. Folke, and R. Costanza, eds. 1994. *Investing in Natural Capital: The Ecological Economics Approach to Sustainability*. Island Press, Washington, DC.

Karr, J. R. 1991. Biological Integrity: A Long Neglected Aspect of Water Resource Management. *Ecological Applications* 1: 66-84.

Karr, J. R. 1993. Protecting Ecological Integrity: An Urgent Societal Goal. *Yale Journal of International Law* 18: 297-306.

Karr, J. R. 1995a. Using Biological Criteria to Protect Ecological Health. In *Evaluating and Monitoring Health of Large-scale Ecosystems*, D. J. Rapport, C. Gaudet, and P. Calow, eds. Springer-Verlag, New York, NY, pp. 137-152.

Karr, J. R. 1995b. Protecting the Integrity of Aquatic Ecosystems: Clean Water is Not Enough. In *Biological Assessment and Criteria: Tools for Water Resource Planning and Decision Making*, W. S. Davis and T. P. Simon, eds. Lewis Publishers, Boca Raton, FL, pp. 7-13.

Karr, J. R. 1995c. Ecological Integrity and Ecological Health Are Not the Same. In *Engineering Within Ecological Constraints*, P. Schulze, ed. National Academy Press, Washington, DC, in press.

Karr, J. R. 1995d. Restoring Wild Salmon: We Must Do Better. *Illahee* 10: 316-319.

Karr, J. R., and D. R. Dudley. 1981. Ecological Perspective on Water Quality Goals. *Environmental Management* 5: 55-68.

Keating, M. 1993. The Earth Summit's Agenda for Change: A Plain-Language Version of Agenda 21 and the Other Rio Agreements. Centre for Our Common Future, Geneva, Switzerland.

Leopold, A. 1949. *A Sand County Almanac*. Oxford University Press, New York.

Litfin, K. T. 1994. *Ozone Discourses: Science and Politics in Global Environmental Cooperation*. Columbia University Press, New York, NY.

Lubchenco, J., and fifteen coauthors. 1991. The Sustainable Biosphere Initiative: An Ecological Research Agenda. *Ecology* 72: 371-412.

Meffe, G. K., and C. R. Carroll. 1994. *Principles of Conservation Biology*. Sinauer Associates, Sunderland, MA.

Myers, N. 1993. *Ultimate Security: The Environmental Basis of Political Stability*. W. W. Norton, New York, NY.

Miller, J. A. 1995. Save the Languages, Both Natural and Invented. *BioScience* 45: 386-387.

Nash, J. A. 1991. *Loving Nature: Ecological Integrity and Christian Responsibility*. Abingdon Press, Nashville, TN.

National Academy of Sciences (U.S.), and Royal Society of London. 1992. Population Growth, Resource Consumption, and a Sustainable World. Joint Statement by the Officers of the Royal Society of London and the U.S. National Academy of Sciences, Washington, DC.

National Research Council. 1993. *Research to Protect, Restore, and Manage the Environment*. National Academy Press, Washington, DC.

Nehlsen, W., J. E. Williams, and J. A. Lichatowich. 1991. Pacific Salmon at the Crossroads: Stocks at Risk from California, Oregon, Idaho, and Washington. *Fisheries* (Bethesda) 16(2): 4-21.

Orr, D. W. 1994. *Earth in Mind: On Education, Environment, and the Human Prospect*. Island Press, Washington, DC.

Pimentel, D., ed. 1993. *World Soil Erosion and Conservation*. Cambridge University Press, New York, NY.

Ponting, C. 1991. *A Green History of the World: The Environment and the Collapse of Great Civilizations*. St. Martin's Press, New York, NY.

Rapport, D. J., C. Gaudet, and P. Calow, eds. 1995. *Evaluating and Monitoring the Health of Large-Scale Ecosystems*. Springer-Verlag, New York, NY.

Richards, D. J., and A. B. Fullerton, eds. 1994. *Industrial Ecology: U.S.-Japan Perspectives*. National Academy Press, Washington, DC.

Sachs, A. 1995. Humboldt's Legacy and the Restoration of Science. *World Watch* 8(2): 28-38.

Schmidheiny, S. 1992. *Changing Course: A Global Business Perspective on Development and the Environment*. MIT Press, Cambridge, MA.

Science Summit. 1993. Population Summit of the World's Scientific Academies. A Joint Statement by Fifty-eight of the World's Scientific Academies. Office of International Affairs, National Research Council, Washington, DC.

Sigma Xi. 1992. Global Change and the Human Prospect: Issues in Population, Science, Technology, and Equity. Forum Proceedings, November 16-18, 1991. Sigma Xi, Research Triangle Park, NC.

Sustainable Seattle. 1993. The Sustainable Seattle 1993 Indicators of Sustainable Community. Sustainable Seattle, Seattle, WA.

Union of Concerned Scientists. 1992. World Scientists' Warning to Humanity. Union of Concerned Scientists, Cambridge, MA.

United Steelworkers of America. 1990. *Our Children's World: Steelworkers and the Environment.* Report of USWA Task Force on Environment. United Steelworkers of America, Washington, DC.

Vitousek, P. M., P. R. Ehrlich, A. H. Ehrlich, and P. Matson. 1986. Human Appropriation of the Products of Photosynthesis. *BioScience* 36: 368-373.

Washington State Department of Ecology. 1995. Washington's Environmental Health 1995: A Summary of Environmental Indicators. Olympia, WA.

Westra, L. 1994. *An Environmental Proposal for Ethics: The Principle of Integrity.* Rowman and Littlefield Publishers, Lanham, MD.

White, R. 1995. *The Organic Machine: The Remaking of the Columbia River.* Hill and Wang, New York, NY.

Wilson, E. O. 1992. *The Diversity of Life.* Harvard University Press, Cambridge, MA.

Woodwell, G. M., ed. 1990. *The Earth in Transition: Patterns and Process of Biotic Impoverishment.* Cambridge University Press, Cambridge, UK.

World Commission on Environment and Development. 1987. *Our Common Future.* Oxford University Press, Oxford, UK.

Chapter 4
EMBRACING COMPLEXITY
THE CHALLENGE OF THE ECOSYSTEM APPROACH

James J. Kay[1]
Eric Schneider[2]

1. Introduction

As environmental degradation and change continues, decision makers and managers feel significant pressure to rectify the situation. Scientists, in turn, find themselves under pressure to set out simple and clear rules for proper ecosystem management. The response has been one of frustration. Michael Soulé and Laurence Slobdokin both loudly complain that ecology is an intractable science, immature and not very helpful. Kristin Shrader-Frechette and Robert Peters reproach ecologists for not producing simple testable hypotheses.[1] Meanwhile policy makers and managers clamour for a measure of ecosystem integrity whose value in different situations can be predicted by simulation models. The question on everyone's mind is "what does ecosystem science identify as the main, simple, basic, universal laws which will allow quantitative prediction of ecosystem behaviour and what are the resulting rules for ecosystem management?"

All of these demands on ecology are predicated on a vision of science which assumes that it can provide firm knowledge, and that the only way of obtaining this knowledge is the scientific method. The standard scientific method works well with billiard balls and pendulums, and other very simple systems. However, systems theory suggests that ecosystems are inherently complex, that there may be no simple answers, and that our traditional managerial approaches, which presume a world of simple rules, are wrong-headed and likely to be dangerous.

In order for the scientific method to work, an artificial situation of consistent reproducibility must be created. This requires simplification of the situation to the point where it is controllable and predictable. But the very nature of this act removes the complexity that leads to emergence of the new phenomena which makes complex systems interesting. If we are going to deal successfully with our biosphere, we are going to have to change how we do science and management. We will have to learn that we don't manage ecosystems, we manage our interaction with them. Furthermore, the search for simple rules of ecosystem behaviour is futile.

Take for example the diversity-stability hypothesis.[2] This is a classic example of the kind of simple rule people are looking for. Students are taught that diversity in ecosystems is important because it maintains their stability. Yet, we know that to obtain an increase in

---

[1]Faculty of Environmental Sciences, University of Waterloo, Waterloo, Ontario N2L 3G1, Canada; [2]Chief scientist (retired) of the U.S. National Oceanic and Atmospheric Administration.

diversity in ecosystems, we need only stress them.[3] Daniel Goodman long ago dissected and refuted this hypothesis and yet we still see it being promoted as a guideline for management and policy.[4] Why? Because we want simple answers to complex questions.

The diversity-stability hypothesis illustrates this nicely. Examination of what is meant by diversity and stability quickly leads us into the quagmire of complexity. Is diversity to be measured by number of species? The relative abundance of species? Their richness? Which species do you include? Big ones that are easy to count? All the micro-organisms in the soil? Very quickly it turns out that there is no one correct way to measure diversity, and in the end, it is an observer-dependent phenomenon, dependent on which species the researcher decides to include.

The notion of stability is even more slippery.[5] The traditional approach developed by M. Lyapunov focuses on some numerical state function and whether that function has a constant value which the system tends towards and returns to when disturbed. But what state function should we measure? Population numbers? Biomass? Productivity?... The list is endless and the problem doesn't end there. We may choose a function to represent the ecosystem and its stability, but we are now discovering that these functions are not stable.[6] Instead ecosystems are dynamic and constantly changing. Stability gives way to the notion of a shifting steady mosaic.[7] Thus, the diversity-stability hypothesis evaporates because the basic concepts of diversity and stability are just too simple to describe the complex reality of ecological phenomena.

The same is true of the notion of "succession," the idea that an ecosystem develops through a series of dominant vegetation types and ultimately reaches a climax community. Robert McIntosh has documented the ongoing debate about succession and identifies six major schools of thought in the US alone.[8] The thinking ranges from succession as an orderly pattern of development, which is reproduced time and time again, to succession as a myth, that is there is nothing but random assemblages of species with no underlying patterns. There is nothing approaching consensus about succession in the ecology community. In fact, ecologists bemoan that there is not one single "law" of the science of ecology. Why? Because we are asking the wrong questions.

There is a group of thinkers who argue that to deal with ecology requires an "ecosystem approach," an approach based on the notions of complex systems theory, the grandchild of Ludwig von Bertalanffy's general systems theory.[9] It is a fundamentally different approach to knowing about the world, and it is, not surprisingly, complex itself. Any effort to study complex systems must look at them in the context of space, time, energy and information. We shall probe, in turn, each of these aspects of ecosystems as complex systems.

2. The Sky is Falling

Part of our trouble is that our conventional notion of science is based on understanding the temporal and sometimes spatial dynamics of systems in the context of their inertia (mass). We see the world as billiard balls from a Newtonian perspective. Ball A strikes ball B causing it to move. All activities of a system can be explained by mechanisms, in terms of the interactions of components, usually in a linear way. Component interactions are sufficient to explain all. So science focuses on establishing which components are responsible for what.[10]

At the turn of the century, several insights changed how scientists look at the physical world. In terms of space and time, it was realized that there is not a preferred observer and that the relationship between observers, at least in four dimensions, is not linear. Space and time are curved. Furthermore, energy is quantized, mass is a form of energy, and we always lose information about things.[11] The world is running down. The sky is falling.

These insights did not affect how we looked at the world on a day-to-day basis. Its direct impact has been on the development of things "nuclear" and things "solid state" (e.g.

computer chips). As for the world running down, we already knew that. So scientific inquiry continued to follow the "scientific method", attempting to explain everything through mechanistic interactions of components. The logical extremes have been the elementary particles of physics, the selfish gene of biology, and "Newtonian ecology".

However, the minute one leaves the physical sciences there is a paradox, a paradox whose resolution ultimately requires us to abandon the hypothesis that the reductionist, mechanistic, scientific method is sufficient for understanding the world. The paradox is that the second law of thermodynamics maintains that the world is running down, but the biological world is not running down. Quite the opposite is happening; life is proliferating. The sky is not falling! The same can be said of the systems studied by the social sciences.

A revolution in science has occurred in the last two decades that is as profound as the one which occurred between 1890 and 1910 with the work of Ludwig Boltzmann, Albert Einstein, Josiah Gibbs, Max Planck, et al. The revolution of the turn of the century was about how we view the microscopic world. It did not change how we look at our world, day-to-day. The current revolution is about how we look at the macro world and it will profoundly affect our day-to-day living, our institutions and our decision making, including decisions on judicial matters.

It is fitting that one cannot put this new set of insights down on paper in a nice linear way. The revolution emerges from the synergism of new insights in several fields. Since the prevalent world view is largely about mechanistic-reductionist predictions about space and time, it seems appropriate to start with the unravelling of this.

3. Space and Time

Catastrophe theory describes the change in systems over time. It predicts that systems will undergo dramatic, sudden changes in a discontinuous way. The classic example is the failure of a structural beam under loading. The choice of the name of the theory is quite unfortunate because it implies abnormal nasty events, when in fact such events are normal and necessary for the continued ordinary functioning of many systems. Your heartbeat is a catastrophic event, as is the emptying of your bladder. Both are necessary for your continuing survival. Both are discontinuous events that occur suddenly.

Furthermore, at the point where a system undergoes a catastrophic change several distinct changes are possible and actually occur—which one is not predictable. For example, dogs (in fact most animals) have a bubble of space around them which is their territory. Enter the space (the catastrophe threshold) and the dog will either retreat or attack, but it is not, a priori, possible to predict with certainty which of the two actions will occur.

The general insight from catastrophe theory is that the world does not always change in a continuous and deterministic way. There are points in any system's development where several possible directions of radical change are open, and it is not possible to predict, with certainty, which one will occur.

Chaos theory takes this one step further by noting that change in any dynamic system is ultimately not predictable, because individually small interactions between components accumulate.[12] This applies even to the balls on a billiard table and the planets in the heavens, those objects whose motion Newtonian mechanics is supposed to predict perfectly. Consequently our ability to forecast and predict is always limited, for example to about five days for weather forecasts, regardless of how sophisticated our computers are and how much information we have.

These two bodies of insight into behaviour in space and time eliminate the possibility of precise, a priori, mechanistic, deterministic predictions of the future. Computers cannot substitute for crystal balls, except for very limited classes of problems that occur over short spatial and temporal dimensions.

4. Thermodynamics and Open Systems

The next insights concern energy, that is thermodynamics. Ilya Prigogine in his Nobel Prize winning work, showed that spontaneous coherent behaviour and organization (e.g. tornadoes) can occur and are completely consistent with thermodynamics.[13] The key to understanding such phenomena is to realize that one is dealing with open systems with a constant flow of high quality energy. In these circumstances, coherent behaviour appears in systems almost magically. Prigogine showed that this occurs because the system reaches a catastrophe threshold and flips into a new coherent behavioural state. (This is evident for example in the vortex which spontaneously appears when draining water from a bathtub.)

Prigogine's work can be taken one step further to explain the energetics of open systems.[14] An open system with high quality energy pumped into it is moved away from equilibrium. But nature resists movement away from equilibrium.[15] So the open system responds with the spontaneous emergence of organized behaviour that uses the high quality energy to maintain its structure, thus dissipating the ability of the high quality energy to move the system away from equilibrium. As more high quality energy is pumped into a system, more organization emerges to dissipate the energy.[16]

This view of the world is radically different from that of a reductionist view which sees the world's workings in terms of mechanical interactions between components of a system. The emergence of organized behaviour, and even life, is now mandated by thermodynamics. This self-organization is characterized by abrupt changes that occur when a new set of interactions and activities emerge among components and the whole system.

The form of expression this self-organization takes is not predictable in advance because the very process of self-organization is by catastrophic (in the catastrophe theory sense) change; it "flips" into new regimes. As noted earlier, one of the characteristics of catastrophic change is that systems may have several possible behavioral pathways available at a catastrophe threshold. Which pathway is followed is largely an accident of circumstances. A reductionist world view, which cannot deal with the reality of emergence and self-organization in non-equilibrium systems, cannot offer sufficient explanation of how the world works.

An important observation about systems that exhibit self-organization is that they exist in a situation where they get enough energy, but not too much. If they do not get sufficient energy of high enough quality (beyond a minimum threshold level), organized structures cannot be supported and self-organization does not occur. If too much energy is supplied, chaos ensues in the system, as the energy overwhelms the dissipative ability of the organized structures and they fall apart. So self-organizing systems exist in a middle ground of enough, but not too much.

Furthermore, these systems do not maximize or minimize their functioning. Rather their functioning represents an optimum, a trade-off among all the forces acting on them. If there is too much development of any one type of structure, the system becomes overextended and brittle. If a structure is not sufficiently developed to take full advantage of the available energy and resources, then some other more optimal (i.e. better adapted) structure will displace it. In sum, these systems represent a fine balancing act. Inevitably then, human management strategies that focus on maximizing or minimizing some aspect of these systems will always fail. Only management strategies which maintain a balance will succeed.

5. Middle Number Systems and Observer Dependence

The description of these self-organizing systems is known as the middle number problem. Small number problems involve a very controlled situation with very few components. (e.g. two billiard balls colliding). Such problems are usually well explained by traditional science. Large number problems involve so many objects interacting that they can be described by statistical means (e.g. the air molecules in a room). This is the domain of classical thermodynamics and statistical mechanics. Middle number problems involve many things interacting in ways that are not random (e.g. most real world problems).[17]

This area of inquiry is the domain of system theorists. There are two important lessons to be learned from the study of middle number systems. First, such systems can only be understood from a hierarchical perspective. Neither a reductionist nor a holistic approach is sufficient. One must look at the system (e.g. a wetland or a woodlot) as a whole and as something composed of subsystems and their components. One must also look at the system in the context of its being a subsystem of a bigger system, which in turn is part of a wider environment. So, study of an animal population without reference to the individuals that make it up, the community it belongs to, and the environment it lives in, is not sufficient. This is not to say that population ecology is useless, but on its own, it cannot explain ecological phenomena.

Another property of these middle number systems is that everything is connected (at least weakly) to everything else. An analyst, in identifying the system to be studied, decides what to include and what to leave out. These decisions, about scale and extent and the hierarchical units to be studied, may be done in a systematic and consistent way, but they are necessarily subjective, and to some extent arbitrary. They reflect the viewpoint of the analyst about which connections are important to the study at hand, and which can be ignored. Thus the notion of a pristine objective scientific observer, is not applicable to the study of self-organizing systems.

It is the observer-dependent nature of the study of self-organizing systems which is the most difficult point for traditional reductionist science to understand. Take for example the notion of an ecosystem. Because the world is made of living and non-living stuff with multitudes of interrelationships, any one defined ecosystem is just one package of stuff and relations. To describe one ecosystem is to take one of many possible perspectives on these entities.[18] An ecosystem can refer to what's happening on our eyelashes, in our gut, or in Lake Ontario, or in the boreal forest. Where one draws the boundaries around an ecosystem depends on the scale and extent from which one needs to observe the whole, given the purpose of the study being undertaken. Different people looking at the same stuff are going to define the ecosystem differently, unless they agree on the inevitably subjective criteria for deciding on scale, extent and hierarchy.

The response of traditional science to this is that ecosystems don't exist, since we cannot come up with an observer-independent way of defining them. One consequence of this logic is that ecosystem research is not considered proper "scientific" research by most North American granting agencies and is not a fit topic in American ecological journals. Luckily, Canadian and European journals do not have this problem. Complex systems theory represents a profound change in the paradigm for doing science, so profound that traditional science rejects it out of hand. The notion of ecosystem is a focal point for the clash between these paradigms of what science is about.

6. Information: The Key to Self-Organization

The notions of observer dependence and hierarchical context lead us to discuss the last player of the space, time, energy and information quartet. The key question is: What information do systems need to self-organize successfully?

All living systems go through cycles of birth, growth, death and renewal. We are all familiar with death and reproduction at the cellular level, and the birth-growth and death of individuals, but it is only recently that Buzz Holling has made us aware that this cycle occurs at many temporal and spatial scales.[19]

Living systems must function within the context of the system and environment of which they are part. If a living system does not conform with the circumstances of the supersystem it is part of, it will be selected against. This process of selection functions at all levels. The supersystem imposes a set of constraints on the behaviour of the system, be it at the level of the cell, individual, population or community. Living systems that are evolutionarily successful have learned what these constraints are and how to live within them. (This is the painful process the human species is now undergoing, assuming it is not selected against).

But this presents a problem. When a new living system is generating after the demise of an earlier one, it would make the self-organization process much more efficient if it were constrained to variations which have a high probability of success. At the level of cells to species, genes play this role. Genes constrain the self-organization process to those options which have a high probability of success. It is not that genes direct or control the process of development, rather they constrain it to forms which will respect the realities of the supersystem and environment. They are a record of successful self-organization. Genes are not the mechanism of development, the mechanism is self-organization. Genes put boundaries on the process of self-organization.[20]

At higher hierarchical levels other devices constrain the self-organization process. For example, some species will kill their young under certain conditions, and many tree species need specific micro-climate conditions to trigger self-organization.[21] In some species, young are taught behaviors and individuals are banished from the group for inappropriate behaviour. Indigenous human cultures have taboos, morals and other cultural mechanisms that constrain behaviors to those which are sustainable in the context of specific ecosystems. Each of these devices acts, at a particular level of organization, as an information database about self-organization strategies that have an historical track record of success. They set out the boundaries of behaviour by self-organizing systems.

Given that living systems go through a constant cycle of birth growth, death and renewal again, at many temporal and spatial scales, a way of preserving information about what works and what doesn't so as to constrain the self-organization process is crucial for the continuance of life. This is the role of the gene. At a larger scale it is the role of biodiversity.

Biodiversity is the information database for ecosystem organization. The ability of an ecosystem to regenerate, as part of the birth, growth and death and renewal cycle, is a function of the species available for the regeneration process. This, of course, is related to the biodiversity of the larger landscape that the ecosystem is part of. Thus preservation of biodiversity is important because we are in effect preserving the library used for regeneration of landscapes.[22]

7. The Ecosystem Approach and Integrity: A New Perspective

So what are the implications of all this? The first is that we need to look at ecosystems from a hierarchical perspective with careful attention to scale and extent. Second, we must examine the spatial, temporal, thermodynamic and information aspects (dynamics) of these systems. This must be done in the context of behaviour which is both emergent and catastrophic. In other words, we must recognize that ecosystems are dynamic, not deterministic, that they have a degree of unpredictability and that they will exhibit phases of rapid change.

This is not to say that ecosystem behaviour is chaotic or random and haphazard. On the contrary, ecosystem behaviour and development is like a large musical piece such as a symphony, which is also dynamic and not predictable and yet includes a sense of flow, of connection between what has been played and what is still to come, the repetition of recognizable themes and a general sense of orderly progression. In pieces such as symphonies or suites we know the stages (allegro, adagio, etc.) that the piece will progress through, even though we don't know the details of the piece. The same is true of ecosystems. Some behave in a very ordered way as does a Baroque suite, while others are full of improvisation as in modern jazz. And yet we know the difference between music and random collections of noise.

Ecosystem self-organization unfolds like a symphony. Our challenge is to understand the rules of composition and the limitations and directions they place on the organization process, as well what makes for the ecological equivalent of a musical masterpiece that stands up to the test of time. However we should not expect to have a science of ecology which allows us to predict what the next note will be.

We must always remember that left alone, living systems are self-organizing, that is they will look after themselves. Our responsibility is to not interfere with this self-organizing process or better yet, to enhance it. Of paramount importance, in this respect, is that we must not destroy the information needed for the regeneration process which is continually ongoing. A damaged ecosystem, left to its own devices, has the capability to regenerate if it has access to the information required for renewal, that is biodiversity; and if the context for the information to be used, that is the biophysical environment, has not been so altered as to make the information meaningless.

Another important thing we need to do is to stop managing ecosystems for some fixed state, whether it be an idealistic pristine climax forest or a corn farm. Ecosystems are not static things, they are dynamic entities made up of self-organizing processes. Management goals that involve maintaining some fixed state in an ecosystem or maximizing some function (biomass, productivity, number of species) or minimizing some other function (pest outbreak) will always lead to disaster at some point, no matter how well meaning they are. We must instead recognize that ecosystems represent a balance, an optimum point of operation, and this balancing is constantly changing to suit a changing environment. And if this isn't radical enough we must bear in mind that all living systems from cells to communities face death and regeneration. This is required by the second law, it is a thermodynamic necessity.

For us, the notion of serving ecological integrity means accepting all of this. If human activities maintain the integrity of the self-organizing entities that we call life, we will be all right. If they don't, we will be selected out of the systems. We have a simple choice, to be stewards of integrity or disrupters of integrity. There is no middle ground.

But what exactly is ecological integrity? For an ecosystem, integrity[23] encompasses three major ecosystem organizational facets.[24] Ecosystem health, the ability to maintain normal operations under normal environmental conditions, is the first requisite for ecosystem integrity. But it alone is not sufficient. To have integrity, an ecosystem must also be able to cope with changes (which can be catastrophic) in environmental conditions; that is, it must be able to cope with stress. As well, an ecosystem which has integrity, must be able to continue the process of self-organization on an ongoing basis. It must be able to continue to evolve, develop, and proceed with the birth, growth, death and renewal cycle. It is these latter two facets of ecosystem integrity that differentiate it from the notion of ecosystem health.

This understanding of the behaviour of complex self-organizing systems provides a framework for the investigation of environmentally induced changes in ecosystem organization and integrity.[25] It establishes that ecosystems can respond to changes in the environment in five qualitatively different ways:

- The system can continue to operate as before, even though its operations may be initially and temporarily unsettled.
- The system can operate at a different level using the same structures it originally had (for example, a reduction or increase in species numbers).
- Some new structures can emerge in the system that replace or augment existing structures (for example, new species or paths in the food web).

- A new ecosystem, made up of quite different structures, can emerge.
- The final, and very rare possibility, is that the ecosystem can collapse completely and no regeneration occurs.

This enumeration of possible ecosystem responses to environmental change is far richer than the simple classical notion, which holds that stress temporarily displaces an ecosystem from its climax community, to which it eventually returns. In fact, an ecosystem has no inherent single preferred state for which it should be managed.

While this identifies the ways in which an ecosystem might re-organize in the face of environmental change, it does not indicate which re-organization constitutes a loss of integrity. It could be argued (and often is) that any environmental change that permanently alters the normal operations of an ecosystem affects its integrity. Ecosystem integrity would then be defined as the ability to absorb environmental change without any permanent ecosystem change. Thus the final four distinct ecosystem responses described above would constitute a loss of integrity, even though all but the last option (collapse) are responses in which the ecosystem reorganizes itself to mitigate the environmental change. However, the reorganized ecosystem is usually just as healthy as the original, even though it may be different. There is no scientific reason that an existing ecosystem should be the only one to have integrity in a situation, just because of its primacy.

At the other extreme, it could also be argued that any ecosystem that can maintain itself without collapsing has integrity. Utter collapses have been rare, desertification being one of the few examples. This definition would encompass almost all ecosystems, including ones whose organization has changed radically in response to major stress.

Neither of these definitions of integrity is operationally useful. The definition which accepts only temporary change is too restrictive in most situations, and reflects a desire to preserve the world as it is currently.[26] This denies the fundamental dynamic nature of ecosystems and leads to disastrous mismanagement (e.g. the complete suppression of forest fires, which eventually results in catastrophic conflagrations). But the latter definition, which accepts all responses except collapse, does not help managers because it restricts loss of integrity to a situation that rarely occurs and that is clearly undesirable. Hence this definition would be trivial.

In between these two extremes of definition lies a third option, which holds that some changes in ecosystems are undesirable, and therefore represent a loss of integrity. This option promises to be the most useful but it embraces many possibilities and requires difficult choices. In particular it requires the value-laden selection of criteria for determining which changes are desirable and which are not. The science of ecology can, in principle, inform us about the kind of ecosystem response or reorganization to expect in a given situation. It does not provide us with a scientific basis for deciding that one change is better than another, except possibly in the two extreme cases just discussed.[27]

Here again the insight into ecological integrity gained from complex systems theory is that the physical and biological sciences can describe and, even to a limited extent, predict human-induced changes in the biosphere, but they alone cannot determine which changes are acceptable. Ultimately, any evaluation of the ecological acceptability of a human activity, will depend on value judgments about whether the resulting changes in the affected ecosystem are acceptable to the human participants.

It should be noted that it is exactly this conclusion that leads classical scientists to reject this whole mode of reasoning as unscientific, soft and useless except as a parlour game. The complaint most often spoken is that such a treatment of ecology is not defensible in court, because there are no black and white answers, no linear causes and effects, no definitive mechanisms and no one person to blame. In short this treatment does not lead to a scientific conclusion that this behaviour is good and that behaviour is bad.

Scientific judgments about right and wrong seemed possible when we viewed the world as a set of billiard balls, and it is this mechanistic, reductionist worldview that our court system assumes. Unfortunately, this worldview with its approach to governance and law does not recognize, and will not help us deal with, the realities of complex systems. And here we have the crux of the issue. If we are truly to use an ecosystem approach, and we must if we are to have sustainability, it means changing in a fundamental way how we govern ourselves, how we design and operate our decision-making processes and institutions, and how we approach the business of environmental science and management.[28] This is the real challenge presented by an ecosystem approach.

8. Acknowledgements

Thanks to Henry Regier, George Francis, and Laura Westra for their support of James Kay's research through their Donner, NSERC and SSHRC grants, Marie Lagimodiere for her extensive search for literature on ecosystem and complex system thinking, and Nina Marie Lister for her work on biodiversity and information.

Reprinted courtesy of *Alternatives*, Canada's environmental quarterly since 1971. Annual subscriptions C$23.50 from *Alternatives*, Faculty of Environmental Studies, University of Waterloo, Waterloo, Ontario N2L 3G1, Canada.

9. References

1. L.B. Slobodkin, "Intellectual Problems of Applied Ecology," *Bioscience*, 38:5 (1988), pp. 337-342.
2. The diversity-stability hypothesis arose from Robert MacArthur, "Fluctuations of Animal Populations and a Measure of Community Stability," *Ecology*, 3 (1955), pp. 533-535) in which he proposed that the diversity of a food web was a measure of community stability. G.E. Hutchinson, "Homage to Santa Rosalia Or Why Are There So Many Kinds of Animals?" *American Naturalist*, 93 (1959), pp. 415-427, mistook this paper as a proof that species diversity explains community stability. Ramon Margalef, "On Certain Unifying Principles in Ecology," *American Naturalist*, 97 (1963), pp. 357-374; and Ramon Margalef, *Perspectives on Ecological Theory* (Chicago: University of Chicago Press, 1968) elaborated a theory of ecosystem development which held that species diversity was the cornerstone of the emergence of a stable system. This hypothesis was "codified" as dogma by the Brookhaven Symposium of 1968 in *Diversity and Stability in Ecological Systems*, G.M. Woodwell, H.H. Smith, eds. (Brookhaven National Laboratories Symposium #22, 1969). It is a very pleasing and simple to understand hypothesis based on the notion that "you don't put all your eggs in one basket." In the early 1970s a number of empirical counter-examples to this hypothesis were presented. Daniel Goodman, "The Theory of Diversity-Stability Relationships in Ecology," *Quarterly Review of Biology*, 50:3 (1975), pp. 237-366, systematically examined the literature and demonstrated clearly that there was no scientific basis for the diversity-stability hypothesis.
3. For example, in southwestern Ontario the most diverse ecosystems can be found in the area between urban development and rural lands. For more discussion see P.S. Petraitis, R.E. Latham, and R.A. Niesenbaum, "The Maintenance of Species Diversity by Disturbance," *Quarterly Review of Biology*, 64:4 (1989), pp. 393-418.
4. See P.J. Burton, *et al.*, "The Value of Managing for Biodiversity," *The Forestry Chronicle*, 68:2 (1992), pp. 225-237, "... the diversity within a biological community confers some measure of stability to that community," p. 229.

5. J.J. Kay, "A Non-Equilibrium Thermodynamic Framework for Discussing Ecosystem Integrity," *Environmental Management,* 15:4 (1991), pp. 483-495.
6. C.S. Holling, "The Resilience of Terrestrial Ecosystems: Local Surprise and Global Change, *Sustainable Development in the Biosphere,* W.M. Clark and R.E. Munn, eds. (Cambridge: Cambridge University Press, 1986), pp. 292-320; C.S. Holling, "Cross-scale Morphology, Geometry, and Dynamics of Ecosystems," *Ecological Monographs,* 62:4 (1992), pp. 447-502; and Kay, "Non-equilibrium" [note 5].
7. F. Bormann, G. Likens, *Pattern and Process in a Forested Ecosystem* (New York: Springer-Verlag, 1979).
8. R.P. McIntosh, "The Relationship between Succession and Recovery Process in Ecosystems," *The Recovery Process in Damaged Ecosystems,* J. Cairns, ed. (Ann Arbor Science, 1980), pp. 11-62.
9. See for example T.F.H. Allen, T.W. Hoekstra, *Toward a Unified Ecology* (New York: Columbia University Press, 1992).
10. This way of looking at the world spills over into our judicial system, where we strive to determine who is responsible, and who is guilty. This is based on the assumption that the observed behaviour can be explained by simple linear interactions between the components. Somebody is responsible for something happening.
11. In the sense that Ludwig Boltzmann spoke of randomization rather than the modern Jaynesian interpretation of information, see E.T. Jaynes, "Where Do We Stand on Maximum Entropy," *The Maximum Entropy Formalism,* R. Levine and M. Tribus, eds. (Cambridge, Massachusetts: MIT Press, 1979), pp. 15-118.
12. In classical analysis, small interactions between components (such as friction), interaction due to spherical imperfections (billiard tables which aren't perfectly flat), etc. are ignored. It turns out that these interactions, after some time, actually determine the system's behaviour as much as anything. But these interactions are essentially noise and unpredictable.
13. G. Nicolis, I. Prigogine, *Exploring Complexity* (New York: W.H. Freeman, 1989). Prigogine showed that such systems do not violate the second law that entropy must increase, even though they increase order or organization.
14. E.D. Schneider and J.J. Kay, "Life as a Manifestation of the Second Law of Thermodynamics," *Mathematical and Computer Modelling* 19:6-8 (1994), pp. 25-48.
15. This is the second law of thermodynamics restated for non-equilibrium situations.
16. More formally, from Schneider and Kay, "Life" [note 14], "the thermodynamic principle which governs the behaviour of self-organizing systems is that, as they are moved away from equilibrium, they will utilize all avenues available to counter applied gradients (high quality energy flows). As an applied gradient increases so does a system's ability to oppose further movement from equilibrium." This seems to be the natural principle behind the emergence of life.
17. See Gerald M. Weinberg, *An Introduction to General Systems Thinking* (New York: John Wiley and Sons, 1975).
18. A.W. King, "Considerations of Scale and Hierarchy," *Ecological Integrity and the Management of Ecosystems,* S. Woodley, J.J. Kay, G. Francis, eds. (Delray, Florida: St. Lucie Press, 1993), pp. 19-46. An ecosystem is a collection of interacting biological entities combined with the physical environment in which they live, which is perceived to act as a whole.
19. Holling, "Resilience" [note 6].
20. J.J. Kay, "Self-Organization in Living Systems" (PhD thesis, Systems Design Engineering, University of Waterloo, 1984), pp. 85-88.
21. For example, jack pine cones require heat from a forest fire to open.

22. N.M. Lister, "Biodiversity: Socio-Cultural and Scientific Perspectives With Reference to Decision Making in the Great Lakes Basin," (unpublished 1994).
23. Ecosystem integrity is about the integrity of ecosystems versus ecological integrity which refers to the integrity of life at all ecological levels including ecosystems. In what follows the focus is on ecosystem integrity.
24. Kay, "Non-equilibrium" [note 5]; and J.J. Kay, "On the Nature of Ecological Integrity: Some Closing Comments," *Ecological Integrity,* Woodley, Kay and Francis [note 18], pp.201-212.
25. Kay, "Non-equilibrium" [note 5].
26. Of course one may wish to preserve an ecosystem as an example or specimen of a specific type.
27. To return to the musical composition analogy, the two extreme cases correspond to the playing of the same piece over and over with minor variations or to no music at all. The third option allows for different compositions, but not all compositions. As in music, the question of taste and need plays an important role in deciding which compositions are acceptable.
28. For an early version of some practical and institutional aspects see H.A. Regier, *A Balanced Science of Renewable Resources* (Seattle: University of Washington Press, 1978).

Chapter 5
ECOLOGICAL INTEGRITY AND SUSTAINABILITY: BUZZWORDS IN CONFLICT?

Reed F. Noss[1]

1. Introduction

The politics of conservation might be seen by an outside observer as a battle of buzzwords. On one side stand the deep ecologists, wilderness advocates, and many biologists proclaiming the values of wildness, naturalness, biodiversity, and ecological integrity. On the other side the bureaucrats, politicians, resource managers, international development planners, and a few biologists chant sustainable development, multiple use, and ecosystem management. Sometimes the distinctions blur as "moderate" conservationists in the middle discuss sustainability, ecosystem management, and biodiversity all in one breath, optimistically believing that development and conservation are interdependent and compatible. A major problem is that everyone seems to have something different--sometimes slightly and other times radically different--in mind when using these terms. What one person sees as sustainable development, another sees as destruction of a priceless heritage. A timber company may justify clearcutting as enhancing biodiversity, while an environmentalist claims that such cutting will lead to loss of old-growth species.

Probably everyone will agree that the battle of buzzwords and slogans is not getting us very far toward solving environmental problems. Although no cause and effect relationship has been demonstrated, the proliferation of buzzwords and the devastation of the global environment show a strong positive correlation. The political use of buzzwords and slogans to rally the public, while at the same time misleading them into complacency and obfuscating real issues, should not be tolerated. But people seem to need an overarching concept they can latch on to. An "umbrella" concept provides a convenient shorthand for complex phenomena and, if nothing else, opens the door to further discussion. The concept can then be made operational by selecting measurable criteria or indicators that provide tangible information about the phenomenon of interest. Unfortunately, taking that next step to make a concept operational is rare; most of us, perhaps because of intellectual laziness, seem to be content with the buzzword.

In this chapter I explore the relationship between two fundamental buzzwords of modern conservation: sustainability and ecological integrity. On the basis of a brief review of the ways these concepts are being applied by conservationists and others, I argue that both concepts have merit but also potential for abuse. The sustainability idea, in particular, seems to have been co-opted by commodity and pro-growth interests in many cases, and is almost always interpreted from an anthropocentric point of view. When development and

---

[1]The Wildlands Project, 7310 NW Acorn Ridge, Corvallis, OR 97330; U.S.A.

conservation are combined as goals, as they often are under the banner of sustainability, development takes priority and biodiversity is usually diminished. The ecological integrity concept, though explicitly non-anthropocentric, has suffered from vagueness and sometimes from a lack of consideration for the inherently dynamic and variable nature of ecosystems. However, some promising attempts have been made to identify indicators by which integrity and its components, such as biodiversity, can be assessed and monitored. I offer examples of some indicators and indices of ecological integrity that may be useful for land-use planning and environmental monitoring. I then conclude with some thoughts on how we might escape the trap of buzzwords and start holding organizations, agencies, and governments accountable for their actions that affect the environment.

2. Sustainability

In a recent essay, Orr (1994) cites Edelman (1962) in reminding us that the purpose of political activity is often not to solve problems but only to appear to solve problems. Government agencies often follow this strategy. Having been employed by two government agencies during my career, I can verify that appearance is everything. Doing good work is seldom rewarded. Rather, making the agency look good to the powers that be is the way to get ahead. And so it is with sustainability. Because it is a "feel good" concept that virtually everyone can agree with, agencies and organizations (including environmental organizations) that claim to be concerned with sustainability gain political power. This seems to be true even when the organization offers no evidence to support its claims. Orr (1994) notes that "the subject of sustainability has become a growth industry. Government and business-sponsored councils, conferences, and public meetings on sustainability proliferate, most of which seem to be symbolic gestures to allay public anxieties, not to get down to root causes."

Environmentalists and developers alike have jumped on the sustainability bandwagon. Government has encouraged consensus-building processes centered around the idea that development and resource extraction are acceptable, even commendable, if they are sustainable. In an editorial in the journal *Conservation Biology*, a U.S. Forest Service scientist lauded sustainability as a "conservation paradigm" (Salwasser 1990) and later proposed that "ecosystem sustainability offers a common ground between the extremes of nature first or people first" (Salwasser 1991). Two Forest Service public relations campaigns, "New Perspectives" and "Ecosystem Management," emphasized sustainable production of commodities and (in theory) non-resource values from national forests. Yet continuing conflicts over public lands management in the United States suggest that little "common ground" has been established so far between those who want public lands devoted to private profit and those who want them managed for broader values such as biodiversity conservation. Furthermore, there is no evidence to suggest that current and proposed timber harvests on national forests—including those sanctioned as New Perspectives or Ecosystem Management projects—can be sustained without degrading ecosystems (Noss 1991, Frissell et al. 1992, Alverson et al. 1994, Noss and Cooperrider 1994). If natural ecosystems are degraded, sustainability has not been achieved and public relations statements to the contrary must be rejected.

On an international front, the 1972 United Nations Stockholm Conference on the Human Environment established the idea of sustainable development as a preeminent world goal and initiated the United Nations Environment Program (UNEP). The concept of sustainable development gained further ground with publication of the *World Conservation Strategy*, an influential report produced by the International Union for the Conservation of Nature (IUCN, now called the World Conservation Union but keeping the original acronym), UNEP, and the World Wildlife Fund (IUCN, UNEP, WWF 1980). In that document, conservation was interpreted to mean actions that "yield the greatest sustainable development to present generations while maintaining its potential to meet the needs and aspirations of future generations." As noted by Robinson (1993), this idea harkens back to the philosophy of Gifford Pinchot, founder of the U.S. Forest Service, who defined conservation as "the greatest good for the greatest number for the longest time." Pinchot, who also wrote that there are just two things in the world—people and natural resources—clearly had the "greatest number" of people, not species, in mind. Similarly, the "needs and aspirations" of present and future generations considered in the *World Conservation Strategy* were of human generations only. According to this philosophy, the needs of nonhuman species can be ignored unless it can be shown that the species in question benefit humans. Several years later but in the same venue, the Brundtland Commission defined sustainable development as "development that seeks to meet the needs and aspirations of the present without compromising the ability to meet those of the future" (World Commission on Environment and Development 1987). Again the focus was exclusively on humans, but the Brundtland report went further by explicitly advocating an expanding world economy fueled by technological innovation and managerial expertise (Irvine 1994).

Should it trouble conservationists that the major international conservation and development organizations have all embraced sustainable development as their primary ambition? At least some scientists and conservationists are worried that the international conservation movement has abandoned its original and more important goal of protecting biodiversity. Robinson (1993) provides a devastating critique of the sustainability paradigm as applied to international conservation and development. Robinson focuses primarily on *Caring for the Earth*, a document produced by the IUCN, UNEP, and WWF in 1991 as a sequel to the *World Conservation Strategy*. Robinson points out that although the goal of *Caring for the Earth* is ostensibly to build a sustainable society by wedding conservation and development, the report fails to recognize that these activities are often incompatible, that is, "human development can lead to species extinction and conservation can limit development" (Robinson 1993: 22). Moreover, the report errs on the side of extinction by emphasizing traditional development goals over conservation. Sustainable uses of natural resources are assumed to be linked to sustainable development, but as Robinson (1993) notes, the report contains no analysis of how sustainability might be achieved. Potential means of achieving sustainability, such as technological advances, a dramatic reduction in the world's human population, or a dramatic reduction in the level of consumption in affluent countries, are not thoughtfully considered or compared. Without such an analysis, *Caring for the Earth*'s claims are hollow and dangerous. They could easily encourage destruction of natural ecosystems and loss of biodiversity in the name of economic development.

Why have international conservationists seemingly abandoned biodiversity in favor of economic development? The reasoning behind the anthropocentric goals of *Caring for the Earth* and its predecessors is that people will not be persuaded to conserve Nature unless their basic needs and desires are met. As stated in the report, "it is important to remember that we are seeking not just survival but a sustainable improvement in the quality of life of several billion people" (IUCN, UNEP, WWF 1991). But because most people worldwide seem to want to live at the same level of wasteful affluence as in the U.S., Canada, western Europe, and Japan, meeting all human desires worldwide while still conserving biodiversity is probably impossible. Increased efficiency in the production and use of resources, recycling and re-use of materials, and a major switch to solar power and other renewable energy sources, can improve or maintain human quality of life while reducing the need for commodity production activities that degrade natural ecosystems (Postel and Ryan 1991, Noss and Cooperrider 1994, Pimentel et al. 1994). But green technologies can be pushed only so far. For 10 billion people to live like the rich Americans they see on television would require a scale of development that would lead to massive extinctions worldwide and to erosion of the life-support systems upon which all species, including humans, depend.

The irresponsible interpretation of the sustainability concept by the major international conservation groups is indeed cause for concern. These groups have the potential to be the leading advocates for biodiversity and ecological integrity at an international level. Their influence on development and conservation policies in developing countries is strong. Yet they have chosen to emphasize economic development over conservation, all in the name of some vague idea of sustainability. We can only hope the leadership of these groups soon changes, or that they come to their senses.

In the meantime, it is encouraging that academic scientists are taking a harder look at sustainability. But they seem to have some difficulty overcoming their traditional infatuation with basic research. For example, the Ecological Society of America has promoted a "Sustainable Biosphere Initiative" (SBI) with three research priorities: global change, biological diversity, and sustainable ecological systems (Lubchenco et al. 1991). However, the SBI emphasizes basic research (as opposed to applied studies or applications) as the foundation for environmental policy, a proposal that some observers find self indulgent and likely to be ineffective (e.g., Schindler 1991, Wilcove 1991). Ludwig et al. (1993) argued that the claim that basic research will promote sustainable use of resources may promote a "false complacency." "We do not require any additional scientific studies before taking action to curb human activities that affect global warming, ozone depletion, pollution, and depletion of fossil fuels," Ludwig et al. (1993) concluded, "Calls for additional research may be mere delaying tactics." A forum on sustainability in the journal *Ecological Applications* (Levin 1993) included some criticism of Ludwig et al. (1993), but most of the invited authors agreed that sustainability is unlikely to be achieved through scientific research alone. Rather, it requires fundamental changes in society. As Meyer and Helfman (1993) stated, "Ecological research is necessary but not sufficient. A better understanding of global ecology will be to no avail without the political will to implement the changes dictated by that understanding." Most authors also agreed with Ludwig et al. (1993) that claims of sustainability should be distrusted and that the burden of proof should be shifted from the public to the parties that stand to gain from resource exploitation (e.g., Costanza 1993).

Whether or not sustainable development is an oxymoron and impossible in principle, the sustainability concept in general seems to have been not only strongly utilitarian, but also more fluff than substance. Despite rhetoric about the rights of future generations, most projects that purport sustainability fail to confront the powerful economic and sociopolitical institutions that make our world less liveable for most species every day.

Among the shortcomings of the sustainability bandwagon noted by Irvine (1994) are: (1) a failure of those who promote sustainability to consider environmental and social limits to growth; (2) an unwillingness to address the unsustainability of the current human population, much less its expected growth; (3) a reluctance to confront the implications of the lifestyles of average citizens of the more affluent societies; (4) an unrealistically optimistic faith in "alternative" technologies, institutional reform, redistribution of wealth, decentralization, and personal empowerment; and (5) a failure to recognize the claims of other species to their share of the planet's resources. This last failure I personally find most troubling. The belief that humans are worth more than other species has no objective basis. It is a prejudice every bit as pernicious as the belief that whites are superior to blacks or males are superior to females (or vice versa).

If people were really concerned about sustainability in the broadest sense, they would challenge the status quo by articulating exactly what must change to make our behaviors and lifestyles sustainable. According to Orr (1994), achieving true sustainability would require three things: (1) raising serious questions about the domination of the economy by large corporations and their immunity from effective public control; (2) reconsidering the present laissez faire direction of technology; and (3)changing the way we think about our responsibilities as citizens to make the world a better place. I would add that making the world a better place requires an overt expression of our appreciation and obligations to other species. There is little evidence that politicians, agencies, or even the major environmental organizations on a national or international level are making an effort to do any of these things. Without leadership at these levels, the public is unlikely to change its ways.

Is the sustainability concept doomed to failure? Perhaps not. If citizens recognize that sustainability is one—but not the only—useful criterion for judging the goodness of a project, strategy, or policy, then it can play a positive role. Humans, just like every other organism, must use resources to survive. Everyone can agree that it is better to use resources sustainably than unsustainably. But the role of the sustainability concept in conservation strategy will be strengthened when we interpret it broadly to include not only sustainable uses of resources, but also sustenance of ecological and evolutionary processes, viable populations of native species, and other non-human qualities of ecosystems, *for their own sakes*. And for every project or policy that makes the claim of sustainability, the public deserves to be informed of how much of these qualities will be sustained, and for how long.

3. Ecological Integrity

> Perhaps we cannot be rigorous about integrity; the idea is soft, visionary, rhetorical, politically and emotionally correct, but philosophically and biologically suspect because it cannot be made operational. Integrity can mean anything you choose it to mean; it has begun to slip around as soon as we start to think about it (Rolston 1994).

Ecological integrity is a concept potentially broader and more biocentric than sustainability. In contrast to the concerns of Rolston (1994), I will argue that we can be rigorous about integrity; the concept can be made operational by selecting measurable indicators that correspond to the ecological qualities we associate with integrity. However, Rolston is correct that the integrity concept remains confusing to many scientists, philosophers, and the public. Like many concepts of great value to people—justice, freedom, love, democracy—integrity is vague and slippery. But these concepts still inspire us; they seem to be fundamentally "right." Indeed, the first reference to integrity in the environmental literature was Aldo Leopold's (1949) famous aphorism in his essay on land ethics: "A thing is right when it tends to preserve the integrity, stability, and beauty of the biotic community. It is wrong when it tends otherwise." Leopold never explained what he meant by integrity, but two generations of biologists and conservationists have quoted this passage with nothing short of reverence.

The next reference to integrity in an environmental context was in the U.S. Water Quality Amendments of 1972, which promoted the restoration and maintenance of "the chemical, physical, and biological integrity of the nation's waters." In 1978 Canada and the United States created the Great Lakes Water Quality Agreement (ratified in 1988), which called on these nations "to restore and maintain the chemical, physical, and biological integrity of the Great Lakes Basin Ecosystem." More recently, the Canadian National Parks Act, amended in 1988, requires that maintenance of ecological integrity be a top priority when developing a park management plan (Woodley 1993).

Definitions of integrity include those of Cairns (1977): "the maintenance of the community structure and function characteristic of a particular locale or deemed satisfactory to society;" Karr and Dudley (1981): "the capability of supporting and maintaining a balanced, integrated, adaptive community of organisms having a species composition, diversity, and functional organization comparable to that of natural habitats of the region;" and one given in reference to Canadian national parks (in Woodley 1993):

> Ecological integrity is defined as a state of ecosystem development that is optimized for its geographic location, including energy input, available water, nutrients and colonization history. For national parks, this optimal state has been referred to by such terms as natural, naturally evolving, pristine and untouched. It implies that ecosystem structures and functions are unimpaired by human-caused stresses and that native species are present at viable population levels.

Characterizing integrity in more precise terms has not been easy, in part because ecosystems are not static. Ecosystems change over time due to purely natural factors and their changes are often chaotic and unpredictable. Recent paleontological evidence, for example, supports the hypothesis of the early American ecologist, Harry Gleason (1926), that most species respond to environmental variation in an individualistic fashion. Environments change both across space (for instance, along elevation or moisture gradients) and across time, such as with changing climate. With both kinds of change, few species coincide precisely in their movements or distributions. During the Pleistocene glacial and interglacial cycles, plant communities did not shift northward and southward as intact entities; rather, individual species of trees dispersed at different rates and along different routes (Davis 1981). Thus, the species composition of communities is continuously changing. Species we see together today may have been separated for most of their evolutionary histories. For example, the Douglas-fir forests of the Pacific Northwest, now famous as the setting for bitter conflicts between the timber industry and conservationists, were first established only about 6,000 years ago, even though Douglas-fir (genus *Pseudotsuga*) and many other species associated with this community have existed in other assemblages for millions of years (Brubaker 1991). If communities are just transient aggregations of species, how can they be said to possess integrity?

Even over much shorter spans of time, natural ecosystems are far from stable and unvarying. Natural disturbances occur at a variety of spatial and temporal scales, so that a landscape is more of a shifting mosaic of patches than a homogeneous vegetation in equilibrium with its physical environment (Bormann and Likens 1979, White 1979, Sousa 1984, Pickett and White 1985). After disturbance, an ecosystem may have multiple potential pathways of successional development. It may also have multiple potential "endpoints," violating the conventional concept of the climax state (Pickett et al. 1992). The type of plant community that develops on a site after a volcanic eruption, fire, flood, clearcut, or agriculture can be predicted only in the most general terms. Many details of species composition, relative abundances, and dominance in biological communities are determined by accidents of history. The entire course of sucession, for example, may be determined by the set of species that just happened to colonize a newly disturbed site first (Egler 1954). And so the question posed above arises in another form: How can a continuously varying, frequently disturbed, unstable, and unpredictable shifting mosaic have integrity?

Reconciling the concept of ecological integrity with what is known scientifically about continuously changing, nonequilibrium ecosystems is difficult but by no means impossible; it requires a "big picture" point of view. Ecologist Dan Botkin's book *Discordant Harmonies* (Botkin 1990) did more than any other recent work to destroy conventional notions of the balance of nature. Botkin demonstrated through many examples that change is intrinsic to natural ecosystems at many scales of space and time. We cannot draw a line around a natural area we want to preserve and expect it to stay forever in the condition in which we found it. Some species, such as fire-adapted pines and the animals dependent on them, require frequent disturbance in order to resist competitive replacement by fire-intolerant species. But accepting that change and disturbance are necessary to keep ecosystems healthy and diverse does not require that we accept all change. As Botkin (1990) emphasized, the rates of change associated with human disturbances are often far beyond what organisms are adapted to coping with. It is interesting that the timber lobbyists and others who find comfort in Botkin's message of continual change, and cite him frequently, fail to mention his caveats about some kinds and rates of change being normal, natural, or desirable, and others not.

Other authors, notably Pickett et al. (1992), have also skillfully overthrown the old equilibrium paradigms about nature without opening the door to the ridiculous assertion that because change is natural, whatever humans do to the change the environment is acceptable. Pickett et al. (1992) pointed out the importance of viewing ecosystems in a functional, evolutionary, and historical context. Nature has functional constraints because organisms have physiological limits to what they can tolerate. Pollution and other human-generated stresses often exceed these functional limits. Evolutionary constraints are also important because species are limited in how quickly they can adapt through natural selection to a changing environment. The rates of environmental change associated with some modern human activities far exceed rates experienced by most species over their evolutionary histories. Finally, each region and site has a unique biogeographic history and a unique set of organisms with their own physiological and evolutionary limits (Pickett et al. 1992). To exceed these limits is to invite extinction. Thus, ecological integrity can be maintained in a dynamic environment only when the changes caused by humans are within the functional, evolutionary, and historical limits characteristic of each ecosystem.

Integrity is the most all-encompassing of all "umbrella" concepts in conservation (Angermeier and Karr 1994). An ecosystem with integrity possesses several qualities—all relative rather than absolute qualities—to a high degree: ecosystem health, biodiversity, stability (in terms of relative resistance and resilience to disturbance), sustainability, naturalness, wildness, and beauty (Noss 1995). Karr (1994) and Westra (1994) discussed some relationships between ecological integrity and ecological health, arguing that health can sometimes be achieved in an impoverished and heavily managed ecosystem such as a tree farm, but that integrity is characteristic only of areas showing little influence from human actions. Angermeier and Karr (1994) compare the concepts of biological integrity and biological diversity, and conclude that integrity is preferable as a management goal because it avoids the pitfalls of assuming that greater diversity is necessarily better (see also Noss 1983).

4. Making Ecological Integrity an Operational Concept

Available definitions of ecological integrity provide a reasonable sense of what integrity means in the context of environmental conservation. One could use these definitions to compare two or more areas and state with some confidence that one area has more integrity than the other. But, definitions by themselves do not allow us to quantify differences in integrity between areas, measure changes in integrity over time, or specify the level of integrity we wish to maintain with preservation efforts or restore with recovery efforts. Thus, available definitions of integrity are not fully operational.

The use of ecological indicators has increased in recent years with the general recognition among scientists and managers that it is a good idea to assess the effects of management activities, and to monitor the state of the environment as a whole. Indicators are required because monitoring the whole environment at any scale directly is impossible; Nature is too complex. The use of ecological indicators in conservation-oriented monitoring and assessment programs is a huge topic beyond the scope of this chapter (for reviews see Cooperrider et al. 1986, Landres et al. 1988, Noss 1990, Angermeier and Karr 1994, Noss and Cooperrider 1994). Indicators may be useful during several phases of conservation work.

Indicators are especially valuable in the inventory and identification phase, where one assesses the distribution of biological and environmental entities and identifies or selects areas that should be conserved; and secondly during the management phase, where one monitors the effects of management treatments, learns from experience, and adjusts management practices accordingly (Holling 1978, Walters 1986).

Some indicators useful for identifying conservation priorities include: (1)locations of roadless, undeveloped, or otherwise essentially wild areas of significant size; (2)concentrated occurrences of rare species, including centers of endemism; (3)areas of unusually high species richness; (4)locations of rare or unusual plant or animal communities, depleted and important seral stages (such as old-growth forest), or animal concentration areas; (5)resource hot spots such as sites of unusually high primary productivity, artesian springs, ice-free bays, outcrops of unusual parent material (e.g., serpentine), mineral licks, etc.; (6)watersheds of high value for anadromous fishes or other elements of aquatic biodiversity; (7)sites of inherent sensitivity to development, such as watersheds with steep slopes or unstable soils, or aquifer recharge areas; (8)sites recognized as important by indigenous peoples; (9)sites that could be added to adjacent protected areas to form larger and more defensible reserves. These indicators were described in more detail by Noss (1995). In any given case, data will be available for some of these indicators but not for others. Also, high values for one or more of these indicators do not automatically suggest that an area has high integrity. For example, an area containing many rare species is not necessarily of higher integrity than an ecosystem with no rare species; both areas may be displaying their inherent biodiversity. But protecting sites that collectively encompass the qualities listed above contributes greatly to the goal of maintaining overall ecological integrity in a region.

The second major phase of conservation where indicators are especially helpful is in monitoring and assessment, either of specific sites such as nature reserves or of the overall landscape (at scales ranging from a local watershed to the biosphere). Some indicators potentially useful for monitoring and assessing the effects of management practices, land uses, or other human activities on the integrity of natural ecosystems are listed in Table 1 (from Noss 1995). Some of these indicators were originally published as potential indicators of biodiversity (Noss 1990); I repeat them here because I believe native biodiversity is one of the best expressions of ecological integrity, though I agree with Angermeier and Karr (1994) that integrity is the broader of the two concepts. Only a few of the indicators listed in Table 1 may be relevant or have data available by which to measure them in any particular case. For instance, amount of late-successional forest interior habitat would not be an appropriate indicator for a prairie landscape, and data on species composition, biomass, and nutrient cycling rates may be unavailable in many cases. However, a multi-parameter approach is essential when dealing with complex ecological systems. No single indicator can be expected to tell us everything we need to know about the ecological integrity of an area (Karr 1981, 1993, Karr et al. 1986, Noss 1990). Other examples of indicators of ecological integrity are discussed by Angermeier and Karr (1994) and in the chapters in Woodley et al. (1993).

The most practical way to use indicators to assess ecological integrity is within a hierarchical framework, proceeding top-down from broad regional assessments to detailed evaluations of particular sites or populations (Noss 1990, 1995). Detailed site-level studies are most needed in areas exceptionally rich in biodiversity (for example, centers of endemism), areas particularly sensitive to human activities, or areas at high risk of losing their ecological integrity in the near future. At the species or population level of organization, genetic assessments are usually suggested only for species at high risk of extinction. Measuring heterozygosity or other genetic indicators for all species in a region, or even all vertebrates and vascular plants, would be extremely costly and would provide little information of immediate value to conservation decisions.

Table 1. Measurable Indicators of Ecological Integrity (from Noss 1995).

LANDSCAPE-REGIONAL LEVEL

Structural measures of patch characteristics

patch size frequency distribution for each seral stage and community type, and across all stages and types

size frequency distribution of late-successional interior forest patches (minus defined edge zone, e.g., 100-200 m)

total amount of late-successional forest interior habitat in all patches and per patch

total amount of patch perimeter and edge zone (also patch perimeter:area ratios, edge zone: interior zone ratios)

patch shape indices (e.g., deviation from roundness)

Structural measures of patch dispersion

patch density

fragmentation and connectivity indices

interpatch distance (mean, median, range) for various patch types

juxtaposition measures (percentage of area within a defined distance from patch occupied by different habitat types, length of patch border adjacent to different habitat types)

structural contrast (magnitude of difference between adjacent habitats, measured for various structural attributes)

Access, flow, and disturbance indicators

frequency, return interval, or rotation period of fires and other natural and anthropogenic disturbances

road density (km/km2) for different classes of road and all road classes combined

percentage of zone in roadless area (for different size thresholds, e.g., 1000 ha and above, 5000 ha and above)

kilometers of roads constructed, reconstructed, and closed (seasonally and permanently) each decade

amount of roadless area restored through permanent road closures and revegetation each decade

density of airstrips, boat landings, and other access points, how frequently they are used, and how many people go in and out per landing

density of livestock (or in some cases, historical density)

energy, nutrient, water, and organism (including human) fluxes between and among habitats or zones

Table 1. Continued on next page.

Table 1. Continued.

COMMUNITY-ECOSYSTEM LEVEL

Structural measures

    frequency distribution of seral stages (age classes) for each community type and across all types

    average and range of tree ages within defined seral stages of forest

    ratio of area of natural habitat to anthropogenic or human-disturbed habitat

    abundance and density of key structural features (e.g., snags and downed logs in various size and decay classes) in either terrestrial or aquatic habitats

    spatial dispersion of structural elements and patches

    physiognomy, including foliage density and layering (profiles), and horizontal diversity of foliage profiles in stand

    canopy density and size and dispersion of canopy openings

    woody stem density in various size (dbh) classes

    diversity of tree ages or sizes in stand

    cover of native graminoids in open forest, grassland, shrub-steppe, and tundra communities

Compositional measures

    identity, relative abundance, frequency, richness, and evenness of species and guilds (in various habitats)

    ratio of exotic species to native species in community (species richness, cover, and biomass)

Functional measures

    frequency, return interval, or rotation period of fires and other natural and anthropogenic disturbances

    areal extent of each disturbance event

    intensity or severity of disturbance events

    seasonality or periodicity of disturbances

    predictability or variability of disturbances

    invasion rates of weedy or opportunistic species (e.g., shrubs or purple loosestrife in wetlands, shrubs in prairies, exotic plants or birds in rangelands, exotic fish in streams or lakes)

    human intrusion rates and intensities

    nutrient cycling rates (for key limiting nutrients)

Table 1. Continued on next page.

Table 1. Continued.

Composite indices

    Karr's index of biotic integrity (IBI), composed of 12 attributes of fish communities in 3 major classes (species richness and composition, trophic composition, and fish abundance and condition), or adaptations to other taxa or habitats

SPECIES LEVEL

Measures of genetic integrity (or lack thereof)

    heterozygosity

    allelic diversity

    presence/absence of rare alleles

    phenotypic polymorphism

    symptoms of inbreeding depression or genetic drift (reduced survivorship or fertility, abnormal sperm, reduced resistance to disease, morphological abnormalities or asymmetries)

    inbreeding/outbreeding rate

    rate of genetic interchange between populations (measured by rate of dispersal and subsequent reproduction of migrants)

Measures of demographic integrity (or lack thereof)

    abundance, density, cover or importance value

    fertility or recruitment rate

    survivorship or mortality rate

    sex ratio and age distribution

    health parameters (fecundity, individual growth rate, body mass, stress hormone levels, etc.)

    population growth and fluctuation trends

    distribution and dispersion of subpopulations or individual home ranges across the region

    trends in habitat components (varies by species)

    trends in threats to species (depends on life history and sensitivity of species; see fragmentation and other landscape indices above)

The only indicator that has been rather thoroughly tested and related explicitly to ecological integrity is Karr's multi-parameter index of biotic integrity (the so-called IBI) (Karr 1981, 1990, 1991, 1993, Karr et al. 1986, Hughes and Gammon 1987, Steedman 1988). The IBI considers the biotic integrity of an aquatic ecosystem to be affected by five classes of environmental factors: water quality, habitat structure, energy source, flow regime, and biotic interactions. Conventional approaches to water management, which focus narrowly on chemical measures of water quality, are not very informative about integrity and can even conflict with integrity objectives. For example, chlorine added in secondary sewage treatment can degrade biotic integrity (Karr et al. 1985). Because they integrate the effects of numerous environmental factors, biotic communities tell more about integrity than chemical measures.

As originally applied, the IBI is based on 12 metrics of fish communities falling into three major classes: species richness and composition, trophic composition, and fish abundance and condition (Karr 1991). The number and identity of fish in various taxonomic and functional groups, as well as their physical condition, reflect the condition of the stream. As human disturbance increases, total species richness, number of intolerant species, and number of trophic specialists usually decline, while the number of trophic generalists increases (Karr 1993). In southern Ontario, the IBI was strongly associated with the condition of the overall watershed (Steedman 1988). IBIs should be constructed individually for each region, and be both designed and interpreted by biologists familiar with the regional fauna. Reference sites (control areas with minimal human disturbance) are critically important as standards for comparison. Although the original IBIs were based on fish, IBIs have been applied to benthic invertebrate communities with informative results (Ohio EPA 1988, Plafkin et al. 1989, Karr and Kerans 1991).

There is a great need to develop multi-parameter indices of integrity for other habitats, including terrestrial and marine ecosystems. So far no terrestrial IBI has been satisfactory, perhaps because so many confounding variables exist in terrestrial ecosystems (Karr 1991, Karr, personal communication). However, many of the components of a potential terrestrial IBI are in place, and include the indicators listed in Table 1. We know that as a landscape is fragmented by human activities, certain kinds of species decline and others increase. Sensitive species include wide-ranging animals (especially large carnivores), non-vagile species, species with specialized habitat or resource requirements, ground-nesting birds, interior (edge- sensitive) species, and species vulnerable to exploitation or persecution. Terrestrial species that tolerate or thrive with human disturbance include opportunistic omnivores, edge species, and other weedy and exotic plants and animals; these patterns are very well documented in many different ecosystems (see review in Noss and Csuti 1994). Landscape-scale indicators of human access are among the best measures of the ecological integrity of wild areas. Road access in particular often leads to invasion by alien species, over-exploitation and persecution of furbearers and large carnivores, increased fire ignitions, loss of snags through firewood collecting, and other problems (Thiel 1985, Schowalter 1988, Noss 1992, Noss and Cooperrider 1994).

Potential indicators and indices of ecological integrity need to be tested in a wide variety of ecosystems. Most importantly, agencies must provide the funding necessary to implement monitoring and assessment programs that apply these indicators. Unfortunately, monitoring is often not taken seriously by environmental agencies, and is one of the first programs to be eliminated when budgets get tight (Noss and Cooperrider 1994). Yet there is no other way to learn about the effects of human activities—both negative such as development and positive such as restoration—except through comprehensive monitoring programs using valid indicators.

5. Conclusion

To move from buzzwords to sincere, substantive efforts to maintain and restore ecological integrity requires accountability. When an agency such as the U.S. Forest Service brazenly claims they have changed their ways and have progressed from commodity-oriented management to ecosystem management, citizens must be able to see the proof. They must be shown the statistics on miles of roads closed, acres of wild habitat protected, number of timber sales and grazing leases cancelled, and the proportion of populations of threatened and endangered species that are recovering. Citizens have a right to this information. When statistics on ecological recovery are not available, it is prudent to conclude that management has not improved and that ecosystems are not recovering, despite public relations propaganda to the contrary.

Because biodiversity is a legacy of the global evolutionary process and cannot fundamentally be "owned," it belongs to everyone and to no one (to paraphrase a statement from Ed Abbey about wilderness). The biosphere is one interconnected, functional system. Road-building, deforestation, and species extinctions in the Amazon Basin or in Oregon are legitimately the concerns of all citizens worldwide. Hopeful propaganda about sustainable development and sustainable use—whether coming from the World Bank, a multinational corporation, or the IUCN—should not be trusted until it can be shown that biodiversity and ecological integrity will not be harmed in the process. The burden of proof must be firmly placed on the developers.

The concept of environmental performance review holds promise for making agencies and governments accountable for their actions (Lykke 1992). As summarized by Stokke (1992):

> Performance review can be loosely defined as an authoritative evaluation of state policy conducted by an external party. The *behaviour* one seeks to influence, however, may ultimately be that of sub-national actors such as private companies; an international review mechanism is founded on the belief that such actors can usefully be addressed indirectly, by influencing the regulatory behaviour of states.

What if Indonesia, Japan, British Columbia, or the U.S. Forest Service were held strictly accountable for the ecological integrity of their lands, airs, and waters (or those of the international commons they influence), and were evaluated by quantifiable indicators on how well they are achieving environmental goals? And what if, failing to maintain or restore ecological integrity, they suffered severe international trade restrictions or other substantive censure? Perhaps they would do more to protect their environment and would not be able to get away with making irresponsible claims about sustainability. Governments and agencies could be evaluated by a series of questions, as suggested by Noss et al. (1992):

> How adequate is the...inventory of biodiversity and what steps are being taken to gain more information? How thoroughly are changes in environmental conditions and biodiversity being monitored at a national scale, if at all? Have destructive land-use practices, pollutant emissions, and other threats to biodiversity been accurately identified? Has the country [or agency] established strong goals to improve its national conservation effort? Is the country implementing programs to protect land, limit pollutant emissions, and take other steps to maintain and restore biodiversity? And finally, is the status of biodiversity in the country declining, stable, or improving?

The indicators and indices reviewed in this chapter will be helpful for addressing some of these questions. The answers can be used to apply pressure on governments and agencies to improve their programs. With enough pressure, they will listen, they will change, and ecosystems will begin to heal. But most important is the conviction, sense of responsibility, and behavior of the individual citizen. Each of us needs to determine—and then improve—the net influence we have on ecological integrity and sustainability everywhere, through our reproduction and use of resources, our votes, our writings and lectures, our teaching and inspiring others, and our constant vigilance and pressure against the powers that be. As Thoreau implored us, "Let your life be a counter friction to stop the machine."

6. References

Alverson, W.S., W. Kuhlmann, and D.M. Waller. 1994. *Wild Forests: Conservation Biology and Public Policy*. Island Press, Washington, DC.

Bormann, F.H., and G.E. Likens. 1979. *Pattern and Process in a Forested Ecosystem*. Springer-Verlag, New York.

Botkin, D.B. 1990. *Discordant Harmonies: A New Ecology for the Twenty-First Century*. Oxford University Press, New York.

Angermeier, P.L., and J.R. Karr. 1994. Biological Integrity Versus Biological Diversity as Policy Directives. *BioScience* 44: 690-697.

Brubaker, L.B. 1991. Climate Change and the Origin of Old-Growth Douglas-Fir Forests in the Puget Sound Lowland. Pages 17-24 in L.F. Ruggiero, K.B. Aubry, A.B. Carey, and M.H. Huff, technical coordinators. *Wildlife and Vegetation of Unmanaged Douglas-Fir Forests*. USDA Forest Service, Pacific Northwest Research Station, Portland, OR.

Cairns, J. 1977. Quantification of Biological Integrity. Pages 171-187 in R.K. Ballentine and L.J. Guarraia, eds. *The Integrity of Water*. U.S. Environmental Protection Agency, Office of Water and Hazardous Materials, Washington, DC.

Cooperrider, A.Y., R.J. Boyd, and H.R. Stuart, eds. 1986. *Inventory and Monitoring of Wildlife Habitat*. USDI Bureau of Land Management, Washington, DC.

Costanza, R. 1993. Developing Ecological Research That is Relevant for Achieving Sustainability. *Ecological Applications* 3: 579-581.

Davis, M.B. 1981. Quaternary History and the Stability of Forest Communities. Pages 132-153 in D.C. West, H.H. Shugart, and D.B. Botkin, eds. *Forest Succession*. Springer-Verlag, New York.

Edelman, M. 1962. *The Symbolic Uses of Politics*. University of Illinois Press, Urbana-Champaign, IL.

Egler, F.E. 1954. Vegetation Science Concepts, I. Initial Floristic Composition, A Factor in Old-Field Vegetation Development. *Vegetatio* 4: 412-417.

Frissell, C.A., R.K. Nawa, and R. Noss. 1992. Is There Any Conservation Biology in "New Perspectives?": A Response to Salwasser. *Conservation Biology* 6: 461-464.

Gleason, H.A. 1926. The Individualistic Concept of the Plant Association. *Bulletin of the Torrey Botanical Club* 43: 463-481.

Holling, C.S., ed. 1978. *Adaptive Environmental Assessment and Management*. John Wiley and Sons, New York.

Hughes, R.M., and J.R. Gammon. 1987. Longitudinal Changes in Fish Assemblages and Water Quality in the Willamette River, Oregon. *Transactions of the American Fisheries Society* 116: 196-209.

Irvine, S. 1994. The Cornucopia Scam: Contradictions of Sustainable Development. Part I: Ignoring the Limits to Growth. *Wild Earth* 4(3): 73-81.

IUCN/UNEP/WWF. 1980. *World Conservation Strategy: Living Resource Conservation for Sustainable Development*. Gland, Switzerland.

IUCN/UNEP/WWF. 1991. *Caring for the Earth: A Strategy for Sustainable Living*. Gland, Switzerland.
Karr, J.R. 1981. Assessment of Biotic Integrity Using Fish Communities. *Fisheries* 6: 21-27.
Karr, J.R. 1990. Biological Integrity and the Goal of Environmental Legislation: Lessons for Conservation Biology. *Conservation Biology* 4: 244-250.
Karr, J.R. 1991. Biological Integrity: A Long-Neglected Aspect of Water Resource Management. *Ecological Applications* 1: 66-84.
Karr, J.R. 1993. Measuring Biological Integrity: Lessons from Streams. Pages 83-104 in S. Woodley, J. Kay, and G. Francis, eds. *Ecological Integrity and the Management of Ecosystems*. St. Lucie Press, Ottawa, Canada.
Karr, J.R. 1994. Ecological Integrity and Ecological Health Are Not the Same. In P. Schulze, R. Frosch, and P. Risser, eds. *Engineering Within Ecological Constraints*. National Academy of Engineering, Washington, DC.
Karr, J.R., and D.R. Dudley. 1981. Ecological Perspective on Water Quality Goals. *Environmental Management* 5: 55-68.
Karr, J.R., K.D. Fausch, P.L. Angermeier, P.R. Yant, and I.J. Schlosser. 1986. *Assessing Biological Integrity in Running Waters: A Method and its Rationale*. Illinois Natural History Survey, Champaign, IL.
Karr, J.R., R.C. Heidinger, and E.H. Helmer. 1985. Sensitivity of an Index of Biotic Integrity to Change in Chlorine and Ammonia Levels from Wastewater Treatment Facilities. *Journal of the Water Pollution Control Federation* 57: 912-915.
Karr, J.R., and B.L. Kerans. 1991. Components of Biological Integrity: Their Definition and Use in Development of an Invertebrate IBI. Pages 1-16 in W.S. Davis and T.P. Simon, eds. Proceedings of the Midwest Pollution Control Biologists Meeting. EPA-905/R-92/003. U.S. Environmental Protection Agency, Chicago, IL.
Landres, P.B., J. Verner, and J.W. Thomas. 1988. Ecological Uses of Vertebrate Indicator Species: A Critique. *Conservation Biology* 2: 316-328.
Leopold, A. 1949. *A Sand County Almanac*. Oxford University Press, New York.
Levin, S.A. 1993. Science and Sustainability. *Ecological Applications* 3: 545-546.
Lubchenco, J., A.M. Olson, L.B. Brubaker, S.R. Carpenter, M.M. Holland, S.P. Hubbell, S.A. Levin, J.A. MacMahon, P.A. Matson, J.M. Melillp, H.A. Mooney, C.H. Peterson, H.R. Pullman, L.A. Real, P.J. Regal, and P.G. Risser. 1991. The Sustainable Biosphere Initiative: An Ecological Research Agenda. *Ecology* 72: 371-412.
Ludwig, D., R. Hilborn, and C. Walters. 1993. Uncertainty, Resource Exploitation, and Conservation: Lessons From History. *Science* 260: 17, 36.
Lykke, E., ed. 1992. *Achieving Environmental Goals: The Concept and Practice of Environmental Performance Review*. Belhaven Press, London, UK.
Meyer, J.L., and G.S. Helfman. 1993. The Ecological Basis of Sustainability. *Ecological Applications* 3: 569-571.
Noss, R.F. 1983. A Regional Landscape Approach to Maintain Diversity. *BioScience* 33: 700-706.
Noss, R.F. 1990. Indicators for Monitoring Biodiversity: A Hierarchical Approach. *Conservation Biology* 4: 355-364.
Noss, R.F. 1991. Sustainability and Wilderness. *Conservation Biology* 5: 120-121.
Noss, R.F. 1992. The Wildlands Project: Land Conservation Strategy. *Wild Earth* (Special Issue): 10-25.
Noss, R.F. 1995. Maintaining Ecological Integrity in Representative Reserve Networks. World Wildlife Fund Canada and World Wildlife Fund-U.S., Toronto, Ontario, and Washington, DC.

Noss, R.F., S.P. Cline, B. Csuti, and J.M. Scott. 1992. Monitoring and Assessing Biodiversity. Pages 67-85 in E. Lykke, ed. *Achieving Environmental Goals: The Concept and Practice of Environmental Performance Review*. Belhaven Press, London, UK.

Noss, R.F., and A. Cooperrider. 1994. *Saving Nature's Legacy: Protecting and Restoring Biodiversity*. Defenders of Wildlife and Island Press, Washington, DC.

Noss, R.F., and B. Csuti. 1994. Habitat Fragmentation. Pages 237-264 in G.K. Meffe and R.C. Carroll, eds. *Principles of Conservation Biology*. Sinauer Associates, Sunderland, MA.

Ohio EPA. 1988. Biological Criteria for the Protection of Aquatic Life. Ohio Environmental Protection Agency, Division of Water Quality Monitoring and Assessment, Columbus, OH.

Orr, D.W. 1994. Twine in the Bailer. *Conservation Biology* 8: 931-933.

Pickett, S.T.A., V.T. Parker, and P.L. Fiedler. 1992. The New Paradigm in Ecology: Implications for Conservation Biology Above the Species Level. Pages 65-88 in P.L. Fiedler and S.K. Jain, eds. *Conservation Biology: The Theory and Practice of Nature Conservation, Preservation, and Management*. Chapman and Hall, New York.

Pickett, S.T.A., and P.S. White. 1985. *The Ecology of Natural Disturbance and Patch Dynamics*. Academic Press, Orlando, FL.

Pimentel, D., G. Rodrigues, T. Wang, R. Abrams, K. Goldberg, H. Staecker, E. Ma, L. Brueckner, L. Trovato, C. Chow, U. Govindarajulu, and S. Boerke. 1994. Renewable Energy: Economic and Environmental Issues. *BioScience* 44: 536-547.

Plafkin, J.L., M.T. Barbour, K.D. Porter, S.K. Gross, and R.M. Hughes. 1989. *Rapid Bioassessment Protocols for Use in Streams and Rivers: Benthic Macroinvertebrates and Fish*. EPA/444/4-89-001. U.S. Environmental Protection Agency, Washington, DC.

Postel, S., and J.C. Ryan. 1991. Reforming Forestry. Pages 74-92 in L. Starke, ed. *State of the World 1991: A Worldwatch Institute Report on Progess Toward a Sustainable Society*. W.W. Norton, New York.

Robinson, J.G. 1993. The Limits to Caring: Sustainable Living and the Loss of Biodiversity. *Conservation Biology* 7: 20-28.

Rolston, H. 1994. Foreword. Pages xi-xiii in L. Westra. *An Environmental Proposal for Ethics: The Principle of Integrity*. Rowman & Littlefield, Lanham, MD.

Salwasser, H. 1990. Sustainability as a Conservation Paradigm. *Conservation Biology* 4: 213-216.

Salwasser, H. 1991. New Perspectives for Sustaining Diversity in U.S. National Forest Ecosystems. *Conservation Biology* 5: 567-569.

Schindler, D.W. 1991. Comments on the Sustainable Biosphere Initiative. *Conservation Biology* 5: 550-551.

Schowalter, T.D. 1988. Forest Pest Management: A Synopsis. *Northwest Environmental Journal* 4: 313-318.

Sousa, W.P. 1984. The Role of Disturbance in Natural Communities. *Annual Review of Ecology and Systematics* 15: 353-391.

Stokke, O.S. 1992. Environmental Performance Review: Concept and Design. Pages 3-24 in E. Lykke, ed. *Achieving Environmental Goals: The Concept and Practice of Environmental Performance Review*. Belhaven Press, London, UK.

Steedman, R.J. 1988. Modification and Assessment of an Index of Biotic Integrity to Quantify Stream Quality in Southern Ontario. *Canadian Journal of Fisheries and Aquatic Sciences* 45: 492-501.

Thiel, R.P. 1985. Relationship Between Road Densities and Wolf Habitat Suitability in Wisconsin. *American Midland Naturalist* 113: 404-407.

Walters, C.J. 1986. *Adaptive Management of Renewable Resources*. McGraw-Hill, New York.
Westra, L. 1994. *An Environmental Proposal for Ethics: The Principle of Integrity*. Rowman & Littlefield, Lanham, MD.
White, P.S. 1979. Pattern, Process, and Natural Disturbance in Vegetation. *Botanical Review* 45: 229-299.
Wilcove, D.S. 1991. Comments on the Sustainable Biosphere Initiative. *Conservation Biology* 5: 553.
Woodley, S. 1993. Monitoring and Measuring Ecosystem Integrity in Canadian National Parks. Pages 155-176 in S. Woodley, J. Kay, and G. Francis. *Ecological Integrity and the Management of Ecosystems*. St. Lucie Press, Ottawa, Canada.
Woodley, S., J. Kay, and G. Francis. 1993. *Ecological Integrity and the Management of Ecosystems*. St. Lucie Press, Ottawa, Canada.
World Commission on Environment and Development. 1987. *Our Common Future*. Oxford University Press, New York.

Chapter 6
ECOSYSTEM INTEGRITY: A CAUSAL NECESSITY

Robert E. Ulanowicz[1]

1. Introduction

Ecosystem integrity has become very topical of late. As is usual with emerging concepts, the bulk of what has been written on integrity deals with progressive definitions of the concept. For example, it has been necessary to distinguish between the integrity of an ecosystem and it's "health"—another popular notion. Ecosystem health was crafted to quantify how well a system is functioning. Costanza (1992), for example, cites three aspects of ecosystem health—vigor, organization and resilience. The first two components refer to the system in its current state. Only the last property addresses the immediate future of the system.

The intention behind ecosystem integrity has been to focus on the well-being of ecosystems over a longer time span. It necessarily follows that integrity subsumes ecosystems health. Hence, Westra (1994) identifies ecosystem integrity, which she labels I_a, and ecosystem health, which she labels I_b. I_a encompasses most of what is included under the rubric of ecosystem health, and reveals not only how the system is functioning at the present moment, but how it might deal with an array of unforeseen circumstances in the future. That is, integrity addresses a system's entire trajectory of past and future configurations (sensu Holling 1992). The direction in which a system is headed is crucial not only to the meaning of integrity, but it also imports a legitimacy to ethical considerations on how society should interact with its natural surroundings (Westra 1994).

Ecosystem health, because it pertains to relatively brief time spans, fits readily into the accepted framework of scientific ideas. Hence, it is not too surprising that considerable progress has been made towards quantifying I_b. The larger notion of integrity I_a, however, entails a secular direction, and endogenous direction (telos) is steadfastly eschewed by the consensus of practicing scientists. Thus, no matter how adroit and succinct we may be in defining integrity, it is obvious from the beginning that the concept does not fit easily into the patterns of thought that have dominated science for the past three centuries. To be more precise, the orthodox, or newtonian worldview is that of a closed universe, wherein only material and mechanical agencies may act. The (Twentieth Century) corollary of this weltanschauung is that if any novelty can arise in the world, it can do so only in the netherworld of atoms and smaller particles or somewhere in the vast reaches of the remote

---

[1]University of Maryland, Chesapeake Biological Laboratory, Solomons, Maryland 20688-0038, U.S.A.

cosmos. From these peripheries of the observable world, causes are propagated in closed fashion to intermediate scales, such as that of the ecosystem.

Against such a background ecosystem integrity simply doesn't stand a chance, no matter how well-articulated it may be. At best it will be regarded either as an epiphenomenon or as pure metaphor. More likely, it will, like other purported "emergent" properties, be derided and rejected as a metaphysical or transcendental construct. Most ecologists will draw analogies with Clements' earlier notion of the ecosystem as "superorganism," or with Lovelock's (1979) vision of a transcendent Gaia and dismiss it forthwith. In brief, the incompatibility of ecosystem integrity with contemporary scientific attitudes is such that coexistence is scarcely possible. One or the other must give. Heretical as it may seem, I wish to suggest that it is the conventional picture of the world that today is at risk. We need to reconsider our conceptions of causality—how things happen in the world (Popper 1990). I submit that in an open world that cannot be contained by our closed models, integrity appears not as some mysterious artifact or addendum, but rather as part of the necessary glue that imparts form and order to the observable world.

We embark, therefore, on a reconsideration of the origins of natural events in the living world. I begin by accepting the radical proposal by Popper to expand the indeterminacy of the quantum realm to all scales—his description of an open universe. The problem with opening up the world to such indeterminacy is that the possibility then arises that everything will simply unravel. To avoid this scenario Popper invokes a form of dynamical cohesion, or what he calls "propensities"—a generalization of the conventional notion of force meant to pertain to a probabilistic theatre of events. Unfortunately, Popper's formulation was quite vague, and it lacks a more concrete image of what might lie behind propensities. For this purpose I am suggesting the oft-used example of positive feedback. Unfortunately, however, positive feedback, or more precisely autocatalysis, is considered by most to be a form of mechanism, wholly consistent with the newtonian formulation. Therefore, we need to elaborate these non-mechanical aspects of autocatalysis that differentiate it from conventional agencies. But if autocatalysis is not mechanism, then in exactly what capacity does it function? To close out the argument, we search back into antiquity to rediscover that Aristotle's notions of formal and final causes remain highly applicable to our contemporary description of positive feedback. This marriage of radically new with ancient ideas brackets the evolutionary narrative quite nicely and provides the context wherein integrity now appears as a legitimate and robust attribute of ecosystems—one that yields a direction along which society may predicate an ethical treatment of the natural world (Westra 1994).

2. An Open World

Ecology, sometimes called "the subversive science," is a fertile breeding ground for revolutionary attitudes about nature. Unfortunately, some purported revolutions have been ill-considered, such as Clements' aforementioned suggestion that ecosystems are superorganisms—a claim for which little evidence exists. More recent hypotheses by Lovelock (1979) about the origins of order in the global ecosystem (the Gaia hypothesis) also follow Clements' mistaken ontology, and thereby invite derision by the majority of scientists. It is only when we discover that Clements had his ordinalities reversed that Lovelock begins to make sense. Ecosystems are not superorganisms; organisms are superecosystems (Depew and Weber 1995).

Interestingly enough, not all radical notions in ecology are immediately attacked. An example of a truly exotic perspective that has received relatively little resistance is that of hierarchy theory (Allen and Starr 1982). Hierarchy theory rejects the Newtonian postulate that causes are propagated universally. That is, in a newtonian world, an event at any scale is ramified to all other scales. In the hierarchical view, however, the consequences of an event at a given scale are attenuated at adjacent levels and become inconsequential at remote scales. This is another way of saying that causality received from other scales is always incomplete (although current descriptions of the theory rarely stress this point). Causal closure begs for something at the focal level to complete the picture. If the focal level happens to be that of the ecosystem, something must arise at that scale to lend integrity to the system. But what?

Popper (1990) suggests why causes do not propagate intact between scales. He cites the "interference" of stochastic and uncontrollable events that have the effect of "opening" the world, causally speaking. That is, he now bids us abandon the notion of a causally closed universe (Popper 1982). To be sure, it is always possible to cite phenomena in relative isolation as examples of strict determinism. But in proportion to the enormous welter of events that make up our world, the strictly mechanical ones comprise but a minor fraction. Of course, mechanisms are useful as ideal limits to which other phenomena conform to greater or lesser degree. The universe in general, however, is open. In accounting for the reasons why some particular event happens, it is often not possible to identify all causes, even if one includes all levels of explanation. There will always remain a small (sometimes infinitesimal) open window that no cause covers. This openness is what drives evolution, and it is only by acknowledging such lacunae that Popper maintains we can embark upon the pathway to a solid "evolutionary theory of knowledge."

Popper's world, though open, is not wholly without form. Those agencies that keep reality from dissolving into total randomness Popper (1990) calls "propensities." In his opinion, we inhabit a "world of propensities." They are the loose glue that keep the world from flying apart. Propensities are the tendencies that certain processes or events might occur within a given context (Ulanowicz 1995a). The subjective "might" connotes a probabilistic aspect to propensities. In fact, Popper's initial example could have come from any textbook on probabilities: Suppose we estimate the probability that a certain individual will survive until 20 years from the present, say to a particular day in 2015. Given the age, health and occupation of that individual, we may use statistics on the survival of past similar individuals to estimate the probability our subject will survive until 2015. As the years pass, however, the probability of survival until the given date does not remain constant. It may increase, if the person remains in good health, decrease if accident or sickness should intervene, or even fall irreversibly to zero in the event of death.

What Popper wishes to convey with this simple example is that there is no such thing as an absolute probability. All probabilities are contingent to a greater or lesser extent upon circumstances and interfering events. While this is manifestly clear in the example just mentioned, it is mostly ignored in classical physics, where one deals largely with events that are nearly isolated. In classical physics events are either independent or rigidly coupled in mechanical fashion—if A occurs then B follows in lock-step fashion. B is forced to follow A.

What, then, in physics are called "forces," Popper regards as the propensities of events in near isolation. The classical (and motivating) example is the mutual attraction of two heavenly masses for each other. The virtual absence of interfering events in this case allows for very precise and accurate predictions. With only a well-defined force at play, the probability of a given effect subsequent to its eliciting force approaches unity. Propensities in the limit of no interfering agencies degenerate (in the mathematical sense of the word) into forces.

Propensities are those agencies that populate the causal realm between the "all" of newtonian forces and the "nothing" of stochastic infinitesima. They can appear spontaneously at any level of observation because of interferences among processes occurring at that level. This circumstance highlights Popper's second difference between propensities and common probabilities. Propensities are not properties of an object; rather they are inherent in a situation. Propensities always exist among, and are mutually defined by, other propensities. There are no isolated propensities in nature, only isolated forces or nothingness.

While it may be permissible to talk about the force of attraction between two heavenly bodies in a context that is almost vacuous, Popper maintains that one cannot apply the same reasoning to the fall of an apple from a tree. "Real apples are emphatically not newtonian apples!," he opines. When an apple will fall depends not only upon its newtonian weight, but also upon the blowing wind, and the whole process is initiated by biochemical events that weaken the stem, etc. Exactly what happens and when is conditional upon any number of other events. For this reason Popper appeals, "We need a calculus of relative or conditional probabilities as opposed to a calculus of absolute probabilities."

3. Dynamical Cohesion

As Popper describes them, his propensities remain mysterious as to their origins. What sort of agencies are these propensities? Can one give an example of how they might arise in a given context?

We begin our search for ordering agencies by considering what sort of interactions might ensue when two processes occur in proximity to each other. There are three qualitative effects the first process could have on the second: it could be beneficial (+); it could be detrimental (-); or it could have no effect (0). The latter process in its turn could have any of these same effects on the former. Whence, there are nine pairs of possible interactions, e.g., (+-), (-0), (--), etc. I wish to argue that one of these combinations gives rise to behavior that is qualitatively very different from the other eight possibilities. In fact, I would go so far as to assert that it has the potential for generating decidedly non-mechanical behaviors and for imparting cohesion or integrity to systems that contain mutualistic dynamics.

Mutualism (++) is a special case of positive feedback. Positive feedback can arise according to any number of scenarios, some of which involve negative interactions (e.g., two negative interactions taken serially can yield a positive overall effect). "Mutualism" we shall take as "positive feedback comprised wholly of positive component interactions." Mutualism need not involve only two processes, and when more than two elements are involved, it becomes "indirect mutualism."

A schematic of indirect mutualism among three processes or members is presented in Figure 1. The plus sign near the end of the arrow from A to B indicates that an increase in the rate of process A has a strong propensity to increase the rate of B. Likewise, growth in process B tends to augment that of C, which in its turn reflects positively back upon process A. In this sense the behavior of the loop is said to be "autocatalytic," a term borrowed from chemistry that means "self-enhancing." An increase in the activity of any member of the triad will tend to increase the activities of the others as well.

In keeping with Popper's idea of an open universe, we do not require that A, B and C be linked together in lock-step fashion, only that the propensities for positive influence be stronger than cumulative decremental interferences. Also, there is the issue of the phasing of the influences. It is conceivable that the timing of sequential positive effects could result in overall negative feedback. Such configurations are simply excluded from our definition of autocatalysis.

Many examples of indirect mutualism in ecology are subtle and require much elaboration. One I discovered by chance personal encounter is somewhat more straightforward (Ulanowicz 1995b). Inhabiting freshwater lakes over much of the world, and especially in subtropical, nutrient-poor lakes and wetlands are various species of aquatic vascular plants belonging to the genus *Utricularia*, or the bladderwort family. Although these plants are sometimes anchored to the lake bottom, they do not possess feeder roots that draw nutrients from the sediments. Rather, they absorb of their sustenance directly from the surrounding water. One may identify the growth of the filamentous stems and leaves of *Utricularia* into the water column with process A in Figure 1.

Upon the leaves of the bladderworts invariably grows a film of bacteria, diatoms and blue-green algae that collectively is known as periphyton. There is evidence that some species of *Utricularia* secrete mucous polysaccharides (complex sugars) to bind algae to the leaf surface and to attract bacteria (Wallace 1978). Bladderworts are never found in the wild without their accouterment of periphyton. Apparently, the only way to raise periphyton without its film of algae is to grow *Utricularia* seeds in a sterile medium (Bosserman 1979). If we identify process B with the growth of the periphyton community, it is clear that, bladderworts provide an areal substrate which the periphyton species (not being well adapted to growing in the pelagic, or free floating mode) need to grow. Some species may even provide other subsidies to the periphyton film.

Enter component C in the form of a community of small, almost microscopic (ca. 0.1mm) motile animals, collectively known as "zoophytes," that feed upon the periphyton film. These zoophytes can be from any number of genera of cladocerae (water fleas), copepods (other microcrustacea), rotifers and ciliates (multi-celled animals with hairlike cilia used in feeding). In the process of feeding upon the periphyton film, these small animals occasionally bump into hairs attached to one end of small bladders, or utrica, that give the macrophyte its family name. When moved, these trigger hairs open a hole in the end of the bladder, the inside of which is maintained by the plant at negative osmotic pressure with respect to the surrounding water. The result is that the animal is sucked into the bladder, and the opening quickly closes behind it. Although the animal is not digested inside the bladder, it does decompose, releasing nutrients that can be absorbed by the surrounding bladder walls. The cycle of Figure 1 is now complete.

It is appropriate at this juncture to ask how prevalent is autocatalysis in the ecological realm? Other familiar examples of indirect mutualism include symbiosis of algae-zoophytes in coral reefs and the homeostatic regulation of nutrients in the euphotic zone of the open oceans by the "microbial loop" of picoplanktonic organisms (Stone and Weisburd 1992). Interestingly, all three examples pertain to oligotrophic, or nutrient poor environments. One might hastily conclude, therefore, that autocatalysis is relegated to the margins of ecosystem behavior. Such judgement ignores, however, the observation that oligotrophy appears to be

Figure 1. *Utricularia* and Indirect Mutualism (see text for discussion).

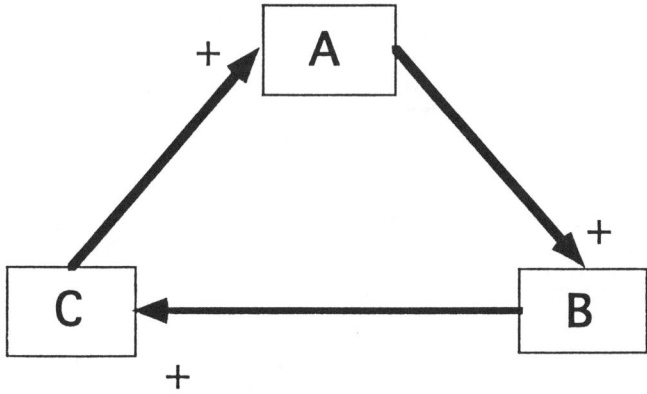

an endpoint for ecological succession, rather than its starting point (Baird et al. 1991). That is, the trend in ecological succession is to sequester as many available resources as possible within the system biomass. Unless sources of nutrients are extremely abundant, this tendency ultimately drives the available abiotic resources towards very low values, i.e., the system eventually becomes oligotrophic. A more correct conclusion, therefore, would be that autocatalysis characterizes the endstate towards which systems, left on their own, naturally converge. One recognizes, therefore, that autocatalysis and system integrity are inextricably entwined as the natural ends for living systems. As such they deserve special consideration whenever we propose to intervene in the natural course of events.

Because the example of indirect mutualism provided by *Utricularia* is so colorful, it becomes too easy to get lost in the mechanical-like details of how it, or any other example of mutualism, operates. For it is important in biological systems that the components of any system maintain some plasticity or indeterminacy. Such is obviously the case with the periphyton and zooplankton communities, as their compositions change with various habitats. Plasticity applies as well over the longer time scale to *Utricularia* itself, which has evolved into numerous species, and even exhibits a degree of polymorphism over rather short intervals (Knight and Frost 1991). Such plasticity or adaptability contrasts with the usual situation in chemistry, where the reactants in any autocatalytic process are fixed, thereby contributing to the stereotypical image of autocatalysis as a "mechanism."

4. Non-Mechanical Attributes

Although autocatalysis as mechanism may well pertain to most chemical examples, I wish to argue that such identification is wholly inappropriate as soon as the elements that constitute the autocatalytic loop become adaptable. In general, autocatalysis is not a mechanism. Taken as a whole, autocatalytic systems exhibit properties that transcend the much-overused metaphor of nature-as-machine (Ulanowicz 1989).

As a first example, autocatalytic configurations, by definition, are growth enhancing. An increment in the activity of any member engenders greater activities in all other elements. The feedback configuration results in an increase in the aggregate activity of all members engaged in autocatalysis over what it would be if the compartments were decoupled.

Of course, even conventional wisdom acknowledges the growth enhancing characteristic of autocatalysis. Far less attention is paid, however, to the selection pressure which the overall autocatalytic form exerts upon its components. For example, if a random change should occur in the behavior of one member that either makes it more sensitive to catalysis by the preceding element or accelerates its catalytic influence upon the next compartment, then the effects of such alteration will return to the starting compartment as a reinforcement of the new behavior. The opposite is also true. Should a change in the behavior of an element either make it less sensitive to catalysis by its instigator or diminish the effect it has upon the next in line, then even less stimulus will be returned via the loop.

Unlike newtonian forces that always act in equal and opposite directions, the selection pressure associated with autocatalysis is inherently asymmetric. Autocatalytic configurations impart a definite sense (direction) to the behaviors of systems in which they appear. They tend to ratchet all participants toward ever greater levels of performance.

Perhaps the most intriguing of all their attributes is the way autocatalytic systems affect the transfers of material and energy between their components and the rest of the world. Figure 1 does not portray such exchanges, which generally include the import of substances with higher exergy (available energy) and the export of degraded compounds and heat. The degradation of exergy is a spontaneous process mandated by the second law of thermodynamics. But it would be a mistake to assume that the autocatalytic loop is itself passive and merely driven by the gradient in exergy. Suppose, for example, that some arbitrary change happens to increase the rate at which materials and exergy are brought into a particular compartment. This event would enhance the ability of that compartment to catalyze the downstream component, and the change eventually would be rewarded. Conversely, any change decreasing the intake of exergy by a participant would ratchet down activity throughout the loop. The same argument applies to every member of the loop, so that the overall effect is one of centripedality, to use the term coined by Sir Isaac Newton. The autocatalytic assemblage behaves as a focus upon which converge increasing amounts of exergy and material that the system draws unto itself.

Taken as a unit, the autocatalytic cycle is not simply acting at the behest of its environment, it actively creates its own domain of influence. Such creative behavior imparts a separate identity and ontological status to the configuration above and beyond the passive elements that surround it. We see in centripedality the most primitive hint of entification, selfhood and id. In the direction toward which the asymmetry of autocatalysis points we see a suggestion of a telos, an intimation of final cause (Rosen 1991). Popper (1990) put it all most delightfully, "Heraclitus was right: We are not things, but flames. Or a little more prosaically, we are, like all cells, processes of metabolism; nets of chemical pathways."

To be sure, autocatalytic systems are contingent upon their material constituents and usually also depend at any given instant upon a complement of embodied mechanisms. But such contingency is not, as strict reductionists would have us believe, entirely a one-way street. Autocatalysis, by its very nature, is prone to induce competition, not merely among different properties of components (as discussed above under selection pressure), but its very material and (where applicable) mechanical constituents are themselves prone to replacement by the active agency of the larger system. For example, suppose A, B, and C are three sequential elements comprising an autocatalytic loop as in Figure 1, and that some new element D: (1)appears by happenstance, (2)is more sensitive to catalysis by A, and (3)provides greater enhancement to the activity of C than does B. Then D either will grow to overshadow B's role in the loop, or will displace it altogether.

In like manner one could argue that C could be replaced by some other component E, and A by F, so that the final configuration D-E-F contains none of the original elements. (Simple induction will extend this argument to an autocatalytic loop of n members.) Important to notice in this case is the fact that the characteristic time (duration) of the larger autocatalytic form is longer than any of its constituents. Persistence of active form beyond present makeup is hardly an unusual phenomenon. One sees it in the survival of corporate bodies beyond the tenure of individual executives or workers; of plays, like those of Shakespeare that endure beyond the lifetimes of individual actors. But it also is at work in organisms as well. One's own body is composed of cells that (with the exception of neurons) did not exist seven years ago. The residencies of most chemical constituencies (even of those comprising the neural synapses by which are recorded long-term memory in the brain) are of even shorter duration. Yet most people would be recognized by close friends they haven't met in the last ten years.

The influence of the overall kinetic form is not exerted only during evolutionary change, but acts also to effect the normal replacement of parts. For example, if one element of the loop should happen to disappear for whatever reason, it is (to use Popper's own words) "always the existing structure of the...pathways that determines what new variations or accretions are possible" to replace the missing member.

The appearance of centripedality and the duration of form beyond that of constituents make it particularly difficult to maintain any hope that a strict reductionist, analytical approach to describing organic systems will succeed in the end. Although the system requires material and mechanical elements, it is evident that some behaviors, especially those on a longer time scale, are, to a degree, autonomous of lower level events. Attempts to predict the course of an autocatalytic configuration by ontological reduction to material constituents and mechanical operation are doomed over the long run to failure.

It is important to note that the autonomy of a system may not be apparent at all scales. If, for example, one's field of view does not include all the members of an autocatalytic loop, the system will appear linear in nature, i.e., one can identify an initial cause and a final result. The subsystem possibly would appear wholly mechanical in its behavior. However, once the observer expands the scale of observation enough to encompass all members of the loop, then autocatalytic behavior with its attendant centripedality, persistence and autonomy emerges as a consequence of this wider vision.

To recapitulate, our study of indirect mutualism has revealed that autocatalytic systems can possess at least seven properties. Autocatalysis induces (1)growth and (2)selection. It exhibits (3)an asymmetry that can give rise to the (4)centripetal amassing of material and available energy. The presence of more than a single autocatalytic pathway in a system presents the potential for (5)competition. Autocatalytic behavior is (6)autonomous to a degree of its microscopic constitution. It (7)emerges whenever the scale of observation becomes large enough.

5. Expanded Causality

In our consideration of autocatalytic systems, we see how agency can arise quite naturally at the very level of observation. This occurs via the relational form that processes bear to one another. Furthermore, there is an assymetry inherent in autocatalytic systems, and a direction is defined by the centripedality they exhibit. Neither of these observations fits well into newtonian descriptions of events. Rather, they are reminiscent of an earlier narrative of how things happen — one made by Aristotle.

Aristotle's image of causality was more complicated than the one subsequently promulgated the by founders of the Enlightenment (Rosen 1985). He taught that a cause could take any of four essential forms: (1)material, (2)efficient, or mechanical, (3)formal, and (4)final. Any event in nature could have as its causes one or more of the four types. The textbook example for parsing causality into the four categories concerns the building of a house. Behind this process, the material causes are obviously the stone, mortar, wood, etc., that are incorporated into the structure, and as well the tools that are used to put these elements together. The workers whose labor brings the material elements together comprise the efficient cause.

The formal cause behind the construction of a house is not as clear-cut as the first two. Aristotle posited abstract forms towards which developing entities naturally progressed. Thus, he thought the form of the adult chicken was immanent in the fertilized egg, and it was this endpoint that attracted all earlier forms of the growing chicken into itself. This notion does not translate well outside the realm of ontogeny, so that the closest one can come to the formal cause for building a house is the image of the completed house in the mind of the architect. Usually, this image takes on material reality as a set of blueprints that orders the construction of the building. The final cause for building the house is the need or desire for shelter on the part of those who will occupy it.

Blueprints or an image in an architect's mind provide rather equivocal examples of formal cause. I have suggested instead (Ulanowicz 1995a) that one consider a military battle, which, despite its unsavory image, nonetheless lends a more appropriate example of formal cause. The material causes of a battle are the weapons and ordnance that individual soldiers use against their enemies. Those soldiers, in turn, are the efficient causes, as it is they who actually swing the sword, or pull the trigger to inflict unspeakable harm upon each other. The officers who are directing the battle concern themselves with the formal elements, such as the juxtaposition of their armies via-a-vis the enemy in the context of the physical landscape. It is these latter forms that give shape to the battle and serve as agents of the third type. In the end, the armies were set against each other for reasons that were economic, social and/or political in nature. Together they provide the final cause or ultimate context in which the battle is waged.

In addition to lending more concrete reality to formal cause, the example of a battle also serves better to highlight the hierarchical nature of Aristotlean causality. All considerations of political or military rank aside, the soldier, officer and head of state all participate in the battle at different scales. It is the officer whose scale of involvement is most commensurate with those of the battle itself. In comparison, the individual soldier usually affects only a subfield of the overall action, whereas the head of state influences events that extend well beyond the time and place of battle.

That is, formal cause should act most frequently at what is called the "focal" level of observation. Efficient causes tend to exert their influence over only a small subfield, although their effects can be propagated up the scale of action. The entire catastrophe transpires under constraints set by the final agents. The three contiguous levels of observation constitute the fundamental triad of causality (Salthe 1993), all three elements of which should be apparent to the observer of any physical events. It is normally (but not universally) assumed that events at any hierarchical level are contingent upon (but not necessarily determined by) material elements at lower levels.

6. The Full Causal Picture

In light of the foregoing, autocatalysis seems to exhibit aspects of both formal and final causes, sensu Aristotle. An expanded view of causality in nature can now be cast in metaphorical terms: One imagines the full suite of natural phenomena as a background over which one tries to place a curtain of causality. Newtonian description provides enough curtain to cover only part of the background. In heeding Rosen, many now believe that formal and final causality will provide enough cloth to cover the remainder. Popper, however, warns us that the cloth is not entirely whole. There exist at all levels small holes in the cover, thru which novel events may emerge unexpectedly. That the curtain doesn't come apart is due to the preponderance of cloth between the holes. We now acknowledge the connecting fabric consists not only of material and mechanical threads, but of formal and final ones as well.

Philosophical imagery is all well and good, but one must also address practical considerations, such as how one can identify an ecosystem with a high degree of integrity. Fortunately, the possibility for such distinction follows from our identification of integrity with autocatalysis and self-organization. That is, an ecosystem with strong integrity is one that is relatively insensitive to its physical inputs. Whenever the response of the ecosystem to changing inputs can be tested without doing irreparable harm to the system, one should be able to assess the relative level of system integrity. For example, one expects that the productivity and structure of the *Utricularia* system is, up to a point, rather insensitive to fluctuations in dissolved abiotic nutrients. The advantage of positive feedback to this macrophytic system waxes and wanes in compensatory fashion as nutrient levels rise and fall within the oligotrophic range (Ulanowicz 1995b). Once a threshold in nutrient level is exceeded, however, the entire system can collapse and be displaced by another biotope (one dominated by *Typha* in the Everglades) that responds more directly to fluctuations in available nutrients.

Such a "black-box" test for integrity tells us about the entire ecosystem, but provides us with little information as to which species could be manipulated or removed from the system without considerable loss of integrity. An empirical search for key species, if at all feasible, would be extremely demanding of time and research resources. One possible shortcut might be to elaborate the full suite of trophic transfers of key nutrients within the ecosystem. The structure of cycling within the resultant network should highlight the major pathways of internal recycling (Ulanowicz 1983), and thereby provide clues as to the major players in the maintenance of system integrity.

The reader should be cautioned that any analysis to identify potentially manipulatible or expendable species, such as the cycle method just cited, is fraught with grave dangers. Given the considerable indeterminacy that is a necessary part of any ecosystem, coupled with the pervasive uncertainty of ecological methods, techniques and information, it is highly unlikely that ecologists will be able to predict with reasonable certainty future ecosystem states. Promoting ecological integrity, then, seems to require a precautionary approach, or a shifting of the burden of proof on to those who propose that their activities will not impair integrity (see e.g., Lemons 1995).

In conclusion, we regard ecological integrity as very much a part of the overall fabric of nature. It is not some mysterious or elusive construct that some are trying to tack on to a clockwork universe. Ecosystems may be defined only insofar as they are capable of behaving as an integrated whole, like the *Utricularia* system (Norton and Ulanowicz 1991). Without the integrity afforded by indirect mutualism, ecosystems would not exist, nor for that matter could life as we know it continue. Integrity is an essential attribute of all living systems. To ignore it could be disastrous, to destroy it wantonly would be immoral.

7. References

Allen, T.F.H. and T.B. Starr. 1982. *Hierarchy*. University of Chicago Press, Chicago.
Baird, D., J.M. McGlade, and R.E. Ulanowicz. 1991. The Comparative Ecology of Six Marine Ecosystems. Phil. Trans. R. Soc. Lond. B 333: 15-19.
Bosserman, R.W. 1979. The Hierarchical Integrity of Utricularia-Periphyton Microecosystems. PhD Dissertation, University of Georgia, Athens, Georgia.
Costanza, R. 1992. Toward an Operational Definition of Ecosystem Health. Ecosystem Health: New Goals for Environmental Management, R. Costanza, B.G. Norton, and B.D. Haskell, eds. Island Press, Washington, pp. 239-256.
Depew, D., and B.H. Weber. 1995. *Darwinism Evolving: Systems Dynamics and the Geneology of Natural Selection*. MIT Press, Cambridge, MA.
Hollings, C.S. 1992. Cross-Scale Morphology, Geometry, and Dynamics of Ecosystems. *Ecol. Monogr.* 62: 447-502.
Knight, S.E., and T.M. Frost. 1991. Bladder Control in *Utricularia* macrorhiza: Lake Specific Variation in Plant Investment in Carnivory. *Ecology* 72: 728.
Lemons, J., ed. 1995. *Scientific Uncertainty and Environmental Problem-Solving*. Blackwell Science, Cambridge, MA.
Lovelock, J.E. 1979. *Gaia: A New Look at Life on Earth*. Oxford University Press, New York.
Norton, B.G., and R.E. Ulanowicz. 1991. Scale and Biodiversity Policy: A Hierarchical Approach. *Ambio* 21: 244-249.
Peters, R.H. 1993. *A Critique for Ecology*. Cambridge University Press, Cambridge.
Popper, K.R. 1990. *A World of Propensities*. Thoemmes, Bristol.
Popper, K.R. 1982. *The Open Universe: An Argument for Indeterminism*. Rowman and Littlefield, Totowa, NJ.
Rosen, R. 1985. Information and Complexity. In *Ecosystem Theory for Biological Oceanography*, R.E. Ulanowicz and T. Platt, eds. Canadian Bulletin of Fisheries and Aquatic Sciences 213, Ottawa, pp. 221-233.
Rosen, R. 1991. *Life Itself: A Comprehensive Inquiry into the Nature, Origin and Foundation of Life*. Columbia University Press, New York.
Salthe, S.N. 1993. *Development and Evolution: Complexity and Change in Biology*. MIT Press, Cambridge, MA.

Stone, L., and R.S.J. Weisburd. 1992. Positive Feedback in Aquatic Systems. *Trends Res. Ecol. Evol.* 7: 263-267.
Ulanowicz, R.E. 1983. Identifying the Structure of Cycling in Ecosystems. *Math. Biosci.* 65: 219-237.
Ulanowicz, R.E. 1989. A Phenomenology of Evolving Networks. *Systems Research* 6: 209-217.
Ulanowicz, R.E. 1995a. The Propensities of Evolving Systems. In *Social and Natural Complexity*, E.L. Khalil and K.E. Boulding, eds. Routledge, London.
Ulanowicz, R.E. 1995b. Utricularia's Secret: The Advantage of Positive Feedback in Oligotrophic Environments. *Ecological Modelling* 79: 49-57.
Wallace, R.L. 1978. Substrate Selection by Larvae of the Sessile Rotifer Ptygura Beauchampi. *Ecology* 59: 221-227.
Westra, L. 1994. *An Environmental Proposal for Ethics: The Principle of Integrity*. Rowman and Littlefield, Lanham, Maryland.

Chapter 7
ECOSYSTEM INTEGRITY IN A CONTEXT OF ECOSTUDIES AS RELATED TO THE GREAT LAKES REGION

Henry A. Regier[1]

1. A Context in Recent History

Western culture apparently crossed a great historical divide some two to three decades ago, with a bit of help from ecostudents. Seers or prophets as gifted people who can see into the present are offering their perceptions about the new realities and their perspectives on the new mores. Ideologues and theologues committed to fundamentals imposed in an earlier age decry what they take to be agnosticism, relativism and manifold heresies. Creative pathfinders of early post-modernism have been discovering how the new culture and the old nature may interact harmoniously ecosystemically. Entrepreneurs of obsolescent mindsets have attempted to subvert the new to discredited old ends, as in the greedy neo-conservative echo of capitalism that metastasized in Western countries in the 1980s.

Each new reality brings its own new bad things as well as good things. Each creative event, whether ultimately judged good or bad on balance, brings with it its own emergent domain of new ignorance, for scientists and scholars to explore. Creative scientists can help to create their own ignorance, which they can then help to dispel; this may be perceived to be a bad vicious circle or good positive feed-back, depending upon one's predisposition about contemporaneous "science."

Certainly the new reality will have its catastrophic phenomena, some good and some bad; a hope is that there will be fewer and less severe bad catastrophes than would have been the case had the old realities continued in their old ways. Perhaps we can help to prevent some bads and to foster some goods, and swing the balance to the goods.

The approach of this paper might be termed "transdisciplinary empirical" in an analytical sense or "post-normal narrative" in a synthetic sense. It has not been written according to any philosophical criterion of how to write such a paper, due in part to my ignorance of such criteria. If there is critical content in the paper, it may be "critical" in the sense of crisis as a crux with respect to a major opportunity. It is a sketch of a worldview that I share and in which I relate to the works of philosophers, and especially to those of Westra (1994).

For convenience, suppose that it was in the year 1968, plus or minus a year or two, that Western Culture crossed that highest divide within recent history (Drucker 1989), or embarked on a new path at a crux that appeared in the old path. Consider some events of about

---

[1]Institute for Environmental Studies, University of Toronto, Toronto, Ontario M5S 1A4, Canada.

1968 as related particularly to ecostudies, as examples of intellectual components of the new mindset.

1.1. ALPBACH SYMPOSIUM—A. KOESTLER, L. VON BERTALANFFY AND OTHERS.

Living nature consists of interactive holarchies (or "hierarchies") as relatively open self-organizing systems in which the elements—holons at different levels—are each also relatively open self-organizing systems. Superficially the etymology of "holon" is that of an oxymoron, i.e. whole-part; but the reality of the holon transcends this apparent contradiction. A holon, including a holarchy as a holon, has a dual Janus-like tendency: to be self-assertive and to persist as a quasi-independent whole through partial internal control of itself and of the nested smaller holons within it; and to integrate and to network participatively and reciprocally with other holons within a larger holarchy of which it is a part. Fully predictable order and stability may occur when these two tendencies are in equilibrium and other conditions are also met; such a state may conceivably occur occasionally and temporarily (see Koestler and Smithies 1969, Koestler 1978). Implicitly this mindset—Koestler referred to it as SOHO for self-organizing, holarchic and open—was consistent with evolving democratic and communitarian traditions as in the United Nations' approach to governance and inconsistent with the presuppositions of hierarchic top-down imperialism, socialism and fascism.

1.2. PARIS SYMPOSIUM ON THE BIOSPHERE—V. KOVDA, M. BATISSE, F. SMITH, J. HUXLEY, W. DILL AND OTHERS

Concepts of a global evolutionary hierarchy—lithospheric, hydrospheric, biospheric and noospheric phenomena of V. Vernadsky and others—were invoked as bases for transcultural cooperation in Unesco's Man and the Biosphere or MAB Program, to resolve severely discordant behavior by culture with respect to nature within our biosphere (see Unesco 1970). MAB, with the ankh as its symbol, was led by Russian and French scientists and may be a kind of echo of interactions in Paris early in the 20th Century between V. Vernadsky, P. Teilhard de Chardin and idealists and intuitionists such as H. Bergson. The American A. Leopold may also have contributed indirectly to this MAB leitmotif through F. Smith, W. Dill and others. Americans, other than L.K. Caldwell, generally have not shown much interest in the MAB approach; political differences with Unesco's leaders, American parochialism or Americans' ideological hubris may be partly to blame. The invoking of Vernadsky's non-Marxist mindset by U.S.S.R. scientists involved transcendence of their country's political ideology (Serafin 1988) at about the time of the Prague spring, - perhaps a harbinger of the collapse of Marxist-Leninist ideology.

1.3. OECD EUTROPHICATION STUDY—R.A. VOLLENWEIDER

A lake was perceived as a Bertalanffian subsystem nested within a landscape; together lake and landscape were relatively open and self-organizing. Eutrophication was understood to involve ecosystem transformation that included non-linear phenomena, as in W. Einsele's inference of positive feedback or "Mitwirkung" in his 1930s work on the Schleinsee. Systemic effects on lakes of certain practices on land, inferred from analytical and comparative work, were described in a way that was useful for formulating corrective measures for bad practices that triggered the eutrophication transformation. This work quickly contributed to a somewhat broader synthesis of ecosystemic responses to cultural interventions (see Vollenweider 1968, Regier and Henderson 1973).

1.4. BROOKHAVEN SYMPOSIUM ON STABILITY AND DIVERSITY—
G.M. WOODWELL, L.K. CALDWELL, C.S. HOLLING AND OTHERS

Systemic consequences of various kinds of natural and cultural phenomena on natural parts of ecosystems were explored from a perspective, more or less explicit, of open self-organizing systems (see Woodwell and Smith 1969). A polarized discussion of whether taxonomic diversity contributed directly to ecosystemic stability was a current distraction, which has since been transcended.

1.5. CYBERNETICS AS AN APPROXIMATION TO A SCIENCE OF THE ECOSYSTEM—R. MARGALEF

The role of information as an attribute of reality (as distinct from "something" that contributes to mere knowledge about reality), as in the substantive content of "noosphere" of Teilhard and Vernadsky (see above), was related to natural evolution, ecological succession and ecosystem transformation as influenced by external influences (see Margalef 1968). Some versions of cybernetics were linearized conceptually within a kind of "deductive" science (Jantsch 1972) and thus were not incorporated into the emerging set of traditions sketched here.

1.6. CARELESS TECHNOLOGY—B. COMMONER, T. FARVAR,
R. DASMANN AND OTHERS

Major consequences to natural systems of progress-oriented technology were explored from a perspective of an early version of ecodevelopment. The notion that a beneficial positive feedback form of development would necessarily follow appropriate capital and technical investment in any culture, as happened in Western Germany with the Marshall Plan following the Second World War, was questioned (see Farvar and Milton 1972). A rancorous debate in the U.S.A. between B. Commoner and P. Ehrlich on whether the global environmental problematique was due more to careless technology than to careless human procreation—a kind of replay of the Malthus-Condorcet debate of two centuries ago (Sen 1994)—was quickly transcended by many reformers (Artin 1973) but still continues to excite sophomores.

1.7. ECONOMICS OF OPEN SYSTEMS—J.H. DALES, H.E. DALY,
J.K. GALBRAITH AND OTHERS

Some roles for a "free market," as part of a larger institutional system for allocating use of natural phenomena, were explored (see Daly 1968, Dales 1968, 1975). In some cultural contexts a partially self-organizing market has institutional advantages over bureaucratic control, especially if the latter evolves into a pathologically self-organizing abomination and the market is prevented from doing so by non-bureaucratic institutional means. Politically astute scholars were noting that people with power could subvert or circumvent both the market and the bureaucracy, as allocative mechanisms; the reification of the relevant abstractions inevitably involves suppression of the weak, directly or indirectly, to the advantage of those who know how to influence the market or the bureaucracy, or both.

1.8. EKISTICS—C.A. DOXIADIS

A systemic science of human settlements was sketched (see Doxiadis 1968). Doxiadis' version was received with some skepticism perhaps because it reflected some attributes of the Germanic "Ordnung" of Berlin in the 1930s. The emergence of megalopolises was generally perceived in a more open ekistical context (Leman and Leman 1976, Papaioannou 1987).

1.9. EXPLODING HUMANITY—I. MACTAGGART COWAN, L. COLE, J.M. STYCOS AND OTHERS

Our species, as a natural and cultural system, was perceived by some to be behaving as an abnormal ecological phenomenon, a pest species with little internal self-control. This view contrasted sharply with that of Vatican-oriented religious fundamentalism in which a god intervened to create humans to gratify himself somehow, and with faith in the near-miraculous capabilities of "economic development", especially when these two were synthesized as by C. Clark of Oxford (see Regier and Falls 1969).

1.10. AN ECOLOGY OF HUMANS AS A SUBVERSIVE SCIENCE—P. SEARS, P. SHEPARD

Americans also generated a largely American version of the eco-worldview which emerged in the late 1960s. Many American biologists were hard at work attempting, rather belatedly, to create a positivistic ecological paradigm, as in the work of some of the regional case studies under the International Biological Program of the International Council of Scientific Unions; this kind of effort had largely petered out by the late 1980s. Concurrently in the 1960s numerous creative ecostudents were pioneering with the "subversive science of an ecology of man." (See Shepard and McKinley [1969], which was widely used in undergraduate courses in ecology.) But relatively few Americans managed to assert enough freedom from their doctrine as God-appointed global emancipators to become fully cosmopolitan as ecostudents. In major UN Conferences, since 1972 in Stockholm, American activists have persisted in operating within the methods and mores of the Washington D.C. worldview and political dynamics. Activists from other countries have had difficulty relating to the Americans.

1.11. MOON ASTRONAUTS

The photograph of our world from space provided an unforgettable image of the biosphere as an open self-organizing system. A sense of aesthetics and of "place" had been broadening from the local as perceived from a hill (as by the 19th century romantics), to the regional as viewed from a high flying aircraft, to the hemispheric and global as viewed from a spacecraft. A global aesthetics fostered a global ethics.

1.12. SUBJUGATION, DOMINATION, EQUITY AND INTEGRITY

"It became apparent during the late 1960s and early 1970s that the multiple practices of domination—race, class, gender and destruction of the earth—were mutually supportive and shared a logic of domination within Western cultural ideology....The global ecological movement comprises mostly women, and ecological concerns are being added to [those of] women's groups, which were previously working on peace and justice issues and participatory democracy. ...Ecofeminism continues to comprise analysis, critique and vision, and it can be broadly defined as the study of and the resistance to the associated exploitation and subjugation of women and the earth" (Eaton 1995) The iniquity of such inequity was intended to be corrected through "integration" toward some state of cultural "integrity" and/or cultural-natural integrity. That none of the conventional 19th century ideologies—

capitalism, socialism, communism, fascism, imperialism—offered an acceptable vision of such integrity or a way towards such a state was widely apparent in the 1960s. The politically active major religions have been fragmented on this complex issue between doctrinaire fundamentalistic "proponents of truth" (generally led by hierarchic males) and more humble "seekers of a path" (generally fostered with female and male partnerships).

1.13. ROLE OF THE MILITARY

It came to be widely appreciated, partly as a result of events in the Vietnam War, that military destruction had progressively come to target women, children, the aged and nature. Pillaging, looting and raping had long been an accepted fringe benefit for unprincipled soldiers; then the weapons came, increasingly and deliberately, to be aimed directly at non-combatants. Though soldiers sometimes combat other soldiers directly, this is not now the accepted way to win a war. For decades the military systems on each side of the Cold War were in a mutualistic systemic relationship, they needed each other, and the tensions within and between them were dissipated in ways that were relatively harmless to themselves, e.g. by "scapegoating" small nations of no great interest to either side. The indiscriminate disintegrative propensity of the modern military became widely apparent in the 1960s.

1.14. POLITICAL REFORMS

Political reforms of various kinds, that bordered on revolution, were initiated in the 1960s. In Western countries the old conventions that fostered inequities and the related "marginalization" of people with respect to race, gender and class were attacked though not fully inactivated. The military lost much of its glory. After the 1968 Prague spring, reform in the Eastern Bloc countries was stalled for two decades, but it did then come, with some good and some bad consequences. Perhaps the Vatican spring of John XXIII, postponed by Humanae Vitae and other developments, will eventually reappear. Environmentally, the concerns about acid rain that Sweden took to the United Nations in 1968 eventually combined with other concerns to lead to the 1972 Stockholm Conference on the Human Environment, with the emergence shortly afterwards of the policy concept of "ecodevelopment" which has since evolved into "sustainable development." Though the human population issue was banned from the formal Stockholm Conference it did appear there in the informal non-governmental activities (Artin 1973) and then, in a broad context, in the Bucharest Conference on the Human Population in 1974.

1.15 DEFLATION OF POSITIVISTIC SCIENCE

The 1960s may also be taken as marking the beginnings of a transition from the earlier dominance of positivistic science and the related fundamentalistic ideology of scientism to a less hubristic, more evolutionary science of open self-organizing systems (Grabow and Heskin 1973). The mindset of the new science, with a renewed emphasis on procedural rationality to complement substantive rationality (Funtowicz and Ravetz 1991, 1993), was broadly consistent with that of the political reformers. The process of integration, effected in part by an empowered "living entity" apparently striving toward a goal of equitable integrity, became a diffuse focus for the scientific elements of the new ecostudies.

2. A Perspective on Ecostudies

The science of the newly perceived emergent realities and of the new perceptions of the old realities is post-positivistic in that an exclusive presupposition of deterministic bottom-up causality is denied. Linear causality is now coming to be accepted only as an approximation to natural phenomena of peripheral evolutionary importance, but sometimes of practical importance. Linear causality relates as an approximation to stereotypical interactions among a few things and where nothing very interesting is happening. To the extent that urban planning methods, engineering practices and clinical medicine relied exclusively on linear rational approaches, they were contra-nature. The new science and practice relate to a kind of complexity in which new phenomena that emerge are not dismissed as unreal or surreal epiphenomena. The new science also transcends the quasi-linearity related to statistical phenomena in which countless numbers of things interact mechanistically, to which the "law of large numbers" can be applied. One focus now is on middle-number complex phenomena (Allen and Starr 1982).

Within a catechism of the old science one set of key questions related to "teleological causes" as a heretical concept. This issue is being transcended—pragmatically and empirically, if not in a rational philosophical sense—within a holarchic perspective on reality. A holon exerts some partial, guiding influence on the behavior of any component sub-holon. This appears to be the case for a holon of any "size," i.e. of any set of spatio-temporal dimensions so long as the holon exhibits identity or integrity. Note that many holons may exist in a number of manifestations with "rules of transition" between different states; identity and integrity then relate to the whole interactive set of such manifestations of a polyphasic holon. The question then arises as to what the largest holon might be, given that the holon may be perceived to exhibit some self-organizing capabilities that help it to persist with a recognizable identity or state of integrity. Some ecoscientists perceive the biosphere to be a holon of this type. If so, then the biosphere might be expected to exert some "existential teleological" guidance on all the nested component holons within it, at whatever level of nesting. Others may perceive our solar system or our galaxy as holons to provide quasi-teleological guidance to our biosphere, etc.

When a paradigm gets to be boring, new scientists cannot easily be recruited to its service (Feyerabend 1988). A paradigm is, after all, a cultural invention and there are many scientists who compete for wide recognition as a paradigm inventor. A current emphasis is on pluralism of perspectives on what may be a single reality but not knowable as such by a scientist; a hope that one monistic paradigm will eventually be perceived to be true has been set aside generally. With pluralism, procedural rationality becomes more important than substantive rationality.

Ecoscience may be perceived to include compatible parts of ecology, economics, ekistics and ecumenics—as four perspectives on different but compatible aspects of the behavior of humans in the contexts of their homes, broadly defined. This kind of schema may also be appropriate for ecoethics and ecoaesthetics, with all three contributing to ecosophy, but this possibility is not explored further here. Some explanation with respect to the ecostudies is offered here, as related to Figure 1.

- Ecologists who felt embarrassed by the apparent vitalism of biology sought to explain ecosystems in terms of old preconceptions of energy acting on mass within space and time all in linear causal ways. As the new physicists have been discovering a reality in "information," ecologists are noting that an interest in such "information," except as linearized in old-fashioned physiology and genetics, used to be dismissed as vitalism, by unperceptive "peers" who enjoyed the Procrustean stretching and cutting of reductionistic "strong science." And the phenomenon of life may be more than some linear interactive linkage of space, time, mass, energy and information (Regier 1992, Woodley et al. 1993, Kay and Schneider 1994).
- Financial accounting, monitoring of market behavior and valuing things in monetary units serves some interests of all economists and becomes a consuming obsession with many who would be little more than acolytes of the arbitrary market. With

94 Henry A. Regier

Figure 1. A Perspective on Ecostudies.

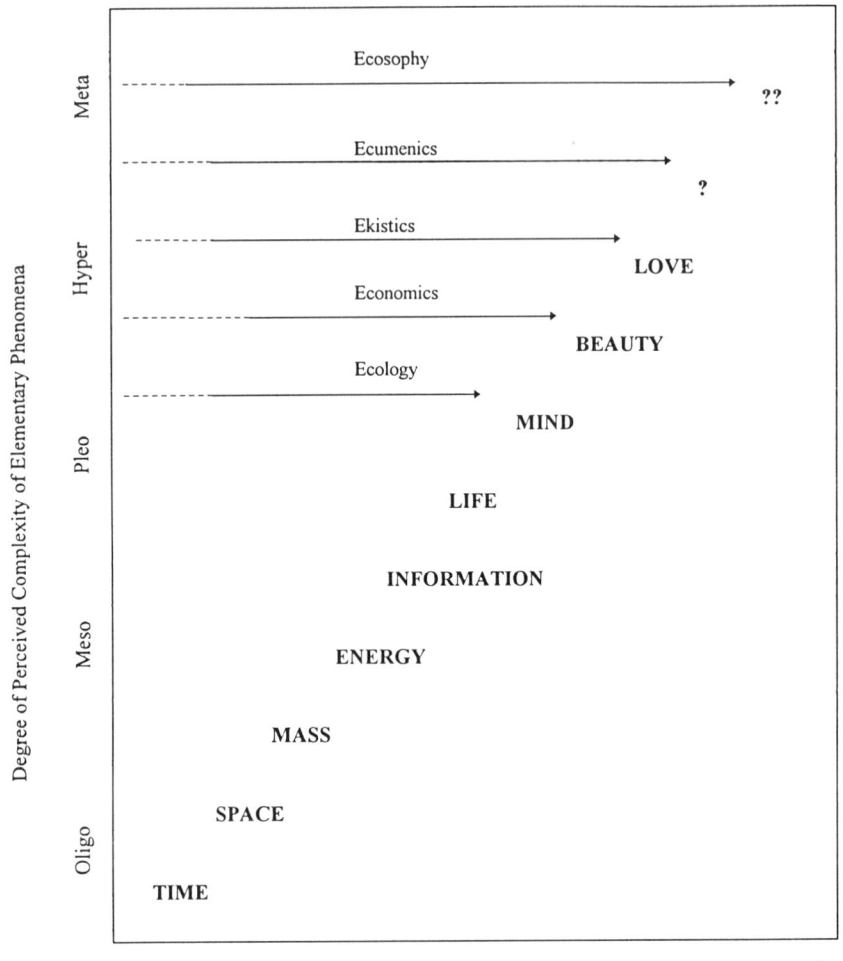

Recency of Emergence of Systemic Studies of Elementary Phenomena,
in an Eco-context

Note that the "things" in bold face type are here taken to be elementary phenomena each of which may be complex in an open self-organizing way. The "field of reality" in which these phenomena are taken to be real is that of practical and political coping, planning and decisionmaking. The continuous part of the arrow for a particular type of ecostudy implies the primary range of directed study while the dashed part is usually implicit or addressed indirectly, at least in reductionistic studies, e.g., in economics time may be addressed through the discount rate and space in the monetary cost of transportation and communication.

Dales (1968, 1973) and Daly (1968), ecoscientists try to leave accounting to accountants and to keep the market in an appropriately subsidiary role, noting that any self-organizing mechanism out of some external control is a catastrophe preparing to happen. For ecoscientists the excessively powerful arbitrary market is a heresy, a small truth that has been permitted to escape its context to become a large falsehood. The subversion of "information" by players in the arbitrary market has attracted increasing attention, and money is basically a phenomenon of the mind or a convenient fiction of the culture. The term "centralized bureaucracy" may be substituted for "market" in preceding sentences. Much of the slanging between the political advocates of the capitalistic market and those of the socialistic or welfare bureaucracy is tendentious or merely sophomoric, ecostudents are transcending such 19th century rhetoric.
- Urbanologists, with an evolutionary systems approach to the emergence of form and function in human settlements, perceived a creative role for "information" far beyond considerations of the efficiency of use of inputs of mass and energy (as resources) or of disposal of outputs of varied wastes. Buckminster Fuller's wondrous geodesic dome was an early distraction, as was the intricate systemic taxonomy of C. Doxiadis. Marshall McLuhan's perception of the global village, with "information" serving as an attractive and cohesive capability, was found difficult to assimilate into ekistics (Doxiadis 1968, Leman and Leman 1976). In a megalopolis with "integrity" the cultural phenomena including the built environment may be perceived as comprising nodes and networks, or cores and corridors, with the natural phenomena also organized in a similar and complementary way, but offset spatially from the cultural cores and corridors. A current challenge is how to design a vibrant co-system of this type. Architecture and planning of cities and landscapes is much influenced by a sense of mosaic beauty.
- Human institutions of many kinds have been perceived as open self-organizing systems in which "information," broadly defined, plays a central role in the emergent reality. A pre-occupation with taxonomy (as in Talcott Parsons' early work) had to be overcome, as it was in social ecology with its self-organizing interactive actor systems (Emery and Trist 1973). What people love or care about is particularly relevant here. To keep a version of "eco" as a prefix, this theme in ecoscience may be termed ecumenics, with a vast broadening of its more usual and narrow church-related connotation.
- A field of philosophy may be coalescing or efflorescing under a banner of ecosophy, as reflected, say, in *The Trumpeter*, a journal on ecosophy edited by A. Drengson (1983-1995). Contributions relate to the "deep ecology" of A. Naess and also to ecofeminism, ecopsychics and ecophilia.
- Some people may be committed to ecotheology of animist, Franciscan Christian, Taoist, Buddhist, Sufi Muslim or other persuasion. Ecofeminists with a commitment to feminine deities may prefer the term ecothealogy (Eaton 1995). This class of ecostudies is not included in Figure 1.

The schema in the figure is in two dimensions, which is clearly simplistic in that respect. It reflects a rough generalization, induced empirically. Any attempt to use it deductively should be cautious. In the conventional methods, based on harsh rigor, a living entity is first killed before it is examined; it is not surprising that no special attributes of life are found with this approach.

Criteria are evolving for what is acceptable as proper ecoscience, in the contexts of ecology, economics, ekistics and ecumenics. The new transdisciplinary mindset is itself being perceived as holarchic, self-organizing and relatively open as an epistemological system. The epistemology of ecoepistemology or the new science of the new science (Funtowicz and Ravetz 1991, 1993), is itself a focus for creative tension. Though unattainable, objectivity is still an abstract ideal for the ecoscientist. In the emergent science as currently practised an onus is on the ecostudent to indicate, if one can, the nature and extent of one's deviations from abject objectivity.

Increasingly and sometimes discordantly, the applied scientist, as a member of a profession, is expected to serve the interests of the client who pays for the professional services, and any prior commitment to disinterested objectivity takes a secondary position to that of responsibility to the client. The financial rewards for idealistic objectivity may be small compared to those that a talented quasi-scientific sycophant can attract.

The following additional comments relate to the contents of Figure 1:
- It used to be held that phenomena to the bottom left were real and could be described and explained with respect to universalistic and deterministic linear causes while phenomena to the top right were ideal, contextual, non-deterministic and could be studied or appreciated subjectively. In post-post-modern times we may be transcending this apparent antithesis.
- By definition an ecostudy must necessarily address phenomena of life and its environment; complex studies of non-living phenomena may be compatible with complex studies of living phenomena, in a holarchic context.
- Systemic studies of sub-life holons, as by I. Prigogine and associates, are providing concepts and methods that are of interest in the more complex conceptual holons. Some may expect that a set of isomorphic concepts may be useful at all levels of this holarchy.
- Complexity implies more than confusing complication; ecostudies relate to complexity, inter alia, in a pluralistic or post-normal way.
- The inferred scientific laws with respect to a particular systemic phenomenon as in Figure 1 may act as partial constraints on phenomena further toward the top and right in the sequence. This would justify a particular version of reductionism that would focus on partial constraints rather than sufficient causes.
- Each of the types of ecostudy are emerging from earlier efforts at understanding within the general mindset of logical positivism, through descriptive and classificational states, towards post-normal pluralistic methods; an exception may be ecosophy which appears to have emerged in a post-normal way or may involve a re-emergence in post-modern thought of earlier philosophies.
- Conventionally, reductionists perceive that the apparently less complex elements strongly constrain and may fully determine the more complex phenomena, while the holists perceive such constraints to be relatively weak. Ecosystems students infer or presuppose that these conceptual perspectives may have comparable but partial explanatory power, and that a transcending synthesis of the two, i.e,. reductionist and holist is preferred as in the holarchic synthesis.
- An appropriate ecotechnology can be developed for any type of ecostudy, and versions for each type are already available and in practice.
- Eco-phenomena are not managed in a control sense but are fostered, guided, husbanded, constrained, administered, adapted to, etc., toward desired or good states, which may be the pristine state in locales in which nearly pristine nature is to be preserved, and away from bad states in settings that have been strongly modified through culture.
- Each of the elementary types of complex elements in Figure 1 may be in different states some of which may be perceived by us to be better than others and some states

may be perceived to be bad or even evil, e.g., exergy vs. entropy, truth vs. lies, life vs. death, beauty vs. ugliness, love vs. hate, peace vs. war.
- Some self-organizing capability toward one or another "normal" but not fully predetermined state of identity may be inherent to all the elementary phenomena of Figure 1. "Integrity" may be a property of well-organized states that we find desirable and "disintegrity" with states, perhaps either organized or disorganized, that we do not prefer. Alternatively open systems which are strongly self-organized but in an unwanted or undesirable state may be considered to reflect "pathological integrity." Many types of holon may manifest themselves in more than one state, one or more of which may be found to be undesirable by some or all humans.

3. An Approach to Ecosystem Integrity in the Great Lakes Basin

The concepts and commitments sketched in the preceding sections have become increasingly more coherent and apparent in a large, loosely-organized network of reformers across all jurisdictions of the Great Lakes-St. Lawrence River Basin (Regier 1992). What is actually shared within the network may be inferred from the Vision Statement of an "Ecosystem Charter for the Great Lakes-St. Lawrence Basin" developed over a period of years by volunteer representatives from a broad spectrum of stakeholder interests in the Basin (Great Lakes Commission 1994):

Our vision is a Great Lakes-St. Lawrence Basin Ecosystem...
- where all people consider and conduct themselves as part of their Ecosystem;
- where all people recognize the fundamental and inextricable link between economic well-being and the health of the Ecosystem;
- in which all beneficial organisms can thrive free from preventable ecological threats to their well-being;
- where environmental degradation is a legacy of the past and a basis for remedial action;
- that exists as an evolving natural and cultural system that can successfully adapt to change;
- in which use of natural resources is compatible with conservation of such resources;
- that maintains the integrity of the Ecosystem and accommodates appropriate development;
- that is a rich mosaic of waters and lands, of natural areas and places of human activity, and of different peoples who govern themselves in various ways;
- that nurtures an abundance and diversity of plant and animal species in their natural communities and habitats as well as in specially protected and rehabilitated sites;
- that embraces the concept of sustainable development by meeting the needs of this generation without compromising the ability of future generations to meet their needs;
- where all people and their governments act as stewards and are committed to informed action and supportive policy decision;
- in which a shared governance process, among diverse and respected traditions, provides an accessible and equitable basis for responsible actions and accountability among all people and their institutions.

In March 1995 I participated as a witness in a hearing on the ecosystem approach in Ottawa, Canada, convened by Charles Caccia, chair of the Parliamentary Committee on Environment and Sustainable Development. On that occasion I submitted a checklist of

guidelines for an approach to ecosystem integrity and have subsequently revised it to read as follows:
- Commitment to a personal, professional and corporate ethic of cultural/natural healthful integrity with quite thorough internalization of the constraints and costs of appropriate husbandry and stewardship.
- Acceptance that humans' rights of ownership or of use of particular natural phenomena are always constrained, because of ecosytemic interconnectedness inter alia; water and air as well as mobile and migratory creatures cannot be "owned" to any major extent and are necessarily part of the "commons" shared by others and to which privileges of shared use apply; and excessive exploitation of one's rights may involve indirect pre-emption of others' rights and amount to theft of others' "property."
- Recognition that cultures of indigenous peoples may reflect desirable co-adaptations to natural phenomena that may be transferable to the ecosystemically less sensitive cultures that are currently dominant; appropriately selected new immigrants from relatively sensitive cultures might help to reform the wasteful and careless dominant culture, rather than acculturate to it.
- Commitment to equity, empowerment and respect within the human family with respect to gender, race and cultures with realization that the people who suffer the effects of conventional inequity and disrespect may often be less inclined to be harmful to nature than are those who benefit directly from the abuses.
- Expecting that neither culture nor nature will act in a fully predictable way and that their interaction will compound some unpredictable features resulting in unprecedented surprises, some good and some bad, so that flexible adaptability is always a necessary capability.
- Awareness that fair allocation of any privileges of use may be achieved through a variety of institutional mechanisms, separately or in combination as appropriate within a culture; none is best a priori.
- Zoning of a kind of human intervention in nature to particular locales with natural, pre-adapted tolerance for that kind of human activity, and preservation of appropriately large areas of all types of ecosystems in a wild state.
- Assurance of sustainability of any legitimate use of nature as related to a priori formal benchmarks of permissible cumulative effects both qualitatively and quantitatively, with intervenors working cooperatively with nature's own techniques of adapting to human culture, as in green technology.
- Assessment of expected effects, direct and cumulative, prior to some legitimate intervention in nature, whether new or corrective of some past misuse, and fair compensation of legitimate "owners and users" that are adversely affected by the intervention; the relevant science should reflect pluralistic procedural as well as substantive rationality.
- Criminalization of egregious abuses of nature, with respect to qualitative and quantitative features, e.g., extinction of species, release of hazardous contaminants and careless genetic engineering.

The set of guidelines above implicitly refers to numerous existing acts of government at different levels, in Canada and the U.S., that each contribute something to an approach to ecosystem integrity. It may be time to create a consolidated federal policy that would clearly formulate the main commitments and would cull numerous contradictions from within the set of relevant legislation. Such a policy would also help to clarify the federal role and disentangle it from responsibilities that are already being devolved, quite appropriately in a holarchic context, to different levels of government.

4. Conclusion

I have sketched a perspective or context and some empirical content, all in recent history, about ecosystem integrity as it relates to contemporary politics and practice especially in the Great Lakes Basin. To emphasize, this is a largely empirical account, and the arguments—pro and con—for a policy and an ethic of ecosystem integrity have been left largely implicit. Thus this essay may complement work by Westra (1994) and others in which such arguments are the focus of attention.

As a general summary, it may be recalled that holarchic versions of the concept of "integrity" were part of the reform terminology of the 1960s. People in Western Countries who had been marginalized in the dominant culture were to be integrated into that culture. Differences in gender, race and economic class had been a basis for such marginalization. The oppressive elements of society, both personal and institutional, were to be disempowered in some respects and the suppressed were to be empowered, all toward an appropriate balance. Rejuvenated and re-invigorated democratic processes were expected to self-organize to realize such reforms. This struggle is still underway, and some partial reforms can be demonstrated to have occurred. Opportunistic scoundrels are ever alert to exploit for personal advantage any opportunity that appears in the course of such a transformation.

The well-integrated military-industrial complex was recognized as a case of pathological integrity, e.g., by C. Wright Mills and others. In the 1960s serious efforts were initiated toward the disintegration of this complex, with particular focus on the Vietnam War and the invasion of Czechoslovakia in both of which the pathological nature of this complex was all too apparent. Some reforms are now underway, especially with the cooling of the Cold War.

A kind of human-nature, mutualistic reciprocity was labeled "integrity" by Aldo Leopold in *The Sand County Almanac*. Rachel Carson's numerous books, including *Silent Spring*, provided scientific insight and aesthetic appreciation for this type of healthy integrity. Eventually George Woodwell, Thomas Jorling and others collaborated to bring the concept of such ecological integrity into U.S. federal environmental law (Edwards and Regier 1990). Again the reforms have been partial and there is much more to do.

With growing acceptance of a holarchic perspective on reality as an open, self-organizing system, a concept of healthy integration will likely have a strong future. An ethic will be necessary to prevent integration toward an undesirable or evil pathological state and to foster integration toward a desirable or good healthy state. Old, objective hard science could not provide a sufficient answer for this problem; post-normal ecosciences can help, but again cannot provide a sufficient answer.

5. Acknowledgements

I have been helped and influenced, with respect to the results of my ruminations as sketched above, through interactions with many colleagues some of whom are listed here: L. Westra, R. Vollenweider, J. Vallentyne, R. Ulanowicz, P. Timmerman, G. Thornburn, G. Stewart, I. Stephanovic, C. Starrs, R. Shimizu, J.S. Rowe, L. Regier, D. Rapport, R. Mason, N.-M. Lister, S. Lerner, J. Kay, C.S. Holling, H.F. Henderson, H.J. Harris, R. Hansell, U. Franklin, G. Francis, A. Dale, R. Cauvokian, L.K. Caldwell, J. Caddy, L. Botts, S. Bocking, B. Bandurski and T. Allen. Perhaps none would endorse fully what is in my paper; no two holons are identical.

6. References

Allen, T.F.H. and T.B. Starr. 1982. *Hierarchy.* Chicago, University of Chicago Press. 310 p.

Artin, T. 1973. *Earth Talk: Independent Voices of the Environment.* New York, Grossman Publishers. vi + 176 pp.
Bocking, S. 1994. Visions of Nature and Society: A History of the Ecosystem Concept. *Alternatives* 20(3): 12-18.
Dales, J.H. 1968. *Pollution, Property and Prices.* Toronto, University of Toronto Press. vii + 111 pp.
Dales, J.H. 1975. Beyond the Market Place. *Canadian Journal of Economics* 8: 483-503.
Daly, H.E. 1968. On Economics as a Life Science. *Journal of Political Economy* 76: 392-405.
Doxiadis, C.A. 1968. *Ekistics: An Introduction to the Science of Human Settlements.* London, U.K., Hutchinson.
Drengson, A., ed. 1983-1995. *The Trumpeter, Journal of Ecosophy.* Volumes 1 to 12. Victoria, B.C. University of Victoria.
Drucker, P.F. 1989. *The New Realities.* New York, Harper & Row. xi + 276 pp.
Eaton, H. 1995. Ecofeminist Spiritualities: Seeking the Wild or the Sacred. *Alternatives* 21(2): 29-31.
Edwards, C.J. and H.A. Regier, eds. 1990. *An Ecosystem Approach to the Integrity of the Great Lakes in Turbulent Times.* Great Lakes Fishery Commission, Ann Arbor, MI. Special Pubication 90-4, vi + 299 pp.
Emery, F.E. and E.L. Trist. 1973. *Towards a Social Ecology: Contextual Appreciation of the Future in the Present.* London, U.K., Plenum Press. xv + 239 pp.
Farvar, M.T. and J.P. Milton, eds. 1972. *The Careless Technology: Ecology and International Development.* Garden City, N.J., Natural History Press.
Feyerabend, P.K. 1988. *Against Method.* London, U.K., Verso. 296 pp.
Funtowicz, S.O. and J. Ravetz. 1991. A New Scientific Methodology for Global Environmental Issues. pp. 137-152 in R. Costanza, ed. *Ecological Economics: The Science and Management of Sustainability.* New York, Columbia University Press.
Funtowicz, S.O. and J. Ravetz. 1993. Science for Post-Normal Age. *Futures* 25(7): 735-755.
Grabow, S and A. Heskin. 1973. Foundations for a Radical Concept of Planning. *American Institute of Planning Journal* 39: 106-114.
Great Lakes Commission. 1994. Ecosystem Charter for the Great Lakes-St. Lawrence Basin. Ann Arbor, MI. 8 pp.
Jantsch, E. 1972. *Technological Planning and Social Futures.* London, U.K., Casell/Associated Business Programmes. xiv + 256 pp.
Kay, J.J. and E. Schneider. 1994. Embracing Complexity: The Challenge of the Ecosystem Approach. *Alternatives* 20(3): 32-39.
Koestler, A. 1978. *Janus: A Summing Up.* London, U.K., Hutchinson, vii + 354 pp.
Koestler, A. and J.R. Smythies, eds. 1969. *Beyond Reductionism.* London, U.K., Hutchinson.
Leman, A.B. and I.A. Leman. 1976. *Great Lakes Megalopolis: From Civilization to Ecumenization.* Ottawa, Supply and Services Canada, Cat. No. SU31-311 1976F. x + 118 p.
Margalef, R. 1968. *Perspectives in Ecological Theory.* Chicago, University of Chicago Press. 111 p.
Papaioannou, J.G. 1987. Ekistics Research: Its Relevance for the Present and the Future. *Ekistics* 54, (325/326/327): 228-242.
Regier, H.A. 1992. Ecosystem Integrity in the Great Lakes Basin: An Historical Sketch of Ideas and Actions. *Journal of Aquatic Ecosystem Health* 1: 25-37.
Regier, H.A. and J.B. Falls, eds. 1969. *Exploding Humanity, the Crisis of Numbers.* Toronto, Anansi. iv + 188 pp.
Regier, H.A. and H.F. Henderson. 1973. Towards an Ecological Model of Fish Communities and Fisheries. *Transactions of the American Fisheries Society* 102: 56-72.

Sen, A. 1994. Population: Delusion and Reality. New York, The New York Review of Books, 41(15) Sept. 22, 1994.

Serafin, R. 1988. Noosphere, Gaia and the Science of the Biosphere. *Environmental Ethics* 10: 121-127.

Shepard, P. and D. McKinley, eds. 1969. *The Subversive Science: Essays Toward an Ecology of Man.* Boston, Houghton Mufflin Co., x + 453 pp.

UNESCO. 1970. Use and Conservation of the Biosphere. Proceedings of the Intergovernmental Conference for Rational Use and Conservation of the Resources of the Biosphere, Paris, September 1968. Paris, United Nations Educational, Scientific and Cultural Organization. 272 pp.

Vollenweider, R.A. 1968. Scientific Fundamentals of the Eutrophication of Lakes and Flowing Waters, With Particular Reference to Nitrogen and Phosphorus as Factors in Eutrophication. Paris, Organization for Economic Cooperation and Development.

Westra, L. 1994. *An Environmental Proposal for Ethics: The Principle of Integrity.* Lanham, Maryland: Rowman & Littlefield, Publ. xxi + 237 pp.

Woodley, S., J.J. Kay and G.R. Francis, eds. 1993. *Ecological Integrity and the Management of Ecosystems.* Delray Beach, Florida: St. Lucie Press. viii + 220 pp.

Woodwell, G.M. and H.H. Smith, eds. 1969. Diversity and Stability in Ecological Systems. New York, Brookhaven Symposium in Biology. Vol. 22, 264 pp.

Chapter 8
UNIVERSAL ENVIRONMENTAL SUSTAINABILITY AND THE PRINCIPLE OF INTEGRITY

Robert Goodland[1]
Herman Daly[2]

1. Introduction

This paper seeks to focus the definition of environmental sustainability (ES), partly by distinguishing ES from social sustainability and from economic sustainability. The challenge to social scientists is to produce their own definition of social sustainability, rather than load social desiderata on to the definition of ES. Similarly with economic sustainability; let economists define it or use previous definitions of economic sustainability. The three types of sustainability—social, environmental and economic—are clearest when kept separate. They are contrasted in Table 1. While there is some overlap among the three in the goals of economic development (Figure 1), and certainly major linkages, the three are best disaggregated and addressed separately by different disciplines. Social scientists are best able to define social sustainability, and environmentalists do not have a major role in that task. The disciplines best able to analyze each type of sustainability are different; each follows different laws and methods. After disaggregating environmental sustainability we show that it is not ecosystem or nation specific, rather it is universal. Furthermore, we show that while all nations and eco-regions may need their own different approaches to ES, it is essentially non-negotiable.

2. Sustainability and Development

Clearly the three different types of sustainability are related in parts. However, this paper focuses on environmental sustainability, rather than on social or economic sustainability. Environmental sustainability is here the goal; sustainable development can be part of the means to approach that goal. The moment the term 'development' is introduced, the discussion becomes quite different and murkier. Although this paper is not focussed on sustainable development, we define it as "development without growth beyond environmental carrying capacity." Our definition of ES thus hinges on distinguishing between growth and development. According to Boutros-Ghali (1994) development is a "fundamental human right" which requires *inter alia* democracy and good governance. "Economic growth is the engine of development....sustained economic growth."

Recent emphases on social development, economic development, development with equity, and development and basic needs suggests that development could become so vague

---

[1] The World Bank, Washington DC 20433, U.S.A.; [2] Public Affairs, University of Maryland, College Park, MD 20742-1821, U.S.A.

Table 1. Comparison of Social, Economic and Environmental Sustainability

Social Sustainability (SS)

ES needs SS—the social scaffolding of people's organizations that empower self-control and self-policing in peoples' management of natural resources (see Cernea 1993). Resources should be used in ways which increase equity and social justice, while reducing social disruptions. SS will emphasize qualitative improvement over quantitative growth; and cradle-to-grave pricing to cover full costs, especially social. SS will be achieved only by strong and systematic community participation or civil society. Social cohesion, cultural identity, institutions, love, commonly accepted standards of honesty, laws, discipline, etc., constitute the part of social capital that is least subject to measurement, but probably most important for SS. This "moral capital," as some have called it, requires maintenance and replenishment by the religious and cultural life of the community. Without this care it will depreciate surely as will physical capital.

Economic Sustainability (EcS)

The widely accepted definition of economic sustainability is "maintenance of capital", or keeping capital intact, and has been used by accountants since the Middle Ages to enable merchant traders to know how much of their sales receipts they and their families could consume. Thus the modern definition of income (Hicks 1946) is already sustainable. But of the four forms of capital (human-made, natural, social and human) economists have scarcely at all been concerned with natural capital (e.g., intact forests, healthy air) because until relatively recently it had not been scarce. Also economics prefers to value things in money terms, so it is having major problems valuing natural capital, intangible, intergenerational, and especially common access resources, such as air etc. In addition, environmental costs used to be "externalized," but are now starting to be internalized through sound environmental policies and valuation techniques. Because people and irreversibles are at stake, economics has to use anticipation and the precautionary principle routinely, and should err on the side of caution in the face of uncertainty and risk. Human capital (investments in education, health, and nutrition of individuals is now accepted in the economic lifestyle (WDR 1990, 1991, 1995), but social capital, as used in SS, is not adequately addressed.

Environmental Sustainability (ES)

Although environmental sustainability is needed by humans and originated because of social concerns, ES itself seeks to improve human welfare by protecting the sources of raw materials used for human needs and ensuring that the sinks for human wastes are not exceeded, in order to prevent harm to humans. Humanity must learn to live within the limitations of the physical environment, both as a provider of inputs ("sources") and as a "sink" for wastes (Serageldin 1993). This translates into holding waste emissions within the assimilative capacity of the environment without impairing it. Also by keeping harvest rates of renewables to within regeneration rates. Quasi-ES can be approached for non-renewables by holding depletion rates equal to the rate at which renewable substitutes can be created (El Serafy 1991).

---

as to require a sanctifying adjective. These should be carefully distinguished and defined anew by others: that is a challenge for development specialists, not for environmentalists. The priorities of development are the reduction of poverty, illiteracy, hunger and disease. While these goals are fundamentally important, they are quite different from the goals of environmental sustainability, namely maintaining human life support system—environmental sink and source capacities—unimpaired. But "environmental sustainability" is legitimized by the latest pronouncement on economic development (Boutros-Ghali 1994) and is the focus of this paper.

The tacit goal of economic development is to narrow the equity gap between the rich and the poor. Almost always this is taken to mean raising the bottom (i.e., enriching the poor), rather than lowering the top or redistribution (Haavelmo and Hansen 1992). Only very recently is it becoming admitted that bringing the low income countries up to the affluence

Figure 1. The Three Goals of Economic Development (After Serageldin 1993a, 1993b; for the Principle of Integrity see Westra 1994).

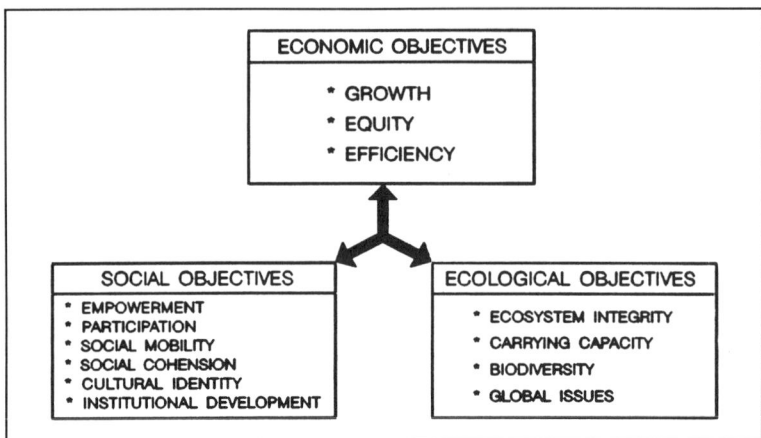

levels in OECD countries, in 40 or even 100 years, is a totally unrealistic goal. We do not want to be accused of attacking a straw man—who ever would be foolish as to claim that global equality at current OECD levels was remotely possible? However, most politicians and most citizens have not yet accepted the unrealistic nature of this goal. Most people would accept that it is desirable for low income countries to be as rich as the North—and then leap to the false conclusion that it must therefore be possible. They are encouraged in this *non sequitur* by the realization that if greater equality cannot be attained by growth alone, then sharing and population stability will be necessary. Politicians find it easier to revert to wishful thinking than to face those two issues. Once we wake up to reality, however, there is no further reason for dwelling on the impossible, and every reason to focus on what is possible.

One can make a persuasive case (see Serageldin 1993, pp. 141-143) that achieving per capita income levels in low income countries of $1,500 to $2,000 (rather than OECD's $21,000 average) is quite possible. Moreover, that level of income may provide 80 percent of the basic welfare provided by a $20,000 income—as measured by life expectancy, nutrition, education, and other measures of social welfare. This tremendously encouraging case remains largely unknown, even in development circles. It needs to be widely debated and accepted as the main goal of development. Its acceptance would greatly facilitate the transition to ES. Colleagues working on Northern overconsumption should address the corollary. Can $21,000/capita countries cut their consumption by a factor of 10 and suffer "only" a 20 percent loss of basic welfare? But to accomplish the possible parts of the imperative of development, we must stop idolizing the impossible. We leave the discussion of development at this point and reinforce the challenge to development specialists to take this important argument deeper.

The paramount importance of sustainability arose partly because, as mentioned, the world is starting to recognize that current patterns of economic development are not generalizable. Present patterns of OECD per capita resource consumption and pollution cannot possibly be generalized to all currently living people, much less to future generations, without liquidating the natural capital on which future economic activity depends. Sustainability thus arose from the recognition that the profligate and inequitable nature of current patterns of development, when projected into the not-too-distant future, lead to biophysical impossibilities. The transition to sustainability is urgent because the deterioration of global life-support systems—the environment, imposes a time limit. We do not have time to dream of creating more living space or more environment, such as colonizing the moon or building cities beneath the sea; we must save the remnants of the only environment we have, and allow time for, and invest in the regeneration of what we have already damaged.

This paper is divided into 9 parts. After distinguishing development from sustainability and from growth, the paper describes the concept of natural capital, and uses the concept to present four alternative definitions of environmental sustainability. The next section presents criteria for analyzing environmental sustainability, and then using the Ehrlich-Holdren (1974) framework in which "Population," "Affluence" and "Technology" (PAT) are examined separately. The seventh section nuances the I = PAT identity. The final section describes how one large development agency, the World Bank, is endeavoring to incorporate these new principles into its operations.

3. Growth Compared With Development

When something grows it gets quantitatively bigger; when it develops it get qualitatively better or at least different. Growth is a physical or material and quantitative expansion; development is qualitative improvement. Quantitative growth and qualitative improvement follow different laws. Our planet develops over time without growing. Our economy, a subsystem of the finite and non-growing earth must eventually adapt to a similar pattern of development without throughput growth. The time for such adaption is now.

It is neither ethical nor helpful to the environment to expect poor countries to cut or arrest their development, which tends to be highly associated with throughput growth. Poor, small, developing economies need both growth and development. Therefore, the rich countries, which are responsible for most of today's environmental damage, and whose material well being can sustain halting or even reversing throughput growth, must take the lead in this respect. Poverty reduction will require considerable growth, as well as development, in developing countries. But global environmental constraints (atmospheric CO_2 accumulation, ozone shield damage, acid rain, etc.) are real, and more growth for the South must be balanced by negative throughput growth for the North, if environmental sustainability is to be achieved. Negative Northern growth need not decrease to the extent $2,000/capita gains 80 percent of basic welfare, as just mentioned. Future Northern growth should be sought from productivity increases in terms of throughput (e.g., reducing energy intensity of production).

Development by the North must be used to free resources (the source and sink functions of the environment) for growth and development so urgently needed by the poorer nations. Large scale transfers to the poorer countries also will be required, especially as the impact of economic stability in Northern countries may depress terms of trade and lower economic activity in developing countries. Higher prices for the exports of poorer countries, as well as debt relief, therefore will be required. Most importantly, population stability is essential to reduce the need for growth everywhere, but especially where population growth impacts the most (i.e., in the Northern high-consuming nations) as well as where population growth is highest (i.e., in the poor, low consuming countries).

4. Natural Capital and Sustainability

4.1. INTERGENERATIONAL AND INTRAGENERATIONAL SUSTAINABILITY

Sustainability in economic terms can be described as the "maintenance of capital," sometimes phrased as "non-declining capital". Historically, at least as early as the Middle

Ages, the merchant traders used the word "capital" to refer to human-made capital. The merchants wanted to know how much of their trading ships cargo sales receipts could be consumed by their families without depleting their capital. Economics Nobelist Sir John Hicks encapsulated the sustainability concept in 1946 when he defined income as "the amount (whether natural or financial capital) one could consume during a period and still be as well off at the end of the period". Solow (1991) is vaguer. Sustainability to Solow is an obligation or injunction "...to conduct ourselves so that we leave to the future the option or the capacity to be as well off as we are...not to satisfy ourselves by impoverishing our successors...."(Solow 1991).

Today's OECD societies have already impoverished much of the world. Most people in the world today are already impoverished or barely above subsistence and can by no stretch of the imagination ever be as well off as the OECD average. Our successors or future generations seem more likely to be more numerous and poorer than today's generation. Sustainability indeed has an element of not harming the future (intergenerational equity), but only addressing the future diverts attention from today's lack of sustainability (intragenerational equity). If the world cannot move towards intragenerational sustainability for this generation it will be greatly more difficult to achieve intergenerational sustainability in the future. This is partly because the world is hurtling away from ES today; environmental source and sink capacities are being impaired. This means the capacity of these environmental services will be lower in the future than they are today. The second reason for tackling intragenerational sustainability first is that world population soars by 100 million new souls each year, some of them OECD over-consumers, most of them poverty stricken. This means achieving intergenerational equity is more difficult for each year that has a bigger human generation.

Of the forms of capital environmental sustainability refers to natural capital. So defining environmental sustainability includes at least two further terms, namely "natural capital" and "maintenance" or at least "non-declining." Natural capital is basically our natural environment, and is defined as the stock of environmentally provided assets (such as soil, atmosphere, forests, water, wetlands), which provide a flow of useful goods or services. The flow of useful goods and services from natural capital can be renewable or non-renewable, and marketed or non-marketed. Sustainability means maintaining environmental assets, or at least not depleting them. "Income" is sustainable by the generally accepted Hicksian definition of economics (Hicks 1946) as mentioned above. Any consumption that is based on the depletion of natural capital should not be counted as income. Prevailing models of economic analysis tend to treat consumption of natural capital as income, and therefore tend to promote patterns of economic activity that are unsustainable. Consumption of natural capital is liquidation, the opposite of capital accumulation.

Natural capital is distinguished from other forms of capital, namely human capital or social capital (people, their capacity levels, institutions, cultural cohesion, education, information, knowledge), and human-made capital (houses, roads, factories, ships). From the mercantilists until very recently capital referred to the form of capital in the shortest supply, namely human-made capital. Investments were made in the limiting factor, such as sawmills and fishing boats, because their natural capital complements—forests and fish— were abundant. That idyllic era has ended.

Now that the environment is so heavily used, the limiting factor for much economic development has become natural capital as much as human-made capital. In some cases, like marine fishing, natural capital has become the limiting factor. Fish have become limiting, rather than fishing boats. Timber is limited by remaining forests, not by saw mills; petroleum is limited by geological deposits and atmospheric capacity to absorb CO_2, not by refining capacity. As natural forests and fish populations become limiting we begin to invest in plantation forests and fish ponds. This introduces an important hybrid category that combines natural and human-made capital—a category we may call "cultivated natural capital." This category is vital to human well-being, accounting for most of the food we eat, and a good deal of the wood and fibers we use. The fact that humanity has the capacity to "cultivate" natural capital dramatically expands the capacity of natural capital to deliver services. But utilities cultivated natural capital (agriculture) is decomposable into human-made capital (e.g., tractors, diesel irrigation pumps, chemical fertilizers) and natural capital (e.g., topsoil, sunlight, water). Eventually the natural capital proves limiting.

4.2. NATURAL CAPITAL IS NOW SCARCE

In an era in which natural capital was considered infinite relative to the scale of human use, it was reasonable not to deduct natural capital consumption from gross receipts in calculating income. That era is now past. The goal of environmental sustainability is thus the conservative effort to maintain the traditional meaning and measure of income in an era in which natural capital is no longer a free good, but is more and more the limiting factor in development. The difficulties in applying the concept arise mainly from operational problems of measurement and valuation of natural capital, as emphasized by Ahmad et al. (1989), Lutz (1993) and El Serafy (1991, 1993).

4.3. FOUR DEGREES OF ENVIRONMENTAL SUSTAINABILITY

Sustainability can be divided into four degrees—weak, intermediate, strong and absurdly strong—depending on how much substitution one thinks there is among types of capital (Daly and Cobb, 1994). We recognize that there are at least four kinds of capital: Human-made (the one usually considered in financial and economic accounts), natural capital (as defined previously, and leaving for the moment the case of cultivated natural capital), human capital (investments in education, health and nutrition of individuals), and social capital (the institutional and cultural basis for a society to function).

Weak sustainability: is maintaining total capital intact without regard to the composition of that capital between the different kinds of capital (natural, human-made, social or human). This would imply that the different kinds of capital are perfect substitutes, at least within the boundaries of current levels of economic activity and resource endowment. Given current gross inefficiencies in resource use, weak sustainability would be a vast improvement as a welcome first step, but would by no means constitute ES. Weak sustainability means we could convert all or most of the world's natural capital into human-made capital or artifacts and still be as well off. Human and social capital is largely lost at death so has to be renewed each generation. We disagree; society would be worse off (fewer choices) because natural and human-made capital are not perfect substitutes. On the contrary, they are complements to a great extent.

Intermediate sustainability: would require that in addition to maintaining the total level of capital intact, some concern should be given to the composition of that capital between natural, human-made, human and social). Thus oil may be depleted as long as the receipts are invested in other capital (e.g., in human capital development, or in renewable energy resources) elsewhere, but that, in addition, efforts should be made to define critical levels of each type of capital, beyond which concerns about substitutability could arise and these should be monitored to ensure that the patterns of development do not promote a total decimation of one kind of capital no matter what is being accumulated in the other forms of capital. This assumes that while human-made and natural capital are substitutable over a sometimes significant but limited margin, they are complementary beyond that limited margin. The full functioning of the system requires at least a mix of the different kinds of capital. Since we do not know exactly where the boundaries of these critical limits for each type of capital lie, it behooves the sensible person to err on the side of caution in depleting resources (especially natural capital) at too fast a rate. Intermediate sustainability is a big improvement over weak sustainability. Its great weakness is that it is difficult, if not impossible, to define critical levels of each type of capital, or rather each type of natural capital which is the limiting factor. We suspect that if the levels of the different types of natural capital become reliably defined, it would approximate strong sustainability.

Strong sustainability: requires maintaining different kinds of capital intact separately. Thus for natural capital, receipts from depleting oil should be invested in ensuring that energy will be available to future generations at least as plentifully as enjoyed by the beneficiaries of today's oil consumption. This assumes that natural and human-made capital are not really substitutes but complements in most production functions. A saw-mill (human-made capital) is worthless without the complementary natural capital of a forest. The same logic would argue that if there are to be reductions in one kind of educational investments they should be offset by other kinds of education, not by investments in roads. Of the four degrees of sustainability, we prefer strong sustainability.

Absurdly strong sustainability: would never deplete anything. Non-renewable resources—absurdly—could not be used at all; for renewables, only net annual growth rates could be harvested in the form of the overmature portion of the stock.

The decision between intermediate and strong sustainability highlights the tradeoffs between human-made capital and natural capital. Economic logic requires us to invest in the limiting factor which now is often natural capital rather than human-made capital, which was limiting yesteryear. Investing in natural capital (non-marketed) is essentially an infrastructure investment on a grand scale, that is the biophysical infrastructure of the entire human niche. Investment in such "infra-infrastructure" maintains the productivity of all previous economic investments in human-made capital, public or private, by rebuilding the natural capital stocks that have come to be limitative. Operationally, this translates into three concrete actions as noted in Table 2.

4.4. CRITERIA FOR ENVIRONMENTAL SUSTAINABILITY

From the above "maintenance of natural capital" approach to environmental sustainability (ES), we can draw practical rules-of-thumb to guide the design of economic development. As a first approximation, the design of investment strategies should be compared with the input/output rules of ES (Table 3) in order to assess the extent to which a project is sustainable. At the next level of detail, specific indicators of environmental sustainability can be used, such as those the World Bank is preparing.

The implications of implementing environmental sustainability are immense. We must learn how to manage the renewable resources for the long term; we have to reduce waste and pollution; we must learn how to use energy and materials with scrupulous efficiency; we must

Table 2. Rebuilding Natural Capital Stocks.

1. REGENERATION: Encouraging the growth of natural capital by reducing our level of current exploitation of it.

2. RELIEVE PRESSURE: Investing in projects to relieve pressure on natural capital stocks by expanding cultivated natural capital, such as tree plantations to relieve pressure on natural forests.

3. EFFICIENCY: Increasing the end-use efficiency of products (such as improved cookstoves, solar cookers, hay-box cookers, wind pumps, solar pumps, manure rather than chemical fertilizer).

Table 3. The Definition of Environmental Sustainability.

1. **OUTPUT RULE**:

   Waste emissions from a project should be within the assimilative capacity of the local environment to absorb without unacceptable degradation of its future waste absorptive capacity or other important services.

2. **INPUT RULE**:

   a. **Renewables**: harvest rates of renewable resource inputs would be within regenerative capacity of the natural system that generates them.

   b. **Non-renewables**: depletion rates of nonrenewable resource inputs should be equal to the rate at which renewable substitutes are developed by human invention and investment. An easily calculable portion of the proceeds from liquidating non-renewables should be allocated to research in pursuit of sustainable substitutes*.

* For a theoretical development of this idea, see El Serafy (1991, 1993).

---

learn how to use solar energy in all it forms; and we must invest in repairing the damage, as much as possible, done to the earth in the past few decades by unthinking industrialization in many parts of the globe. Environmental sustainability needs enabling conditions which are not integral parts of environmental sustainability: not only economic and social sustainability (Table 1 and Figure 1) but democracy, human resource development, empowerment of women, and much more investment in human capital than common today (i.e., increased literacy, especially ecoliteracy, Orr 1992).

As can be seen by the World Bank's recent extraordinary expansion of its traditional investments in human capital and its concerns with governance and promotion of the civil society (forms of investment in social capital), and its growing concern with natural capital and its maintenance, the Bank is increasingly tackling an agenda that tends towards promoting environmental sustainability. The World Bank considers environmental sustainability to be an urgent priority.

The sooner we start to approach environmental sustainability the easier it will become. For example, the demographic transition took a century in Europe, but only a decade in Taiwan: technology and education make big differences. But the longer we delay, the worse the eventual quality of life (e.g., fewer choices, fewer species, more damage, more risk), especially for the poor who do not have the means to insulate themselves from the negative effects of environmental degradation.

Many writers have expressed concern that the world is hurtling away from environmental sustainability at present (Simonis 1990, Goodland 1992, Meadows et al. 1992, Hardin 1993, Brown et al. 1994), although consensus has not yet been reached. But what is not contestable is that the current modes of production prevailing in most parts of the global economy are causing the exhaustion and dispersion of a one-time inheritance of natural capital, such as topsoil, groundwater, tropical forests, fisheries, and biodiversity. The rapid depletion of these essential resources, coupled with the degradation of land and atmospheric quality show that the human economy, as currently configured, is already inflicting serious damage on global supporting ecosystems and is probably reducing future potential biophysical carrying capacities by depleting essential natural capital stocks (Daily and Ehrlich 1992).

Yet, what is galling, is that in spite of spending capital inheritance rather than just income, most of the world consumes at barely subsistence levels. Can humanity attain a more equitable standard of living which does not exceed the carrying capacity of the planet? The transition to environmental sustainability will inevitably occur. However, whether nations will have the wisdom and foresight to plan for an orderly and equitable transition to environmental sustainability, rather than allowing biophysical limits to dictate the timing and course of this transition, remains in doubt.

It is obvious that if pollution and environmental degradation were to grow at the same rate as economic activity, or even population growth, the damage to ecological and human health would be appalling, and the growth itself would be undermined and even self-defeating. Fortunately, this is not necessary. A transition to sustainability is possible, although it will require changes in policies and the way we humans value things. The key to the improvement of the well-being of millions of people lies in the increase of the added value of output after properly netting out all the environmental costs and benefits and after differentiating between the stock and flow aspects of the use of natural resources. In our view, this is the key to sustainable development. Without this needed adjustment in thinking and measurement, the pursuit of economic growth that does not account for natural capital and counts depletion of natural capital as an income stream will not lead to a sustainable development path.

The global ecosystem, which is the source of all the resources needed for the economic subsystem, is finite and has now reached a stage where its regenerative and assimilative capacities have become very strained. It looks inevitable that the next century will witness double the number of people in the human economy, depleting sources and filling sinks with their increasing wastes. If we emphasize the latter, it is because human experience seems to indicate that we have tended to overestimate the environment's capacity to cope with our wastes, even more than we overestimated the "limitless" bounty of such resources as the fish in the sea.

5. Basic Conditions for Environmental Sustainability

The fundamental definition of environmental sustainability is the input/output rule (Table 3). Building on the economic definition of sustainability as "non-declining wealth per capita," and as wealth is so difficult to measure, environmental sustainability is now defined by the two fundamental environmental services—the source and sink functions—that must be maintained unimpaired during the period over which sustainability is required. This general definition is robust and irrespective of country, sector or epoch; that is why we call it universal and non-negotiable. Even so it can in turn be disaggregated (Table 4).

The emphasis on maintenance is to be expected, first for intergenerational equity. Our descendents should have as much choice as we have. Second, as scale increases or matures, production is no longer for growth but for maintenance. Production is the maintenance cost of the stock and should be minimized (Daly 1994). Sustainability demands that production and consumption be equal so that we maintain capital stocks. Efficiency demands that the maintenance cost (production equal to consumption) be minimized, given the capital stock.

To stop throughput of matter and energy from growing or to hold throughput constant (we leave until later the need actually to reduce throughput) means stabilizing population on the demand side, and improving resource productivity or "dematerializing" the economy on the supply side. Resource productivity has increased already, although more progress is possible and needed. These include improvements in energy efficiency, more production with less energy and fewer materials, tight recycling, repair, re-use and decarbonization, another name of the transition to renewables such as wind, photovoltaics and the hydrogen economy.

Table 4. The Basic Conditions for Environmental Sustainability.

1. Maintenance of human-made capital (eg: artifacts, infrastructure) per capita.

2. Maintenance of renewable natural capital (eg: healthy air, natural forests, oceanic fish stocks) per capita.

3. Maintenance of non-renewable substitutable natural capital per capita, with capital values based on the value of the services of the present stock of natural capital. This means that if the cost of supplying energy substitutes rises, sufficient capital must be accumulated to maintain these services.

4. Maintenance of non-substitutable, non-renewable natural resources (eg: waste absorption by environmental sink services). No depletion or deterioration of non-substitutable non-renewable natural capital. This derives from *Figure 5* output rule: no net increases in waste emissions beyond absorptive capacity.

5. All economic consumption should be priced to reflect full cost of all capital depletion, including waste creation, the cost of which is equal to the cost of reducing an equivalent amount of that particular waste.

6. Stating the conditions in per capita terms calls attention to the importance of stopping population growth. Theoretically, the per capita stock of all kinds of capital could remain constant as long as the stocks grew at the same rate as population. But in actuality the rate of growth of population and stocks of physical wealth must move toward zero.

6. Some Common Misconceptions About Environmental Sustainability (ES)

6.1. IS ES THE SAME AS SUSTAINED YIELD?

No, clearly not. There is a lively debate, especially in forestry and fishery circles, whether ES is "sustained yield" (S-Y), in the form of timber removals from forest for example. Clearly ES includes, but certainly is far from limited to, sustained yield. ES is more akin to the simultaneous S-Y of many interrelated populations in an ecosystem. S-Y is often used in forestry and fisheries to determine the optimal—most profitable—extraction rate of trees or fish. ES counts all the natural services of the sustained resource. S-Y counts only the service of the product extracted, and ignores all other natural services. S-Y forestry counts only the timber value extracted; ES forestry counts all services. These include protecting vulnerable ethnic minority forest dwellers, biodiversity, genetic values, intrinsic as well as instrumental values, climatic, wildlife, carbon balance, water source and water moderation values, ecosystem integrity in general (Westra 1994) and, of course, timber extracted. The relation between the two is that if S-Y is actually achieved, then the stock resource (e.g., the forest) will be nearer sustainability than if S-Y is not achieved. S-Y in tropical forestry is doubtful now (Ludwig 1993), and will be more doubtful in the future, as human population pressures intensify. But even were S-Y to be achieved, that resource is unlikely to have also attained environmental sustainability. The optimal solution for a single variable, such as S-Y, usually (possibly inevitably) results in declining utility or declining natural capital sometime in the future, therefore is not sustainable.

6.2. IS ES A VARIABLE, OR A CONSTANT?

ES is a variable, but it changes so slowly that it is probably best to assume it is constant as a first approximation. If humans evolve lungs that can use hitherto unbreathably polluted air, or if we carry cylinders of oxygen on our backs, then that part of ES could be construed as a variable. On the output side, in general, assimilative capacity cannot be substantially increased. As "waste is our fastest growing resource" this is significant. On the non-renewable input side, non-renewables can be used slower or more efficiently, or more ores and substitutes can be found, but the stock of non-renewables is fixed and cannot be increased. Technology and efficiency squeeze more utility out of inputs, but do not increase the stock. It is difficult to get renewables to regenerate faster! Even well fertilized and irrigated trees in the US, for example, grow slower than *laisser faire* trees in Costa Rica, which has a short winter. Light is often more limiting than water and nutrients. Human-made capital such as pond fish, intensive agriculture such as sugarcane or hydroponic laboratory greenhouse crops have reached high levels of productivity, but the ability to get the whole of humanity to produce effectively and in an environmentally sensitive fashion and match performances achievable on experimental farms is another question. So again, ES appears to be more constant than variable—i.e., a very slowly changing variable. This is why we suggest ES is universal.

6.3. IS ES MORE OF A CONCERN FOR DEVELOPING COUNTRIES?

Not really, no. ES is even more relevant to industrial countries than to developing countries. The big difference is in burden sharing. The North is responsible for the overwhelming share of global environmental damage today, and it is unlikely that poor countries will want to move towards sustainability if the North doesn't do so first. The North can contribute more to decreasing the global warming risks by reducing Greenhouse gas emissions, and the release of substances such as CFC's which damage the ozone shield. The North has to adapt to ES more than the South, and arguably before the South. The North can afford to exert leadership on itself. But because developing economies depend to a much greater extent than OECD economies on natural resources, especially renewables, the South has much to gain from reaching ES. In addition, because much tropical environmental damage is irreversible, it is either impossible or more expensive to rehabilitate than temperate environments so the South will gain from a preventive approach, rather than emulating the short-sighted and expensive curative approach and similar mistakes of the North.

6.4. DOES ES IMPLY REVERSION TO AUTARKY OR THE STONE AGE?

Certainly not; ES is not sacrifice. On the contrary, ES increases welfare. The message that affluence and overconsumption do not increase welfare is being acted on by a few people. Much more education is needed for overconsumers to realize that limousine rides are often slower and more polluting than metro, and eating three steaks a day reduces health. As the diseases of overconsumption increase (heart attack, stroke), this message will spread. The concept of sufficiency (doing more with less) needs dissemination. Education is needed that love, pleasure, fulfillment, enjoyment and other rewards do not depend on overconsumption, in fact is decreased by it.

A single measure—population times per capita consumption of natural capital—encapsulates an essential dimension of the relationship between economic activity and environmental sustainability. This scale of the growing human economic subsystem is judged, whether large or small, relative to the finite global ecosystem on which it so totally depends, and of which it is a part. The global ecosystem is the source of all material inputs feeding the economic subsystem, and is the sink for all its wastes. Population times per capita consumption of natural capital is the total flow—throughput—of resources from the global ecosystem to the economic subsystem, then back to the global ecosystem as waste, (as dramatized in Figure 2). In the long gone "empty world" case, the scale of the human

Figure 2. Relationship Between the Economic and the Environmental Systems: Three Views.

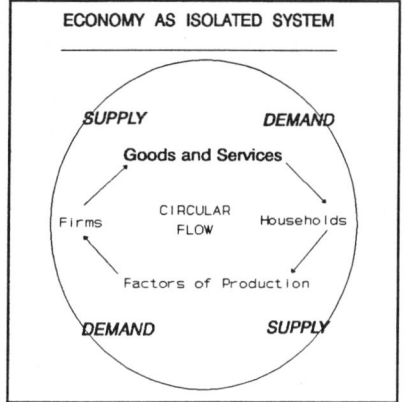

A. The Economy as an Isolated System

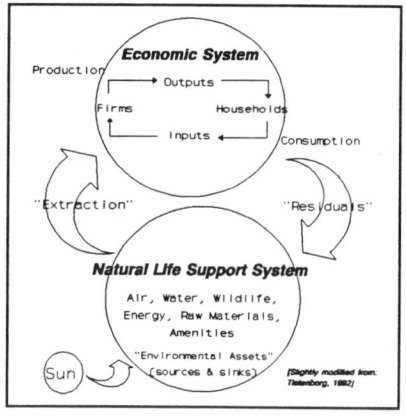

B. Linking the Economic and Environmental Systems

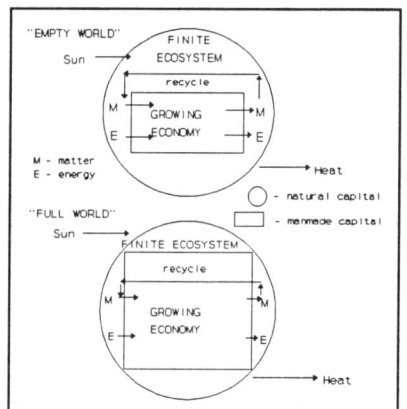

C. The Economy Dependent on the Environment

economic subsystem is small relative to the large, but non-growing global ecosystem. In the lower diagram, the "full world" case, the scale of the human economic subsystem is large and still growing, relative to the finite global ecosystem. In the full world case, the economic subsystem has already started to interfere with global ecosystemic processes, such as altering the composition of the atmosphere (Greenhouse warming), or the now nearly global damage to the ozone shield.

7. Population, Affluence and Technology

Carrying capacity is a measure of the amount of renewable resources in the environment in units of the number of organisms these resources can support. It is thus a function of the area and the organism: a given area could support more lizards than birds with the same body mass. Carrying capacity is difficult to estimate for humans because of major differences in affluence, behavior, and technology. An undesirable "factory-farm" approach could support a large human population at the lowest standards of living: certainly the maximum number of people is not the optimum. The higher the throughput of matter and energy, or the higher the consumption of environmental sources and sinks, the fewer the number of people that can enjoy it.

Ehrlich and Holdren (1974) encapsulate the basic elements of this concept in the forceful identity: The impact (I) of any population or nation upon environmental sources and sinks is a product of its population (P), its level of affluence (A), and the damage done by the particular technologies (T) that support that affluence.

$$I = P \times A \times T$$
$$(I = P \times Y/P \times I/Y)$$

Population (P) refers to human numbers.

Affluence (Y/P) is output (Y) per capita.

Technology (I/Y) refers to environmental impact per unit of output, i.e., a dollar's worth of solar heating stresses the environment less than a dollar's worth of heat from a lignite-fired thermal power plant.

There are a number of ways of reducing environmental impacts of human activities upon the environment. These include changing the structure of production and demand (i.e., more high-value, low-throughput production and service industries) and investing in environmental protection (e.g., for the amenity value, if for nothing else). Reducing environmental impacts of human activities upon the environment can be achieved only by change in the three variables in the equation. Reducing impact (I) means either (1) limiting population growth; or (2) limiting affluence; or (3) improving technology, thereby reducing throughput intensity of production. There is much to be done to limit the impact of human activities upon the environment, although so far many of the measures have proven politically unpopular and difficult to achieve. The changes in variables—population, affluence, technology—through which "I" can be limited, are each examined in more detail below.

7.1. POPULATION

Population stability is fundamental to environmental sustainability. Today's 5.5 billion people are increasing by nearly 100 million a year. Just the basic maintenance of 100 million extra people per year needs an irreducible minimum of throughput in the form of clothing, housing, food, and fuel. There is so much momentum in population growth that even under the United Nations most optimistic scenario, the world's population may level off at 11.6 billion in 2150! Since under current inequitable patterns of production, consumption and distribution, we have not provided adequately for one-fifth of humanity at today's relatively low population, the prospects for being either able or willing to provide better for double that number of people look grim indeed, unless major changes in attitudes and practices were to happen. We do not want to cast a political problem (willingness to share) as a biophysical problem (encountering limits to total product). We urge much greater sharing. However, we do not want to make the opposite error of suggesting that more equitable sharing will permit us completely to avoid the issue of biophysical limits to total production in the face of mounting population pressure. Responsible stewardship of the earth requires that we redouble efforts to slow down population growth, especially in the poorest and most vulnerable countries, where population is currently growing fastest and people are suffering most.

The human population is totally dependent on energy from the sun, fixed by green plants, for all food, practically all fiber (cotton, wool, paper), most building materials (wood), and most of the cooking and heating fuels in many developing countries (fuelwood). The human economic subsystem now appropriates 40 percent of all that energy, according to Vitousek et al. (1986). Yet the sun provides enough energy to cover 6,500 times the total commercial energy consumption of the world. Instead of harnessing this massive source of clean and renewable energy, the bulk of energy research funds are still going to nuclear energy. This speaks poorly of the priorities of energy research worldwide and is a measure of how far we still have to go to get the concept of sustainability thoroughly incorporated into the priorities of those allocating the energy research dollars.

Whether the issue will be joined over the energy fixed by photosynthesis or not, there are reasons to be concerned. Several factors are all working in the same direction to reduce irreversibly the energy available globally through plants. Greenhouse warming, damage to the ozone shield, and less predictable, unstable climates seem inescapable and may have started. Depending on the models used, these will reduce agricultural, forest, fisheries, rangeland and other yields. The increases in UVb light reaching the earth through the damaged ozone shield may decrease the carbon-fixing rates of marine plankton, one of the biggest current carbon sinks. In addition, UVb light may damage young or germinating crops. According to some reports, tiny temperature elevations have already begun to increase the decomposition rates of the vast global deposits of peats, soil organic matter, and muskeg, thus releasing stored carbon. Only in mid-1992 did the circumboreal muskeg and tundra become net global carbon sources (instead of being net C-sinks). Some atmospheric scientists claim that an immediate 50 percent reduction in global fossil fuel use is necessary to stabilize atmospheric composition. Whether one accepts this estimate or not, it dramatizes the gravity of the situation.

7.2. AFFLUENCE

Overconsumption by the OECD countries contributes more to some forms of global unsustainability than does population growth in low income countries (Mies 1991, Parikh and Parikh 1991). If energy consumption is used as a crude surrogate for environmental impact on the earth's life support systems, (crude since the type of energy used is not taken into account), then "A baby born in the United States represents twice the impact on the Earth as one born in Sweden, three times one born in Italy, 13 times one born in Brazil, 35 times one in India, 140 times one in Bangladesh or Kenya, and 280 times one born in Chad, Rwanda, Haiti or Nepal" (Ehrlich and Ehrlich 1989a, 1989b). Although Switzerland, Japan and Scandinavia, for example, have recently made great progress in reducing the energy intensity of production, the key question is: can humans lower their per-capita impact (mainly in OECD countries) at a rate sufficiently high to counterbalance their explosive increases in population (mainly in low income countries)? The affluent are reluctant to acknowledge the concept of sufficiency—to begin emphasizing quality and non-material satisfactions. Redistribution from rich to poor on any significant scale is, at present, felt to be politically impossible. But the questions of increasing equity in sharing the earth's resources and its bounty must be forcefully be put on the table.

Increased affluence, especially of the poor, thus need not inevitably hurt the environment. Indeed, used wisely, economic growth can provide the resources needed to protect and enhance the environment in the poorest developing countries where environmental damage is caused as much by the lack of resources as it is by rapid industrialization. Indeed if poor countries are to have any hope of protecting their natural capital, accelerated economic and human development is imperative. Thus, there is a nexus of problems linking poverty, environmental degradation, and rapid population growth (see e.g., Table 5). Breaking this nexus of problems is essential if the poor are not to continue to be the victims, as well as the unwitting cause, of environmental degradation.

7.3. TECHNOLOGY

Technology continues to play a vital role in driving a wedge between economic activity and environmental damage. Illustrations of this occur in virtually every field of human activity.

In energy, for example, the introduction of mechanical and electrical devices in power generation over the past four decades has reduced particulate emissions per unit of energy generated by up to 99 percent and newer technologies, such as flue gas desulfurization and fluidized bed combustion are dramatically reducing emissions of sulfur and nitrogen oxides. But it will be the transition to non-fossil based sources that will make the permanent difference. Here, technological progress has been remarkable—with costs of solar generation of electricity falling by 95 percent in the past two decades—but not yet enough. Renewable energy continues to receive much too small a share of public research funds.

Technological innovation and application has also done much to make agriculture more sustainable. New technologies (many developed throughout the CGIAR system) have enabled a doubling of food production in the world in just 25 years, with more than 90 percent of this growth deriving from yield increases and less than 10 percent from area expansion. More recently, the dissemination of Integrated Pest Management approaches has enabled pesticide application to be cut dramatically with no loss of productivity.

Despite such remarkable progress, it is a mistake to place too much optimism in technological change. New technology is often adopted in order to improve labor produc-

Table 5. Priorities to Approach Environmental Sustainability.

The main means to accelerate the two crucial transitions: population stability and renewable energy are:

1. **Human capital formation**:
   Education and training, employment creation, particularly for girls equivalent to that for boys; meeting unmet family planning demand.

2. **Technological transfer**:
   For the South and East to leapfrog the North's environmentally damaging stage of economic evolution. For the developing countries, this requires creating an incentive framework conducive to efficient investment. For industrial countries, this requires an open trade regime and adequate investment in new cleaner technologies.

3. **Direct poverty alleviation**:
   Including social safety nets, and targeted aid (*inter alia* see World Development Report 1990, Goodland and Daly 1993a, b).

tivity, which in turn can raise material standards of living, but without adequate attention to the environmental impacts of the manner in which the improvements are reached. The impact of a particular technology depends on the nature of the technology, the size of the population deploying it, and the population's level of affluence. The World Bank, along with others, is increasing investments in more sustainable technologies, such as on wind and solar energy, which have limited or benign impacts on the relations of humanity to the ecosystem that supports us all.

But the level of affluence currently enjoyed by the citizens of the OECD countries cannot be generalized to the rest of the world's current population, much less the massively larger population of the developing countries 40 years from now, no matter what the improvements in technology are likely to be.

The contribution to approaching global environmental sustainability differs markedly in three geographic regions. OECD's main contribution to environmental sustainability should surely be to cease its long history of environmental damage from overconsumption and pollution (corollaries to affluence under today's technology), such as greenhouse warming and ozone shield damage. The contribution of low income countries lies in stabilizing the human population. The former centrally planned economies' contribution seems to be more in accelerating the modernization of its technology, to reduce acid rain by removing subsidies on dirty coal, stop poisoning the land, *waldesterben* forest death, and nuclear risks. It is in OECD's self-interest to accelerate technology transfer to the former centrally planned economies and to the low income countries. It is possible that with current types of technology and production systems the global economy has already exceeded the sustainable limits of the global ecosystem and that manifold expansion of anything remotely resembling the present global economy would not simply speed us from today's long run unsustainability to imminent collapse. We believe that in conflicts between political feasibility and biophysical realities, the former must eventually give way to the latter, although we cannot specify exactly how long "eventually" will be.

8. A Dynamic Formulation: From Sustainability to Sustainable Development

The foregoing discussion of population, affluence and technology (PAT), based on the Ehrlich and Holdren (1974) generalization can be nuanced. The inter-relationships among the three factors, and their links with shifts in the structure of the economy should be further disaggregated. Three trends need to be accelerated as we struggle towards sustainability.

First, given the political unreality of a voluntary decline in the overall affluence of industrial countries, how is the "pattern" of this affluence shifting? Specifically, is the economic structure of the economy shifting away from environmentally damaging activities (e.g., heavy and toxic industries) and towards less "natural capital-depleting" sectors (e.g., services). This trend is to be encouraged although some services deplete much natural capital (e.g., hospitals, air travel, hotels). Furthermore, while affluent Northern nations may becoming less capital-depleting by evolving into the service sector, at the same time much industry and other natural capital-depleting activities are being transferred to developing countries. This is not a net gain for the sole global ecosystem and may be a loss if developing countries environmental standards are weaker than those whence the industries originated. Japan's huge success in using less input per unit of economic output, such as by de-linking energy and production, is based partly on the fact that most of Japan's aluminum, for example, is smelted overseas. Similar, Japan's forest natural capital is almost entirely intact; practically all timber is imported. Recent research shows that structural shifts can have powerful impacts on natural resource consumption.

Second, the clear trend in the consumption of natural resources per unit of output is improving in OECD and in places in developing countries. Two mechanisms need to be monitored here: improvements in economic efficiency (inputs per unit of output), and the degree of substitution away from environmentally critical inputs. Policy instruments, including taxes and user charges, can help promote such transitions, especially when the environment costs are not captured in the marketplace.

Third, the pollution impact per unit of economic activity delining is declining in places; less so in others. Here it is important to distinguish between the innovation of new technologies, and their dissemination and application. Many of the most profound forms of environmental damage in today's world (soil erosion, lack of clean water, municipal waste, etc.) do not require new technologies, but simply the application of existing ones. This in turn requires (1)that decision-makers are persuaded that the benefits of using such technologies exceed the costs, and (2)that resources are available for putting them in place. Public policies can be targeted towards meeting both conditions.

The availability of resources to implement more sustainable strategies was a central argument in the findings of the Brundtland Commission's "Our Common Future," the United Nations Rio Earth Summit's "Agenda 21," and the World Bank's World Development Report 1992. These documents and Goodland and Daly (1993a, 1993b) argue strongly that the reduction of poverty (i.e., empowering the poor with human and financial resources) is a *sine qua non* of sustainability. The point here is a crucial one: The universal and non-negotiable goal of environmental sustainability needs the fuzzy process of environmentally sustainable development. The challenge is to distinguish between environmentally sustainable and unsustainable development.

Interactions among the driving factors—scale, structure, efficiency, technology and investment in environmental protection—together with the key feedback loops between economic activity and human behavior—such as the powerful impact of income on fertility—explain why in some situations economic growth and technological progress will increase environmental damage and sometimes lessen it. For effective policy-making, it is essential that these various paths be disentangled so that policies may be targeted in a manner that induces changed behavior away from environmentally damaging and inequitable growth, and towards accelerated sustainable poverty reduction.

9. World Bank Progress Towards Sustainable Development

The World Bank's latest position on sustainable development has just been published (Serageldin 1993a) so it will not be repeated in this conceptual paper. To summarize: an entire Vice Presidency for Environmentally Sustainable Development (ESD) was created in January 1993, and substantial numbers of staff (c.200) are now focussing on sustainable development. ESD has started to integrate the viewpoints of the three relevant disciplines: economics, ecology and sociology against which any technical/engineering proposals should be evaluated. ESD focuses on improved valuation of environmental concerns, building sustainability into national accounts, and how to value the future. This is amplified by Serageldin (1993a).

But beyond these conceptual concerns, the World Bank is actively pursuing a four-pronged agenda to promote ESD worldwide, through all its activities. This agenda comprises:
1. Assisting borrowing countries in promoting environmental stewardship.
2. Assessing and mitigating whatever adverse impacts are associated with Bank-financed projects.
3. Building on the positive synergies between development and the environment (often called "win-win" strategies).
4. Addressing the global environmental challenges.

9.1. PROMOTING ENVIRONMENTAL STEWARDSHIP

When it comes to assisting countries in environmental stewardship, the Bank is helping in the definition of strategies and is providing funds for environmental management as well as for projects. This last year the Bank committed $173 million in support of environmental management, bringing up the portfolio of environmental management projects and project components financed so far up to about $500 million. This compares with a total portfolio of environmental loans amounting to about $5 billion, of which about $2 billion were committed in this last year alone. Some more details are provided in Table 6. In addition, the Bank is trying to help in the expansion and dissemination of knowledge by promoting the sharing of experiences between the decision-makers in the various countries. But sound environmental stewardship is rooted in sound development and environmental strategies, which must be based on properly identifying the right priorities, and these are very much country specific.

The key point is that environmental priorities will vary from country to country, and the Bank should stand ready to assist each country with the design and implementation of its own environmentally sustainable development strategy. Each will have to address particular problems. Air pollution may be the prevalent issue in Mexico city (where the Bank is helping with a $280 million loan), but there are other forms of pollution that could be a major priority in some other cities of the developing world. Toxic wastes are the most urgent problem in parts of the former Soviet Union. In Niger, it could very well be the problem of overgrazing. But whatever it is, the formulation of these national strategies, we believe, should be the result of a consultative participatory process in the countries themselves. That is how we hope that the National Environmental Action Plans (NEAPs) which are now being promoted in many countries will be done.

9.2. ASSESSING AND MITIGATING ADVERSE IMPACTS

The second part of the four point agenda is assessing and mitigating unavoidable adverse impacts of projects that the Bank agrees to finance. This requires a subjecting every proposal to a rigorous environmental assessment, as well as the traditional technical and economic assessments. Furthermore, the Bank is now trying to introduce social assessment as well. The Bank has published much on environmental assessment procedures (e.g., World Book 1992), so we will not go into detail on this point here.

9.3. BUILDING ON "WIN-WIN" STRATEGIES

Conversely, we should dwell a bit more on the third part of the four pronged agenda: the question of building synergies between development and the environment. The key here is that by adopting the conceptual framework of environmental sustainability, proper development helps environmental protection and vice versa. That is the so called "win-win" strategy. It is to our mind, the most promising area to focus on. There are two parts to this item of our four pronged agenda: investing in people and promoting the efficient use of resources.

Investing in people is particularly important. It is the poor who suffer the most from environmental degradation, especially women. When drought hits, it is the poor who suffer. Women are responsible for getting water, just as they have to gather fuel wood from farther and farther afield all the time, and naturally they also care for the children. The solutions to better natural resource management as well as lowering fertility all involve empowering women. That means that investing in people, in human resource development, must pay special attention to girls' education. This is probably the single most important measure that

Table 6. World Bank Lending for Environment (World Bank 1993a).

Country	Project	Loan/Credit Amount
New Commitments, approved in fiscal 1993		millions of dollars
Brazil	Water Quality and Pollution Control - Sao Paulo/Parana	245.0
Brazil	Minas Gerais Water Quality and Pollution Control	145.0
China	Southern Jiangsu Environmental Protection	250.0
India	Renewable Resources Development	190.0
Korea, Republic of	Kqangju and Seoul Sewerage	110.0
Mexico	Transport Air Quality Management	220.0
Turkey	Bursa Water and Sanitation	129.5
Total		**1,289.5**
Projects under implementation, approved in fiscal 1989-92		
Angola	Lobito-Benguela Urban Environmental Rehabilitation (92)	46.0
Brazil	National Industrial Pollution Control	50.0
Chile	Second Valparaiso Water Supply and Sewerage (91)	50.0
China	Ship Waste Disposal (92)	15.0
China	Beijing Environmental (92)	125.0
China	Tianjin Urban Development and Environment (92)	100.0
Côte d'Ivoire	Abidjan Lagoon Environment Protection (90)	21.9
Czech and Slovak Federal Republics	Power and Environmental Improvement (92)	246.0
India	Industrial Pollution Control (91)	155.6
Korea, Republic of	Pusan and Taejon Sewerage (92)	40.0
Poland	Energy Resources Development (90)	250.0
Poland	Heat Supply Restructuring and Conservation (91)	340.0
Total		**1,439.5**
Portfolio Total		**2,729.0**

we can adopt both for development and for the promotion of sound environmental policy over time.

Investment in people must also include population programs to recognize the pressure that the global population is putting on all of us, and these must be accompanied by the provision of maternal and infant health care.

The efficient management of resources is the second leg of the win-win strategy. Just how inefficient the current management of resources actually is, can be quite striking. Sadly, a large part of this mismanagement is currently induced by government policy. Energy subsidies in the developing world account for $230 billion a year. That is about four to five times the total volume of Official Development Assistance (ODA) going from the North to the South. That is environmentally unsound, economically unsound and wasteful of resources that could be going towards other uses.

Likewise, many of the subsidies that exist today are, in fact, for extractive and destructive industries. In the case of logging, for example, average stumpage fees are a fraction of the cost of reforestation. Among African countries sampled in 1988, the best was less than a quarter of the cost of reforestation, while the worst was running at about one percent of the cost. So subsidies were going to private loggers whereas, in fact, the full restitution to the public commons was not taking place.

If we have focussed here on the developing countries problems, that is because this is where the World Bank lends. Nevertheless, the Bank is equally concerned about the inequitable use of resources world-wide and the wasteful and destructive practices being pursued by northern consumption and pollution patterns and is trying through reports and discussions to help the requisite awareness in the North.

9.4. ADDRESSING GLOBAL CHALLENGES

The fourth part of our four-pronged agenda is addressing global challenges. Here, we include national activities that have global payoffs. These are areas where much can be done to promote the global agenda from a national sovereign decision-making framework. There are, of course, global issues where the costs are local and the benefits are global. For these activities special instruments like the Global Environmental Facility (GEF) have a crucial role to play. The GEF has also been designated as the interim funding mechanism for the two conventions on Biodiversity and Climate Change.

In parallel to working on these global population challenges, the World Bank is also concerned with consumption in the north, and how to address the disparities between the North and the South. It is important to remind ourselves as the UNDP's Human Development Report did so eloquently in the now famous "champagne glass" graph, that the richest 20 percent of the world receive about 83 percent of the world's income. The poorest 20 percent of the world receive 1.4 percent. And that disparity means a huge disparity, both in terms of consumption patterns and in terms of pollution. This argues for sound strategies for people in the South because they have so few degrees of freedom, but it also certainly argues for looking again at the consumption patterns in the North. And by that, we do not mean going back to the horse and buggy days. Switzerland, which by no stretch of the imagination is a deprived country, has a water consumption per capita that is about one-fifth that of the United States. On energy consumption levels, the difference between Switzerland (or Japan for that matter) and the United States is also about one-half. The per capita consumption of energy in India or China is still a very small fraction of that in Switzerland or Japan. So the per capita consumption issues have to be looked at, and these argue for changes in the northern patterns, as much as they argue for sound practices in the South.

The same is true in terms of the global commons, and the contribution on the debit side, in terms of pollution and the use of the environment as a "sink." The contribution in terms of CO_2 emissions, or in terms of global waste production and pollution show the same types of disparities. They are also very large. India's per capita contribution of average annual tons of carbon emitted into the atmosphere is very small compared to Canada or the United States, and this is true of most developing countries, except for the former USSR, where levels are relatively high because of the nature of their industrial activities.

Such disparities encourage one to think in terms of tradeable permits. Low income countries with a large population could trade permits based on proportional population rights to use environmental services (both to consume and to pollute) with some of the richer countries. While this is not currently on the agenda of international negotiations, there is something there for all of us to reflect on.

In addition to the agenda alluded to above, the Bank has substantially increased direct environmental investments (Table 6): a record $2 billion for 23 projects to assist developing countries in improved environmental management. This represents a doubling over one year ago, and a thirty-five-fold increase over lending five years ago. Financing for the enabling conditions of sustainability also soared: $180 million for population, $2 billion for education, $5 billion for poverty alleviation. In addition, more than 30 countries have prepared national environmental action plans with assistance from the Bank. The Bank is convinced that environmental sustainability is essential and costs less than unsustainable development.

A major obstacle to promoting policies that foster sustainability to date has been the incomplete measurement of income and investment, particularly the failure to reflect the use or deterioration of natural capital (Steer and Lutz 1993). To correct this failure, the Bank is promoting improvements in National Income Accounts (SNA) (Ahmad et al. 1989, Lutz et al. 1993, El Serafy 1993). Environmentally adjusted SNA has massive policy implications for most developing countries. Without environmentally adjusted SNA for example, we cannot judge if an economy is genuinely growing or merely living unsustainably on asset liquidation beyond its true income; whether the balance of payments is in surplus or deficit on current account, or whether the exchange rate needs to be changed.

10. Conclusion

This paper represents our current views on the concept of environmental sustainability. This is ongoing work, part of which is being done in the World Bank, much of it being done by concerned scholars around the world. Our aim has been to make a modest contribution to the debate on the essence of environmental sustainability. In particular, we have suggested that ES is a clear concept and that it is universal and non-negotiable. While the many paths leading to ES in each country or sector will differ, the goal remains constant. We fully expect the concept to be refined in the coming months and years. By its actions, it is clear that the World Bank is taking environmental sustainability seriously indeed. But this conceptualization is far from an academic exercise. The monumental challenge of ensuring that possibly ten billion people are decently fed and housed within less than two human generations—without damaging the environment on which we all depend—means that the goal of environmental sustainability must be reached as soon as humanly possible.

11. References

Ahmad, Y., S. El Serafy, and E. Lutz, eds. 1989. *Environmental Accounting*. The World Bank, Washington, DC.
Boutros-Ghali, B. 1994. *An Agenda for Development*. New York, United Nations. NGO Liaison Service 46: 20.
Brown, L.B. et al. 1994. *State of the World: 1994*. Worldwatch Institute, Washington, DC.
Cernea, M. 1993. The Sociologist's Approach to Sustainable Development. *Finance and Development* 30(4): 11-13.
Daily, G.C. and P.R. Ehrlich. 1992. Population, Sustainability and the Earth's Carrying Capacity. *BioScience* 42(10): 761-771.
Daly, H.E. 1991. Sustainable Growth: A Bad Oxymoron. *Grassroots Development* 15(3): 39.
Daly, H.E. and J. Cobb. 1994. *For the Common Good*. Beacon Press, Boston.
Daly, H.E. 1994. *Consumption: Value Added, Physical Transformation, and Welfare* (Draft). Public Affairs, University of Maryland, College Park, MD 20742-1821.
Dasgupta, P. and C. Heal. 1979. *Economic Theory and Exhaustible Resources*. Cambridge University Press, Cambridge.
Ehrlich, P. and A. Ehrlich. 1989a. Too Many Rich Folks. *Populi* 16(3): 3-29.
Ehrlich, P. and A. Ehrlich. 1989b. How the Rich Can Save the Poor and Themselves. *Pacific and Asian Journal of Energy* 3: 53-63.
Ehrlich, P.R. and J.P. Holdren. 1974. Impact of Population Growth. *Science* 171: 1212-1217.
El Serafy, S. 1991. The Environment as Capital. In *Ecological Economics*, R. Costanza, ed. Columbia University Press, New York, pp. 168-175.

El Serafy, S. 1993. *Country Macroeconomic Work and Natural Resources*. The World Bank, Washington, DC. Environment Working Paper No. 58.

Goodland, R., and H.E. Daly 1993a. Why Northern Income Growth Is Not the Solution to Southern Poverty. *Ecological Economics* 8: 85-101.

Goodland, R. and H.E. Daly. 1993b. Poverty Alleviation is Essential for Environmental Sustainability. The World Bank, Washington, DC. Environment Working Paper No. 42.

Goodland, R. 1992. The Case That the World Has Reached Limits. *Population and Environment* 13(2):167-182.

Haavelmo, T. and S. Hansen. 1992. On the Strategy of Trying to Reduce Economic Inequality by Expanding the Scale of Human Activity. In *Population Technology Lifestyle: The Transition to Sustainability*, R. Goodland, H. Daly, and S. El Serafy, eds. Island Press, Washington, DC.

Hardin, G. 1993. *Living Within Limits*. Oxford University Press, New York.

Hicks, Sir J.R. 1946. *Value and Capital*. Clarendon Press, Oxford.

Ludwig, D. 1993. Uncertainty, Resource Exploitation and Conservation: Lessons From History. *Science* 260(2 April): 17, 36.

Lutz, E., ed. 1993. *Toward Improved Accounting for the Environment*. An UNSTAT-World Bank Symposium. The World Bank, Washington, DC.

Meadows, D., D. Meadows, and J. Randers. 1992. *Beyond the Limits*. Chelsea Green Publishing, Post Mills, VT.

Mies, M. 1991. Consumption Patterns of the North: The Cause of Environmental Destruction and Poverty in the South: Women and Children First. UNCED, UNICEF and UNFPA, Geneva.

Orr, D.W. 1992. *Environmental Literacy: Education and the Transition to a Postmodern World*. State University of New York, Albany.

Parikh, J. and K. Parikh. 1991. *Consumption Patterns: The Driving Force of Environmental Stress*. UNCED, Geneva.

Putnam, R.D. (with R. Leonardi and R.Y. Nanetti). 1993b. *Making Democracy Work: Civic Traditions in Modern Italy*. Princeton University Press, Princeton, NJ.

Putnam, R.D. 1993a. Social Capital and Public Affairs. *The American Prospect*. 13(Spring):1-8.

Serageldin, I., H. Daly, and R. Goodland. 1994. The Concept of Environmental Sustainability. In *Towards Sustainable National Income*, W. van Deen, ed. IMSA, Amsterdam, pp. 71-95.

Serageldin, I. 1993a. Making Development Sustainable. *Finance and Development* 30(4): 6-10.

Serageldin, I. 1993b. *Development Partners: Aid and Cooperation in the 1990s*. SIDA, Stockholm.

Simonis, U.E. 1990. *Beyond Growth: Elements of Sustainable Development*. Edition Sigma, Berlin.

Solow, R. 1974. The Economics of Resources or the Resources of Economics. *American Economic Review* (May): 1-14.

Solow, R.M. 1991. *Sustainability: An Economists Perspective*. The Woods Hole Oceanographic Institution: The Eighteenth Seward Johnson Lecture (June 14), Woods Hole, MA.

Solow, R.M. 1992. An Almost Practical Step Toward Sustainability. Resources for the Future (40th Anniversary Lecture), Washington, DC.

Steer, A. and V. Thomas. (forthcoming). *Promoting Development That Lasts*.

Steer, A. and E. Lutz. 1993. Measuring Environmentally Sustainable Development. *Finance and Development* 30(4):20-23.

Tietenberg, T. 1990. The Poverty Connection to Environmental Policy. *Challenge* 33(5): 26-3.
Tietenberg, T. 1992. *Environmental and Natural Resource Economics.* Harper Collins, New York.
Tinbergen, J. and R. Hueting. 1991. GNP and Market Prices: Wrong Signals For Sustainable Economic Success That Mask Environmental Destruction. In *Environmentally Sustainable Economic Development: Building on Brundtland*, R. Goodland, H. Daly and S. El Serafy, eds. The World Bank, Washington, DC. Environment Paper 36: 36-42.
Vitousek, P.M., P. Ehrlich, A. Ehrlich, and P. Matson. 1986. Human Appropriation of the Products of Photosynthesis. *BioScience* 36: 368-373.
Westra, L. 1994. *An Environmental Proposal for Ethics: The Principle of Integrity.* Rowan & Littlefield, Lanham MD.
World Bank. 1990. *World Development Report.* The World Bank, Washington DC.
World Bank. 1991. *World Development Report: Poverty Alleviation.* The World Bank, Washington DC.
World Bank. 1992. *World Development Report 1992: Development and the Environment.* Oxford University Press, New York.
World Bank. 1992. *Environmental Assessment Sourcebook.* The World Bank, Washington, DC.
World Bank. 1993a. *The World Bank and the Environment 1993.* The World Bank. Washington, DC.
World Bank. 1993b. *World Development Report 1993: Investing in Health.* Oxford University Press, New York.
World Bank. 1995. *World Development Report. Infrastructure For Development.* The World Bank, Washington DC.

Chapter 9
HARD ECOLOGY, SOFT ECOLOGY, AND ECOSYSTEM INTEGRITY

Kristin Shrader-Frechette[1]

1. Introduction

What you take as your starting point depends on where you want to go. If you want to sail due south to the Dry Tortugas, then you start with plenty of fresh water, some food, a good navigational system, and arguably a ship-to-shore radio for the long trip. But if you want to sail due west to nearby John's Pass, then you might need some fresh water, but no food, no sophisticated navigational system, and no radio. How you begin a journey depends on where you want to go. So it is with environmental ethics.

How you begin your environmental ethics depends on where you want to go with them. On the one hand, if you want to develop a system of ideals to guide attitudes toward the biosphere, then your environmental ethics and its scientific foundations can be general and qualitative. They can provide principles that are inspirational to backpackers and birders, ecologists and scuba divers. Such ethics, despite their motivational and perhaps poetic character, would be of little practical use in resolving environmental controversies because they are imprecise and have no second- or higher-order ethical principles. They are too soft. On the other hand, if you want to develop a system of norms that withstands courtroom challenges involving wetlands protection or development rights, a system of environmental ethics that is capable of supporting precise, often controversial, environmental policies, then general, qualitative principles will not work. They are more useful in preaching to the converted. In controversial situations, environmental ethics need to be precise and complex enough to handle controversy, and the underlying science must be predictive and explanatory.

Many soft environmental ethics—such as those of Callicott (1989), Leopold (1968), Naess (1973), Rolston (1988), and Westra (1994), have great heuristic and inspirational power, but they are incomplete in being qualitative and general. They mainly preach to the converted—people who already accept environmental values—but fail to provide precise second- and higher-order principles that would make them operationalizable. For those who are less interested in preaching than in policy, less concerned with the philosophical ivory tower than with embattled wetlands or forests, environmental ethics must be grounded in precise science. They must also provide complex, rationally defended principles that are capable of clear, specific, practical applications. Although both general (soft) and specific (practical) approaches are necessary to environmental ethics, too many persons think that the

---

[1]Environmental Sciences and Policy Program and Department of Philosophy, University of South Florida, Tampa, FL 33620-5550, U.S.A.

former is sufficient, not merely necessary, for solving environmental problems. In so doing, they follow "soft ecology" and append to it ethics based on inspiration rather than argument, preaching rather than careful case studies, and intuition rather than second- and higher-order ethical analyses that are capable of adjudicating environmental controversies. Or they follow "hard ecology" and search for general, deductive ecological theories rather than modest rules and concepts that are operationalizable and applicable. My argument is that the science necessary to undergird sound environmental ethics requires that we avoid the extremes of either soft or hard ecology. Sound environmental ethics, at least at present, requires a "practical ecology" based primarily on case studies, natural history, and rules of thumb.

A practical scientific foundation for environmental ethics must chart a middle course between the "hard," hypothetico-deductive ecology of persons like Robert Henry Peters (1991) and the "soft," largely qualitative, ecology espoused by persons—such as Laura Westra or Holmes Rolston—who propose the concepts of ecosystem, community, integrity, or stability as the foundation for environmental policymaking (Rolston 1988, 1987, pp. 246-274; 1986; Westra 1994; Shrader-Frechette and McCoy 1993, ch. 3). The problem with using the ecosystem or integrity concept, as an alleged scientific foundation for environmental ethics, is that such concepts underestimate ecological uncertainty and thus demand too little of ecology. Likewise, the more deductive concepts of Peters overestimate ecological uncertainty and thus demand too much of ecology. We show where both hard and soft ecology go wrong, and we chart a middle course, "practical ecology." First, let's examine some of the difficulties in Peters' "hard ecology."

2. Peters and the Problems of "Hard Ecology"

In an analysis that is both tough-minded and controversial, Peters argues that ecology is a "weak science" (Peters 1991, p. 11). He claims that the primary way to correct this weakness is to judge every ecological theory "on the basis of its ability to predict" (Peters 1991, p. 290). Peters' argument, that the main criterion for ecological theorizing ought to be its predictive power, is somewhat correct in at least two senses. Prediction often is needed for applying ecology to environmental problemsolving. Peters also is right to emphasize prediction because, if scientists did not seek this goal, at least in some cases, they likely would foreclose the possibility of ever having any predictive scientific theories.

Despite the value of prediction in science, Peters' argument is misguided in at least four ways (Shrader-Frechette and McCoy 1993, pp. 106-111). For one thing, he is wrong to use prediction as a criterion for, rather than a goal of, ecological theorizing. Not all sciences are equally predictive. Economics and sociology, for example, are both more explanatory than predictive, yet it is not obvious that they are non-scientific by virtue of being so. Likewise, many geological phenomena—such as whether a given rock formation will be intact in 100,000 years—are not susceptible to precise, long-term prediction. We conclude from this predictive imprecision neither that geology is unscientific nor that we should reject the goal of precise geological prediction, but rather that geology probably deals with long-term phenomena that are less deterministic than those in other sciences. In overemphasizing the importance of prediction in ecology and science generally, Peters has erred in underemphasizing the role of explanation.

Peters' overemphasis on prediction and hypothesis deduction is also highly questionable in the light of the last three decades of research in philosophy of science, much of which has identified fundamental flaws in the positivistic, hypothetico-deductive paradigm for science. Kuhn (1970)—and other critics of the positivist paradigm—have argued that science is likely based more on retroduction and good reasons than on deduction alone. One of the fundamental reasons that no sciences can be perfectly deductive in method is that they depend on methodological value judgments—about whether certain data are sufficient, about whether a given model fits the data, about whether nontestable predictions are reliable, and so on. Because such value judgments render strict deduction impossible, falsification and confirmation of hypotheses are always questionable, at least to some degree. Moreover, although all sciences depend on such value judgments, this dependence is particularly acute for ecology, because ecology is more empirically and theoretically underdetermined than many other sciences. For example, in island biogeography there are many areas of underdetermination that require one to make choices among different methodological value judgments. These choices concern how to interpret data, how to practice good science, and how to apply theory in given situations, such as determining the best design for nature reserves. Such choices are evaluative because they are never wholly determined by the data.

Consider the nature reserve case (see Shrader-Frechette and McCoy 1993, pp. 88-94). Ecologists must decide whether ethical and conservation priorities require protecting an individual species, an ecosystem, or biodiversity, when not all can be protected at once. Different design choices usually are required to protect a particular species of interest, as opposed to preserving a specific ecosystem or biotic diversity (Margules, et al. 1982, p. 116; Soulé and Simberloff 1986, pp. 19-40; Williamson 1987, p. 367; Zimmerman and Bierregaard 1986, p. 134). Also, ecologists often must choose between maximizing present and future biodiversity. Currently they are able only to determine which types of reserves, for example, contain the most species at present, not which ones will contain the most over the long term (Soulé and Simberloff 1986, pp. 24ff.). Moreover, in the absence of adequate empirical data on particular taxa and their specific autecology, ecologists frequently must decide how to evaluate the worth of general ecological theory in dictating a preferred reserve design for a particular case (Boecklen and Simberloff 1987, pp. 247-276, pp. 250-252, 272; Margules, et al. 1982, p. 124; Simberloff and Cox 1987, pp. 63-71; Soulé and Simberloff 1986, pp. 25ff.; see Zimmerman and Bierregaard 1986, p. 135). They also are often forced to assess subjectively the value of different reserve shapes. Besides, reserve shape, as such, may not explain variation in species number (Blouin and Connor 1985, pp. 277-288). Ecologists likewise must frequently rely on subjective estimates and methodological value judgments whenever the "minimum viable population" size is not known in a precise area (Boecklen and Simberloff 1987, pp. 252-255; Orians, et al. 1986, p. 231; Soulé and Simberloff 1986, pp. 26-32). One of the most fundamental sources of value judgments in ecology is the fact that the island-biogeographical theory underlying current paradigms regarding reserve design has rarely been tested (Margules, et al. 1982, p. 117; Zimmerman and Bierregaard 1986, p. 134) and is dependent primarily on ornithological data (Zimmerman and Bierregaard 1986, pp. 130-139), on correlations rather than causal explanations (Zimmerman and Bierregaard 1986, pp. 130-139), on assumptions about homogeneous habitats (Margules, et al. 1982, p. 117), and on unsubstantiated turnover rates and extinction rates (Boecklen and Simberloff 1987, pp. 248-249, 257). Hence, whenever ecologists apply the theory, they must make a variety of methodological value judgments. Some of these value judgments concern the importance of factors other than those dominant in island biogeography (e.g., maximum breeding habitat), especially because such factors often have been shown to be superior predictors of species number (Boecklen and Simberloff 1987, p. 272; Margules, et al. 1982, p. 120; Simberloff and Cox 1987, pp. 63-71; Zimmerman and Bierregaard 1986, pp. 136ff.). Making value judgments regarding reserve design is also difficult because corridors (an essential part of island biogeographic theory) have questionable overall value for species preservation (Orians, et al. 1986, p. 32; Simberloff and Cox 1987, pp. 63-71; Salwasser 1986, pp. 227-247). Recommending use of corridors thus requires ecologists to subjectively evaluate their effectiveness in particular situations. Also, owing to the large variance about species-area relationships (Boeklen and Simberloff 1987, pp. 261-272; Connor and McCoy 1979, pp. 791-833; McCoy 1983, pp. 53-61; 1982, pp. 217-227), those who use island biogeographical theory are often forced to make subjective evaluations of non-testable predictions. Some of these subjective evaluations arise because islands are disanalogous in important ways with nature reserves (Margules, et al. 1982, p. 118). As a result, ecologists who apply data about islands to problems of reserve design must make a number of value judgments about the representativeness and importance of their particular data.

Because of the empirical and theoretical underdetermination illustrated by cases like island biogeography, and because of the resultant methodological value judgments necessary to interpret and apply it in specific cases, ecology does not appear to be fully amenable to hypothesis deduction. The included value judgments break the deductive connections of ecological theory. Of course, there are rough generalizations that can aid problem solving in specific ecological situations, as a prominent National Academy of Sciences Committee recognized (Orians, et al. 1986). Nevertheless, it is unlikely that we shall be able to find many (if any) simple, general, hypothetico-deductive (H-D) laws that we can easily apply to a variety of particular communities or species. A second reason—in addition to the underdetermined, value-laden theory—that such laws are unlikely is that fundamental ecological terms (like "community" and "stability") are imprecise and vague, and therefore unable to support precise empirical laws (Shrader-Frechette and McCoy 1993, ch. 2). Likewise, although the term "species" has a commonly accepted meaning, and although evolutionary theory gives a precise technical sense to the term, there is no general agreement in biology on an explicit definition of "species." There is consensus neither on what counts as causally sufficient or necessary conditions for a set of organisms to be a species, nor on whether species are individuals. Phenetic taxonomy has failed to generate a workable taxonomy, perhaps because species are not natural kinds and because facts cannot be carved up and rearranged in accord with the hopes of numerical taxonomists (Hull 1988, pp. 102ff.; Rosenberg 1985, pp. 182-187; Sokal and Sneath 1963).

Simple, general, hypothetico-deductive laws are also unlikely in ecology because of the uniqueness of ecological phenomena. If an event is unique, it is typically difficult to specify the relevant initial conditions for it and to know what counts as relevant behavior. Often one must have extensive historical information in order to do so (see Fetzer 1975, pp. 7-97; Kiester 1982, pp. 355ff.). Hence, from an empirical point of view, complexity and uniqueness hamper the elaboration of a simple, general set of hypothetico-deductive laws to explain most or all ecological phenomena. And if so, then the "hard ecology" of Peters is not a reasonable foundation for environmental ethics because it does not appear achievable. H-D may be an important ideal but, at present, it appears to demand too much of ecology and to overestimate its potential for certainty.

3. Problems with "Integrity" and Soft Ecology

At the other extreme of proposed scientific foundations for environmental policymaking, concepts like "integrity" demand too little of ecology because they are qualitative, unclear, and vague. They underestimate the ecological uncertainty associated with such fuzzy terms. Arne Naess recognized this point when he claimed that the normative foundations provided by ecology are "basic intuitions" (Naess 1973, pp. 95-100). The problem with intuitions is not only that they are vague and qualitative but also that one either has them or does not. They are not the sort of things amenable even to intelligent debate, much less to scientific confirmation or falsification. Hence intuitions ask too little of ecology; their uncertainty causes us to come up short when ecologists need to defend their conclusions in an environmental courtroom. To illustrate the difficulties with "soft ecology," consider some of the problems associated both with the scientific foundations of the concept of ecosystemic integrity and with its philosophical applications. Much of the scientific and ethical interest in integrity arose as a result of Aldo Leopold's famous 1949 environmental precept: "A thing is right when it tends to preserve the integrity, stability, and beauty of the biotic community.

Ch. 9. Hard Ecology, Soft Ecology, and Ecosystem Integrity

It is wrong when it tends otherwise" (Leopold 1968, pp. 224-5). Numerous persons—such as Callicott (1989), Heffernan (1982, pp. 235-247), Leopold (1968, pp. 224-255), Rolston (1975, pp. 103-109), Sagoff (1985a, pp. 99-116), and Westra (1989, pp. 91-124)—have done insightful analyses of the philosophical concept of integrity, but unfortunately these studies rely on problematic science or soft ecology. One of the major problems with the scientific concept of integrity is that, as late as 1992, one of the leading experts on integrity, Henry Regier, noted that the term has been explicated in a variety of ways: to refer to open-system thermodynamics, to networks, to Bertalanffian general systems, to trophic systems, to hierarchical organizations, to harmonic communities, and so on (Regier 1992a, pp. 25-37). Obviously, a clear, operational scientific concept cannot be explicable in a multiplicity of ways, some of which are mutually incompatible, if one expects the concept to do explanatory and predictive duty for field ecologists and therefore philosophical and political duty for attorneys, policymakers, and citizens involved in environmental controversies.

A second problem with the integrity concept is that when people attempt to define it precisely, often the best they can provide is necessary conditions, such as "indicator species" for ecosystem integrity. For example, the 1987 Protocol to the 1978 Great Lakes Water Quality Agreement formally specified lake trout as an indicator of a desired state of oligotrophy (Regier 1992a, pp. 25-37). One difficulty—with using such species to indicate environmental integrity—is in part that tracking the presence or absence of an indicator species is imprecise and inadequately quantitative. A better idea might be to track the change in species number or taxonomic composition. Another recognized problem is that the presence or absence of an indicator species, alone, presumably is not sufficient to characterize everything that might be meant by "integrity"; otherwise, persons would not speak of "ecosystem integrity" but merely of "ecosystem presence of lake trout." Hence, although the meaning of "integrity" is not clear, defining the term via several indicator species appears both crude and inadequately attentive to the underlying processes likely contributing to the presence or absence of certain species and to the larger processes presumably possessing integrity.

Methodologically, the definition of integrity is also suspect because it is based merely on opinions rather than on confirmed ecological theories or empirical generalizations. As Regier admits, although the aggregated form of the Index of Biological Integrity (IBI), avoids reliance on a single indicator species, it provides an arbitrary definition of integrity. It

> does not relate directly to anything that is observable by the nonexpert, nor to any encompassing theoretical or empirical synthesis. As a conceptual mixture put together according to judgment of knowledgeable observers, it is not 'understandable' in a theoretical sense. It is conceptually opaque in that it provides only a number on a scale; this number is then interpreted as bad or good according to practical considerations (Regier 1992a, pp. 25-37).

Indeed, the whole concept of ecosystem integrity seems to be conceptually opaque and vague. Regier admits, for example, that "general, qualitative, developmental tendencies of healthy organic systems...provide a basis for practical understanding, measurement, and management of ecosystem integrity" (Regier 1992b, p. 191).

If general, qualitative judgments provide the basis for understanding ecosystem integrity, then it is arguable that they are likely to be insufficiently precise and quantitative to do the environmental work required of them if they are challenged in court by developers, polluters, or citizens asked to pay for cleanup. Also, if only experts can recognize integrity, and if "integrity" is not tied to any publicly recognizable criteria, then the term seems incapable of uncontroversial operationalization. Hence, concepts like ecosystem integrity may be closer to "soft ecology"—or as Dan Simberloff would put it, "theological ecology"—

that preaches only to the converted (Regier 1992b, p. 194). Soft ecology may be too uncertain to provide a firm foundation for the precise norms often required in environmental ethics and policy.

Admittedly, at least one branch of theory regarding ecosystems integrity—Kay's and Schneider's nonequilibrium thermodynamic account—is not "soft ecology" in the sense that it is not general, qualitative, and vague. Rather, it is specific, quantitative, and precise. It also yields a number of insights about ecosystem behavior. This thermodynamic version, however, assuming it might be "the" definition of integrity, appears to be "soft ecology" in several other damaging senses. For one thing, the account is based on defining ecological phenomena in terms of a thermodynamic model rather than on discovering, confirming, or falsifying specific hypotheses about the phenomena. Because this account relies on definition rather than discovery, and because it does not show how at least two independent avenues function in advancing our explanation of ecological phenomena, the thermodynamic account appears to provide merely a stipulative definition, rather than a causal explanation, of ecological phenomena.

The thermodynamic account of integrity is also definitionally problematic in a second sense. On the thermodynamic model, ecosystem organization tends to increase degradation of energy, and measures of this organization rely in part on measures of energy utilization in the food web. Yet, because it is often difficult to assign organisms to a particular trophic level, it is difficult to measure accurately ecosystem organization. Linking the integrity of an ecosystem to its ability to maintain its organization, Kay and Schneider (Kay 1993, pp. 201-214; 1991, pp 483-495; Kay and Schneider 1992, pp. 159-182; Schneider and Kay 1993; Victor, et al. 1991) argue that there are certain situations in which an ecosystem would not maintain its organization. One such example is an ecosystem that is stressed by exposure to a 6 degree C increase in temperature of the water effluent from a nuclear power station (Ulanowicz 1985, pp. 23-47). If, as a result of this thermal stress, the size of the ecosystem were diminished, its trophic levels were decreased, it recycled less, and it leaked nutrients and energy, then Kay and Schneider claim that the ecosystem would not have maintained its organization. They claim that such effects are signs of "disorganization and a step backward in development" (Schneider and Kay 1993, p. 21).

One problem with their argument is that there are many ecosystem responses to stress, and complex systems have multiple steady states. After stress, (1)the system could eventually continue to operate as before, or (2)it could operate with a reduction or increase in species number, or (3)it could exhibit new paths in the food web, or (4)it could take on a largely different structure with different species and food webs (Kay 1993, pp. 201-214; 1991, pp 483-495; Kay and Schneider 1992, pp. 159-182; Schneider and Kay 1993; Victor, et al. 1991). Because of the multiple steady states of complex ecosystems, a third definitional problem is that the thermodynamic model, as Kay and Schneider recognize (Kay 1993, pp. 201-214; 1991, pp 483-495; Kay and Schneider 1992, pp. 159-182; Schneider and Kay 1993; Victor, et al. 1991), does not indicate which (if any) of these four changes is more or less natural or acceptable. Hence, the thermodynamic account, in itself, indicates different ways in which ecosystems respond to stress, but not which responses constitute a lack of integrity. And here is the rub. Either we must say, first, that any system maintaining itself at any optimum operating point has integrity—with the consequence that virtually any environmental change anywhere any time is said to be consistent with integrity. This first position likely would cause environmental catastrophe and would delight many developers and polluters. Or we must say, second, that a system has integrity if it resists permanent ecosystem change—a position that doesn't fit the facts of dynamic and evolutionary ecosystems. This second position is inapplicable to the real world. Or, third, we must define, independent of the thermodynamic account, some type of change as a loss of integrity. Hence, the thermodynamic model of integrity reduces us, when using it for environmental policymaking, to using science that is either (1) incapable of defining integrity in an environmentally sound way, or (2) irrelevant, or (3) dependent on some nonthermodynamic (arguably subjective) account of community structure. Thus the thermodynamic model, despite its heuristic power, is definitional in at least three senses. It does not provide two independent avenues for explanation. It assigns organisms to trophic levels in a question-begging way. And it requires one to stipulate some change as a loss of integrity. For all these reasons, the account is fundamentally uncertain. It is not an adequate ecological basis for environmental policymaking. At best, it provides necessary, but not sufficient, scientific grounds for environmental ethics and policy. Insofar as it is uncertain and requires us to fill in our knowledge gaps with subjective judgments, it leads to incomplete and soft ecology.

The objection, of course, is not to philosophical or ethical concepts of integrity which obviously may have heuristic, philosophical, and political power. Rather, the argument is that philosophers and soft ecologists do not call a spade a spade. They do not call soft science "soft" when it is soft, and they appear not to realize that soft science, in the absence of an environmental political consensus, is unlikely to be robust enough to support precise environmental ethics and policy decisions. When a consensus supports particular environmental values, then soft ecology is obviously valuable and heuristically useful. But situations of consensus regarding environmental values are not those in which we most need ecology.

4. Soft Ecology and Westra's Account of Integrity

In the absence of consensus on environmental values, neither hard nor soft ecology is robust enough to provide a firm foundation for a philosophical concept of integrity. Thus, despite the ethical, metaphysical, and heuristic importance of philosophical accounts of integrity, they are adequate to support neither precise and strong policymaking nor the adjudication of environmental disputes. To understand some of the assets and liabilities of philosophical discussions of integrity, consider the recent work of philosopher Laura Westra. Her analysis of integrity, based on the work of ecologists such as James Kay (Westra 1994, pp. 42-50), is significant because it presents the most detailed and carefully argued philosophical account, to date, of the ecosystems approach to environmental ethics.

Using an Aristotelian framework, Westra argues that integrity is an ultimate value, the basis for a new moral imperative, the principle of integrity (PI). The PI, she says, prohibits harming the life-support systems of the earth, but otherwise allows choices of social/ethical/ individual priorities, provided they do not conflict with the duty to avoid harming these life-support systems (Westra 1994, p. 104). Westra argues that only by understanding properly the functioning of ecosystems and their temporal and spatial scales will we know what role culture and humans ought to play on the planet (Westra 1994, p. 179). Offering an insightful metaphor, she suggests that we proceed, in our interactions with the environment, as we "would in a vehicle without certified, guaranteed brakes, in bad weather, with fogged windows, and uncertain roads (Westra 1994, p. 179). In other words, she says we ought to proceed, in environmental decisionmaking, with extreme caution and prudence. Going farther, Westra argues that the PI requires us to practice "environmental triage" (Westra 1994, p. 202), that is, to put preservation of earth's life-support systems ahead of all other social causes such as eliminating racism or sexism.

On the one hand, Westra's account is to be commended. It provides an important discussion of the direction in which Aldo Leopold's views might lead and, as such, is of central importance to environmental ethics. Her treatment also is philosophically significant for its careful arguments and responses to objections. Likewise, from Aristotle, Mill, and Rawls—through Callicott, Gert, and Taylor—Westra displays a careful and comprehensive grasp of the philosophical literature relevant to metaphysics, moral philosophy, and environmental ethics. She also modestly admits the need for continued fine-tuning of both the concept of integrity and her analyses of it (Westra 1994, p. xxi). For example, she is explicit about the problems that her account faces (Westra 1994, p. xvii). She correctly recognizes that the PI—as a general, first-order principle directing our actions toward the environment—does not specify second-order ethical principles, rules for what to do in a variety of situations in which following the PI is impossible because of conflicts between environmental and human interests (Westra 1994, p. 183). Because Westra does not specify explicit second-order principles, however, her holistic position is not completely clear, and this lack of clarity suggests that her PI is not yet operationalizable. To her credit, however, Westra recognizes that her account of holism needs more development along the lines of second-order principles (Westra 1994, p. 183). Another important philosophical asset of Westra's treatment of integrity is that it provides an example of a genuinely Aristotelian theory of environmental ethics. As such, it argues for teleology in nature (Westra 1994, p. 137) and for a common-sensical ethics about how we ought to behave according to the "laws of nature" (Westra 1994, p. 92). Westra's discussion of the principle of integrity is especially valuable because it is motivated by actual policy concerns, such as the 1978 Great Lakes Water Quality Agreement. This agreement called on decisionmakers to "restore integrity" to the Great Lakes Basin Ecosystem. Because she has taken an important step in helping environmental philosophy respond to actual policy dictates, Westra deserves praise for her eminently practical work on ethics. She is concerned not only with the philosophical underpinnings of environmental policy but also with its practical consequences. Indeed, her concerns are a model for all applied philosophers. Westra speaks both to the interests of policymakers and to those of concerned citizens.

On the other hand, Westra's account of integrity faces problems similar to those of other theories built on soft ecology. At least four such difficulties are (1)the metaphorical, scientifically-problematic account of organic teleology, (2)the reliance on ecosystem approaches, (3)the stipulative definition of "integrity," and (4)the lack of clarity about the priority among human and environmental concerns. The first difficulty, the metaphorical claim that ecosystems are "live" (Westra 1994, p. 137), is scientifically problematic because ecosystems include the physical environment of biological communities as well as the communities themselves. Hence ecosystems are not alive in any scientific sense. Claiming that they are alive ties Westra's account of integrity to outmoded, often-criticized organismic claims about ecology (see Shrader-Frechette and McCoy 1993, pp. 11-67). While such theses about living ecosystems may be philosophically meaningful, they threaten the scientific credibility of the integrity enterprise. Likewise, Westra's Aristotelian argument for "organic teleology," for ecosystems' "irreducible potential for form" (Westra 1994, p. 137), may be more metaphorical than factual. Ecosystems are not agents in any sense, and populations (not ecosystems) adapt. Adaptation is restricted to heritable characteristics; no alleged knowledge of the past operates in natural selection, and the individual which is better adapted to the present environment is the one which leaves more offspring and hence transmits its traits. Whole ecosystems do not adapt. They do not maximize anything. Natural selection does not operate to produce phenomena of a particular kind because their presence gives rise to certain effects. According to neo-Darwinian theory, an organism possesses heritable traits because of the genes it carries; which genes it transmits to its progeny is determined by random processes taking place during division and fertilization of the sex cells, not by the effects that the genes produce in the organism carrying them or in its offspring. Because mutation of genes as well as which genes are transmitted to offspring are determined by random processes, natural selection does not operate so as to give rise to certain effects. Although genes mutate as a consequence of environmental changes to which the organism responds, both the specific mutation involved as well as which genes mutate, are independent of the effects which the mutation may have on successive generations. One reason that natural selection cannot perform the task—which Westra and others suggest it has—is that it is not an explanatory principle and thus cannot explain evolution. If anything, our contemporary understanding of evolution is fundamentally anti-teleological because, in the study of evolution, there are neither completely stable and unchanging individuals nor a stable and unchanging goal of the whole process. Given neo-Darwinian theory, it is questionable for Westra and other holists to suggest that there is selection at the level of the ecosystem (see Shrader-Frechette 1986, pp. 84-85).

Westra's reliance on ecosystem approaches presents a second set of problems because there are other scales, methods, and schools in ecology than the ecosystems approach. Those who study individuals, species, communities, and populations, for example, may ask about the ethical grounding for scientific accounts not based on ecosystems ecology. Moreover, even ecologists who model ecosystems, such as Kay, admit that they are "complex, dynamic systems, that cannot be clearly defined, perhaps not even in principle" (Westra 1994, p. 45). Although ecosystems have obvious heuristic value, often it is not possible to specify the conditions under which one ecosystems account, rather than another, might be falsified. Without testability, the applicability of ecosystems theories in helping to preserve the environment is limited. As Schindler noted near the end of his classic study of eutrophication, because the carbon hypothesis was not tested, it was not falsified, and because it was not falsified, much-needed phosphorus control in numerous dying lakes was delayed (Shrader-Frechette 1986, p. 83).

Holistic, ecosystems approaches are also problematic in that they may cause scientists to underestimate the importance of studies of individual species, populations, and communities. They beg the question of what will eventually turn out to be an adequate explanation of ecological phenomena. As many ecological risk assessors have noted, stress and pollution in a particular habitat or environment typically appears first as a problem with a particular species, usually the most sensitive one—like the canary in the coal mine. To define integrity or any other environmental concept in terms only of ecosystems approaches could cause us to ignore the canary in the coal mine. Moreover, as one botanist pointed out recently, if everyone had listened to the so-called holists in chemistry and had argued that certain structures of nature were so complex that scientists would have to be content merely to measure their functions as if they were whole organisms, then we would never have learned about the structure of proteins (Shrader-Frechette 1986, pp. 83-84).

There are also more general philosophical difficulties associated with holistic appeals to ecosystems. For one thing, to presuppose that ecosystems are holistic units which maximize their well-being, as Rolston and Callicott claim, or to presuppose that humans are bound to maximize ecosystemic well being (Calhoun 1963; Jolly 1980; Milton and Farvar 1972), is to attribute interests to ecosystems. Yet, within the accepted philosophical tradition, "interests" logically presuppose desires, aims, or wants (see Feinberg 1977, pp. 49-51; Frankena 1979, p. 11). And ecosystems do not have desires, aims, or wants (see Rodman 1977, p. 91; Taylor 1986, p. 18). Moreover, the capacity for suffering or enjoyment is presumably a prerequisite for having interests (Heffernan 1982, pp. 235-247; see Callicott 1980, pp. 311-336; Singer 1977, p. 8). If it were not, then we would be forced to say, for example, that water had an interest in not being polluted (see Singer 1979, pp. 194-195ff.). This means that attributing interests to ecosystems (via a first-order ethical principle) is incomplete and problematic (because of the multiplication of entities said to have interests), unless one likewise formulates clear and precise second-order ethical principles for how to adjudicate conflicts among beings whose interests differ. One might argue, for example, that when there are conflicts of interest among different beings, human interests ought to receive top priority only in cases in which human lives are at stake; otherwise, environmental interests ought to have top priority, even when property rights are at stake (Shrader-Frechette and McCoy 1993; see Shrader-Frechette 1991a, pp. 311-321; 1991c, pp. 25-31). In the absence of such second-order principles—that specify priorities in cases of conflict—it does not seem reasonable to attribute interests to ecosystems, because there would be no way to adjudicate conflicts among different interests or different beings (Caldwell and Shrader-Frechette 1993, pp. 214).

Another ethical problem with defining "integrity" in terms of holistic ecosystem notions is that, if ecosystems were organic wholes having a good, then it would be very difficult for us, as moral agents, to know what that good is. They cannot tell us what their good is, and there is no general, predictive theory in ecology that can tell us, uncontroversially, what their good is (Shrader-Frechette and McCoy 1993, pp. 1-79). Moreover, it is not clear (apart from human interests) how we ought to define the good for ecosystems because they cannot experience pleasure or pain. Obviously, the good for an organism is life, nourishment, and so on. Given that the ecosystem is a whole comprised of both living and nonliving components, it is difficult to say what its precise good is, without appealing to purely stipulative definitions. Obviously there are purely anthropocentric grounds for condemning wanton destruction or misuse of the environment, because such behavior manifests selfishness or greed. However, it would be very difficult, in every case, to specify purely ecological, or solely nonanthropocentric, criteria for praising or blaming moral actions that have effects on ecosystems (Caldwell and Shrader-Frechette 1993, pp. 214-215).

Moreover, within a holistic ethics, there is a dilemma. Either we humans are on a par with other creatures on the planet, or we are not. Either we humans are equal members of the biotic community and therefore have no special responsibilities—contrary to what all our ethical traditions have taught—or we humans are not equal members of the biotic community. If we are not, because of our moral primacy, then in many cases we have no obligations to any non-human entity whenever its basic welfare conflicts with our own needs for bodily security. Following the consequences of this dilemma, the presupposition—of defining integrity in terms of ecosystemic holism—appears to create a dilemma. Either it leads to actions (such as murder) which are heinous (in the case of human equality with other beings). Or it is inconsistent with the prescriptions of most environmentalists (in the case of human superiority over other beings) (see Fritzell 1987; Taylor 1986, pp. 45-46, 225-226, 246, 259, 281-282). Hence, defining integrity in terms of ecosystemic holism is questionable (Caldwell and Shrader-Frechette 1993, p. 215).

As we have already argued, scientific versions of ecosystemic holism are also problematic. Obviously there are communities of different species, as well as interactions and interdependencies among the abiotic and biotic elements of the environment. Nevertheless, there is no precise, empirically confirmed ecological whole, although pragmatic, holistic thinking clearly has heuristic value. For one thing, most well known ecologists have either remained agnostic or rejected the Gaia hypothesis, the basis of many accounts of holism. They regard it as possibly correct, but at present only unproved speculation. Of course, they admit the ecological fact of interconnectedness and convolution on a small scale. Moreover, an ecosystem, as the same collection of individuals, species, and relationships, certainly does not persist through time. Hence any notion of the "dynamic stability" of an ecosystemic whole is somewhat imprecise and unclear (Goodman 1975, p. 239; MacArthur 1955; see Norton 1987, ch. 4, section 2; Shrader-Frechette 1986, pp. 77-92; Shrader-Frechette and McCoy 1993, ch. 2). Also, the selection of the "ecosystem" as the unit which is or ought to be maximized is peculiar. Why not choose, as the unit, the community, or the association (see McIntosh 1985, pp. 44, 79, 107), or the trophic level? Clements said that the community is an organism (McIntosh 1985, pp. 228, 252-256); if so, then why is the ecosystem also an organism? Which is it, and what are the criteria for a holistic organism? Or, if one is a holist, why not choose the collection of ecosystems, the biosphere, as that which is maximized in nature and which we are morally enjoined to optimize? Once one abandons an individualistic ethics, then how, from a scientific point of view, does one choose among alternative non-individual units to be maximized (see McIntosh 1985, pp. 69-146, 263-267, 146-193, 193-242; Norton 1982, pp. 17-36)? Such questions suggest that ethical or scientific holism—organicism—despite its apparent heuristic power, is an arbitrary and imprecise notion, akin more to metaphysics than to empirical science or to practical ethics. As an empirical notion, ecosystemic holism is further undercut by the current reductionist dispute in ecology among Gleasonian individualists and Clementsian holists. Their controversy indicates that the "levels problem" has not been solved in ecology. Admittedly various ecological conclusions are valid within particular spatial and temporal scales. Nevertheless, a given ecological conclusion (regarding balance) typically holds for some (but not other) "wholes" (e.g., populations, species, communities). For example, there may be some sort of stability or balance for a given species within a certain spatial scale, but not for other species, or not within another such scale. Ecologists cannot optimize the well being of all these different wholes (having different spatial and temporal scales) at the same time. Because they cannot, there is no general level at which ecological problem-solving takes place, and no general temporal or spatial scale within which a stable "whole" is exhibited. Likewise, because of the absence of a universal ecological theory that can be appealed to, in defining the "whole" that is balanced, ecologists are forced to work on a case-by-case basis. They recognize that there is no universal level (across species, populations, or communities) at which some balanced or stable whole exists. In part this is because numerous alleged "wholes", e.g., populations, exhibit density vagueness rather than density dependence, while other "wholes" do not (McIntosh 1985, pp. 126ff., 157ff., 181-182ff., 252; see Shrader-Frechette 1986, pp. 77-92; Strong 1986). This suggests, therefore, both that there is no universal level at which a balanced or stable whole is evident, and that there is a "levels problem" in ecology. But if so, then there is no clear, precise, universal sense of ecosystemic holism to which environmental ethicists can appeal, despite its apparent heuristic value (Caldwell and Shrader-Frechette 1993, pp. 216-217).

All these difficulties suggest that, despite their heuristic power, many ecosystemic or holistic explanations are neither falsifiable nor even testable (McIntosh 1985, p. 193). There is a clear definition neither of what it is to maximize some pattern of excellence, e.g., based on interspecific competition, nor of the ecosystem which is the subject of this alleged excellence. Theorists also do not agree on the underlying processes that structure communities and ecosystems (Cody and Diamond 1975; especially Gilpin and Diamond 1984, pp. 298-315; see Lewin 1983, pp. 636-639; Shrader-Frechette 1990, pp. 47-61). Hence, despite their apparent heuristic value, holistic/ecosystemic notions cannot always contribute practical and precise accounts necessary for resolving environmental conflicts (Caldwell and Shrader-Frechette 1993, p. 217).

A number of people, however, might object to these arguments (that question the presupposition of ecosystemic holism underlying many accounts of integrity). They might object that such arguments ignore the fact that we humans need balanced, holistic thinking, both for our own welfare and for that of the entire biosphere. Admittedly, at a practical level, notions of integrity and ecosystemic holism have heuristic power. However, to posit the existence of something, e.g., ecosystem integrity, just because we need it, would be wishful thinking of the sort that Freud condemned in his *The Future of an Illusion*. Moreover, just because we need something does not mean that it exists. We may need intelligence, for example, but that does not mean that we have it. The limits of reality are not determined by our desires, but by what exists and what is defensible. In particular, we may need holistic thinking, but this does not mean that we can provide a rationally defensible account of holism, an account robust enough to undergird environmental policymaking and likely objections to it. The earlier analysis indicated some of the fundamental conceptual difficulties besetting holistic/organismic ecology and ethics. If this account is correct, then although a given individual may be able to accept some sort of holism as a pragmatic or heuristic hypothesis at the personal level, it is unlikely that scientific or ethical holism is defensible at the level of predictive ecological theory, because of the conceptual problems already noted. Of course, one individual may decide what "whole" to maximize, on the basis of her personal beliefs or stipulative definitions. For holism to undergird a societal approach to policymaking, however, requires that it be free of the conceptual difficulties already noted, be precisely defined, be rationally defensible, and therefore be acceptable to many people. This suggests that, although we may need a holistic way of thinking, and although it has immense heuristic power, we may not be able to ground societal environmental policy on an ethical or scientific framework based on holism, even though holism is often pragmatically desirable. Nevertheless, we may be able to adopt ecosystemic holism at the level of personal morals or at the level of a working hypothesis (Caldwell and Shrader-Frechette 1993, pp. 223-224).

The purpose of this analysis of ecosystemic holism and integrity has not been to leave environmental ethics without a foundation. Rather, the goal has been merely to point out that the foundation is not as simple as many persons currently suppose. As a recent U.S. National Academy of Sciences report on ecology noted:

> the point of discussing the many obstacles to making accurate predictions is not to argue the futility of trying, but to show that the process of prediction must be viewed as complex and probabilistic. An appropriate approach to managing ecological systems recognizes the random component of population dynamics.... Environmental manipulations will always be experimental to some extent, and our most promising course is to structure each one so that we can learn as much as possible from it (Orians, et al. 1986, pp. 91-92).

If the Academy report is correct, then notions such as "integrity" and "ecosystem" are not so much foundations on which to build an ethics, as they are heuristic tools, working hypotheses, and idealizations that provide a useful context for learning more about the environmental perturbations and fluctuations that sometimes preclude precise prediction in ecology.

Because stipulative definition prevents authentic prediction and falsification in science, the stipulative definition of "integrity" presents a third difficulty for philosophers and scientists interested in this concept. Many scholars in environmental ethics, including Westra, speak of "integrity," "health," "wholeness," "stability," "balance," and "harmony" (Westra 1994, p. 22). Yet, the scientific community agrees neither on what each of these terms means, nor on precisely how each of them differs from the others, nor on whether they refer to anything real. Because of the partially metaphorical, stipulative, and controversial nature of each of these terms, it is difficult both to distinguish them from each other and to use them to ground environmental ethics. For example, what makes an ecosystem have integrity—rather than health or wholeness—amounts to a largely stipulative definition. There is no scientifically or philosophically consistent usage, and merely stipulating some difference between them does not solve the problem. Whenever philosophical assessments employ evaluative endpoints (such as "integrity" or "health") that are stipulatively defined and that are far removed from actual empirical data (on species, ecological structures, and biotic processes), the assessments are controversial. Between the empirical data and such stipulatively defined endpoints, there is a large gap that different persons fill differently. Also, why should thermodynamic models, species numbers, or something else define "integrity," rather than "wholeness," "harmony," or some other concept? Why not skip the stipulative definitions and just rely on the empirical data and the assessment endpoints that are very close to the empirical data? Why arbitrarily invent one concept, rather than another, to label a particular set of data? Although Westra deserves praise for attempting to bridge the gap between empirical data and philosophical endpoints, especially when the science is not clear, defining "integrity" in a stipulative way is problematic. Later, in the next section of this chapter, we suggest ways of protecting the environment that are closer to empirical data and that rely less on stipulative definitions.

A fourth difficulty with "integrity" in Westra's account is that the priority ranking of human and environmental concerns, in following the PI, is not clear. Sometimes Westra says that "the 'principle of integrity' is an imperative which must be obeyed before other human moral considerations are taken into account" (Westra 1994, pp. 6, 202). At other times, she says that she "allows choices of social/ethical/individual priorities, provided they do not conflict with the absoluteness of the 'harm principle' [the principle of avoiding harming life-support systems]" (Westra 1994, p. 104). In other words, at times Westra appears to make PI a first-priority principle, while at other times she speaks of it merely as a constraint on standard, anthropocentric, ethical priorities. Although such problems suggest that "integrity" discussions need further development, perhaps along the lines of second-order principles, as Westra clearly recognizes, clarification of philosophical priorities will not resolve the scientific difficulties besetting the concept. Hence, although it makes sense to attempt to improve scientific accounts of integrity, there are alternative ecological approaches (to be discussed in the next section) that might provide more direct, less controversial ways to address environmental ethics and policymaking.

5. Ecological Foundations for Environmental Policymaking

Despite the scientific problems besetting ecosystemic holism, ecology is nevertheless important—for pragmatic, philosophical, and heuristic reasons—in environmental theorizing. What sorts of ecological theorizing might be able to ground environmental ethics? Arthur Cooper, in his 1982 Presidential Address to the Ecological Society of America, argued that ecological findings about the value of wetlands provided "the most direct example of ecological influence on public policy...." (Sagoff 1985a, p. 104). Although Cooper cites the wetlands example as a victory for environmental policy, it really appears to be a case in which policymakers accepted untested, highly doubtful beliefs of ecologists (Sagoff 1985a, p. 107; 1985b, esp. pp. 5ff.; see Nixon 1979 pp. 437-525). Indeed, the acclaimed theoretical ecologist, John Maynard Smith, noted that "ecology is still a branch of science in which it is usually better to rely on the judgment of an experienced practitioner than on the predictions of a theorist" (McIntosh 1985, p. 321). As a consequence of this reliance, the battlefields of environmental policy are littered with the carcasses of untested, now rejected hypotheses (like DDT biomagnification) (see Shrader-Frechette 1991b, pp. 294-301) that were once used as ecological "facts" to support arguments for environmental protection (Caldwell and Shrader-Frechette 1993, pp. 217-218; see Carpenter 1983, pp. 573-595; Cooper 1982; Federal Register 1979; Hunt and Bischoff 1960, pp. 91-106; Levin and Harwell 1985, p. 15; Moriarty 1983, pp. 135-154; Murdoch and Connell 1971, p. 57; Sagoff 1986a; 1985a, p. 110; Suter 1981).

In the U.S.'s longest legal conflict over environmental policy, for example, general ecological theory and notions—such as integrity and ecosystemic holism—were of little help. The controversy began in 1964 and was between the U.S. Environmental Protection Agency (EPA) and five New York utility companies. The basic problem was that the disputants disagreed over the effect of water withdrawals (by the utilities) on the Hudson-River striped-bass population. After spending tens of millions of dollars, scientists could still not estimate the precise ecological effects of the water withdrawals. In other words, they knew, at the level of a first-order ethical principle, that they wished to avoid serious harm to the striped-bass population. Because of the inadequacy of ecological theory, however, they were unable to specify some second-order principle for adjudicating the dispute between those attempting to protect the utility and those attempting to protect the bass (see Barnthouse, et al. 1984, pp. 17-18). This controversy (between the utility and the EPA) suggests a number of reasons that it is difficult for ecologists to get a hold on fundamental processes that might support ecosystemic holism or integrity. For one thing, important ecological problems, such as the causes and consequences of global CO_2 or acid rain, involve many parameters and a high degree of complexity and uncertainty. As a result, there is too much "going on" in natural communities to be captured by any model, e.g., Lotka-Volterra (see Sagoff 1986b, p. 17), and ecologists are not certain which factors are the significant ones (Cooper 1982, p. 350; see McIntosh 1985, pp. 247, 249, 268, 273-74, 278, 284). Also, data bases in ecology are still so limited that they do not provide enough information for making land policy or environmental policy (see Cooper 1982, pp. 350-351). Because of inadequate data bases, different ecologists often claim evidential support for inconsistent hypotheses (this example is from Suter 1981, p. 186), and they are encumbered with masses of untested hypotheses (see McIntosh 1985, pp. 249, 269-270, 273, 284; Poole 1977, pp. 210-213; Simberloff 1981). Just by mere dint of repetition, often these hypotheses achieve the status of facts (see Sagoff 1985a, pp. 110-111). Many of them are not testable in the first place (see Simberloff 1983, pp. 626-635), some are mere tautologies (see Peters 1991; 1976, pp. 1-12), and most are not evaluated against null models (Simberloff 1983, pp. 626-635). As a consequence, ecologists often advocate overly simple theories about ecosystem response because empirical data are hard to obtain. Such simple theories (e.g., regarding linear relationships between two parameters) are easily challenged, even though more complex ones are difficult to establish (Levins 1966, pp. 421-431; McIntosh 1985, pp. 244, 268; see Suter 1981, p. 186). Part of the problem is that ecologists are forced to examine and understand ecosystems that are constantly changing in ways that are not always predictable or uniform. In other words, the natural-selection foundations of ecology may undercut any uncontroversial notion of ecosystemic holism, equilibrium, or integrity (see Desmond 1979; and Johnson 1981). Moreover, even if ecologists could arrive at some noncontroversial notion of ecosystemic balance or integrity, it would not be very useful, for two reasons. Unlike the mathematical models used to portray them, natural ecosystems are not typically at equilibrium; if they were, they would have far fewer species (Lewin 1986, pp. 1072). Also, it is not clear that adverse environmental effects come from loss of system stability rather than from direct impacts (Ricklefs 1987, p. 171). Still another ecological difficulty is that virtually every ecological situation can be said to be so unique that there are no obvious "state variables" and few similarities across cases. Hence there may be no general theoretical laws in ecology because the diversity of the biological community often fails to converge under similar physical conditions (Ricklefs 1987, p. 167; see Macfadyen 1975, p. 351). Often scientists cannot even make ecological measurements, e.g., for r (the intrinsic rate of natural increase of a population), as fine as legitimate use of proposed equations might require (Sagoff 1986b, p. 18). Also, ecologists typically need to know how to optimize a situation involving many individual entities, species, communities, and populations. An analogous problem arises in economics: how to develop a theory of social choice that represents the interests of each person but makes the good of the entire group paramount. We have solved the problem, so far, in neither economics nor ecology (Caldwell and Shrader-Frechette 1993, pp. 218-219).

Although the difficulties just listed mean that ecology often cannot give us fundamental, predictive, theoretical laws capable of informing particular environmental decisions, they suggest both some useful methodological rules (Suter 1981, p. 189; see Pielou 1981, pp. 17-31) and some insights regarding what ecology can tell us (Cooper 1982, p. 351). It can tell us very general things and can give us first-order ethical principles, such as: "preserve integrity," "behave as if everything is connected to everything else", or "do not exceed the carrying capacity of the area or the planet." But none of these generalizations is very helpful in practical, environmental decisionmaking, especially when we need either specific answers or a second-order ethical principle that tells us how to adjudicate disputes. Ecologists can often tell us, for example, what interventions in ecosystems are likely to reduce species diversity. If we define "balance" in terms of species diversity, for example, then indeed ecologists can give us some help in environmental ethics and policymaking. That is, given the end of maximizing species diversity, ecology can tell us about the means of attaining it. Ecologists, however, cannot provide us with a general definition of an end or goal of ecosystemic activity, but they can often reveal the best means of attaining some goal, once it is specified. This is in part because, as a recent U.S. National Academy of Sciences report noted, there is typically no general, predictive, ecological "theory" that can be applied to solve environmental problems, even though lower level theories and particular ecological facts, gained from specific cases, have often been useful in environmental policymaking (Sagoff 1985a, p. 101; see Orians, et al. 1986, esp. p. 1). In other words, ecologists can rarely tell us how to protect entire ecosystems or how to define such protection, although they can often help us manage particular areas, once we can define our goals (Caldwell and Shrader-Frechette 1993, p. 220; Sagoff 1985a, p. 103).

6. A Middle Path: Practical Ecology

Given widespread controversy over environmental ethics and policy, soft ecology is unable to ground biocentric ethics on mere stipulative definition, just as hard ecology is unable to provide hypothetico-deductive theories to resolve environmental controversies. Because both types of ecology are uncertain, anyone who does environmental ethics needs both (1) a procedure for making ethical decisions under conditions of ecological uncertainty and (2) a method for using ecology, in a practical sense, to direct environmental policy.

One procedure for dealing with ecological uncertainty, a procedure defended elsewhere (Shrader-Frechette and McCoy 1993; 1992, pp. 96-99), is to minimize type II, rather than type I, statistical errors when both cannot be avoided. Contrary to current scientific norms, this rule of thumb places the burden of proof not on anyone who posits an effect, but on anyone who argues that there will be no damaging effect from a particular environmental action. One can defend this rule, despite its reversal of the norms of statistical practice, on straightforward grounds of protecting human welfare (Shrader-Frechette and McCoy 1993; 1992, pp. 96-99). Because of the uncertainty of both soft and hard ecology, one rarely has the luxury of using them to ground controversial environmental decisions.

Another means of avoiding the scientific uncertainty of both soft and hard ecology is to develop a more reliable middle path, "practical ecology." Based neither on stipulatively defined concepts nor on general theories lacking precise predictive power, practical ecology is grounded on rules of thumb (like the norm regarding types I and II error), on rough generalizations, and on case studies about individual organisms. A recent National Academy of Sciences (NAS) committee illustrated how case-specific, empirical, ecological knowledge, rather than an uncertain general ecological theory or model, might be used in environmental problem solving (Orians, et al. 1986, pp. 1, 5). According to the U.S. National Academy committee, ecology's greatest predictive successes occur in cases that involve only one or two species, perhaps because ecological generalizations are most fully developed for relatively simple systems. This is why, for example, ecological management of game and fish populations through regulation of hunting and fishing can often be successful (Orians, et al. 1986, p. 8). Applying this insight to our discussion, ecology might be most helpful in undergirding environmental ethics and policymaking when it does not try to predict complex interactions among many species, but instead avoids the uncertainties of both soft and hard ecology and attempts to predict what will happen for only one or two taxa in a particular case. Predictions for one or two taxa are often successful because, despite the problems with general ecological theory, there are numerous lower-level theories in ecology that provide reliable predictions. Application of lower-level theory about the evolution of cooperative breeding, for example, has provided many successes in managing red-cockaded woodpeckers (Walters 1991, p. 518). In this case, successful management and predictions appear to have come from natural-history information, such as data about the presence of cavities in trees that serve as habitat (Walters 1991, pp. 506ff.).

Examples like that of the woodpecker suggest that, if the case studies used in the National Academy committee report are representative, then some of the most successful ecological applications arise when (and because) scientists have a great deal of knowledge about the natural history of the specific organisms investigated in a particular case study (Orians, et al. 1986, p. 13). As the authors of the National Academy report put it, "the success of the cases described...depended on such [natural-history] information" (Orians, et al. 1986, p. 16; Shrader-Frechette and McCoy 1993, pp. 119-120). The vampire-bat case study, for instance, is an excellent example of the value of specific natural-history information when ecologists are interested in practical environmental problemsolving (Orians, et al. 1986, p. 28). The goal in the bat study was to find a control agent that affected only the "pest" species of concern, the vampire bat. The specific natural-history information that was useful in finding and using a control, diphenadione, included the facts that the bats are much more susceptible than cattle to the action of anticoagulants; that they roost extremely closely to each other; that they groom each other; that their rate of reproduction is low; that they do not migrate; and that they forage only in the absence of moonlight (Mitchell 1986, pp. 151-164). Using this natural-history information, ecologists were able to provide a firm foundation for policy about controlling vampire bats and for the ethics of doing so. Rather than attempting to apply some general ecological theory, "top down," they scrutinized a particular case, "bottom up," in order to gain explanatory insights. Their explanation was local or "bottom up" in the sense that it showed how particular occurrences come about. It explained particular phenomena in terms of collections of causal processes and interactions (Shrader-Frechette and McCoy 1994, pp. 45-70). Their explanations do not mean, however, that general laws play no role in ecological explanations, because the mechanisms discussed in the vampire-bat study operate in accord with general laws of nature. Nor do they mean that all explanations are of particular occurrences, because we can often provide causal accounts of regularities. Rather, their explanations, like the accounts of practical ecology that we wish to emphasize, are more inductive or "bottom-up" in that they appeal to the underlying microstructure of the phenomena being explained. They avoid both the hard ecology of more deductive or "top-down" explanation—that appeals to the construction of a coherent world picture and to fitting particular facts into a unified picture (Salmon 1989, pp. 3-219)—as well as the soft ecology based on stipulative definition (see Shrader-Frechette and McCoy 1993).

7. Conclusion

The success of the NAS case study, with its "bottom-up" approach to scientific explanation, suggests that—whenever ecology is needed to resolve environmental controversies—ecological method needs to avoid soft ecology's uncertain, grand concepts like integrity and stability. It also needs to avoid the equally uncertain grand theories of hard ecology. Reliable environmental actions seem to require case studies, natural-history knowledge, autecology, and humans making value judgments about the merits of their actions. Such a recipe for environmental ethics and welfare, however, provides no basis for purely biocentric concepts, laws, or theories. Rather, the modest practical ecology for which we have argued appears to rely on the practice of ecologists and on individual cases that are not separable from human judgments about environmental welfare. A genuinely biocentric ethics seems to require more certainty and more grandiose concepts and theories than are currently available in the modest rules of thumb characterizing case studies and practical ecology. Practical ecology is particularly needed in unique situations, like most of those in community ecology, where we cannot replicate singular events. If we can use the vampire-bat study as a model for future ecological research, and if the National Academy Committee is correct, then both suggest that accounts of ecological method might do well to focus on practical applications and on unavoidably human judgments about environmental management. Moreover, if ecology turns out to be a science of case studies, practical applications, and human-directed environmental management, it is not obvious that this is a defect. Ecology may not be flawed because it must sacrifice universality for utility and practicality, or because it must sacrifice generality for the precision gained through case studies (see Shrader-Frechette and McCoy 1993).

Even with its natural-history knowledge and its subjective human judgments, ecology often can provide the insights necessary for sound preservation and environmental policy. This practical and precise knowledge of natural history, coupled with conceptual and methodological analysis, is a critical departure from the hypothetical deductive and general mathematical models of hard ecology and the untestable, definitional, or incomplete principles of soft ecology. Both soft ecology and hard ecology seem to fail to address the uniqueness, particularity, and historicity of many ecological phenomena. As a consequence, it likely will be difficult for either of them to provide clear directions for how to preserve the environment or how to guide environmental ethics and policy. For this we need a middle path — dictated in part by humans, not merely by biocentric theory. We need the practical ecology of case studies and natural history.

8. References

Barnthouse, L.W., et al. 1984. Population Biology in the Courtroom: The Hudson River Controversy. *BioScience* 34(1): 17-18.

Blouin, M., and E. Connor. 1985. Is There a Best Shape for Nature Reserves? *Biological Conservation* 32: 277-288.

Boecklen, W.J., and D. Simberloff. 1987. Area-Based Extinction Models in Conservation. In *Dynamics of Extinction*, D. Elliot, ed., John Wiley, New York, pp. 247-276, pp. 250-252, 272.

Caldwell, L.K., and K.S. Shrader-Frechette. 1993. *Policy for Land: Ethics and Law*. Rowman and Littlefield, Savage, MD.

Callicott, J.B. 1989. *In Defense of the Land Ethic*. State University Press of New York, Albany.

Callicott, J.B. 1980. Animal Liberation. *Environmental Ethics* 2(4): 311-336.

Calhoun, J.B. 1963. *The Ecology and Sociology of the Norway Rat*. U.S. Department of Health, Education and Welfare, U.S. Public Health Service.

Carpenter, R. 1983. Ecology in Court. *Natural Resources Lawyer* 15(3): 573-595.

Cody, M., and J. Diamond, eds. 1975. *Ecology and the Evolution of Communities*. Harvard University Press, Cambridge.

Connor, E.F., and E.D. McCoy, 1979. The Statistics and Biology of the Species-Area Relationship. *American Naturalist* 113: 791-833.

Cooper, A. 1982. Why Doesn't Anyone Listen To Ecologists—and What Can ESA Do About It? *Bulletin on the Ecological Society of America* 63(4) (December): 348-356.

Desmond, A. 1979. *The Ape's Reflexion*. James Wade, New York.

Federal Register. 1979. 44, 71456 (December 11).

Feinberg, J. 1977. The Rights of Animals and Unborn Generations. In *Philosophy and Environmental Crisis*, W.T. Blackstone, ed., University of Georgia Press, Athens, pp. 49-51.

Fetzer, J.H. 1975. On the Historical Explanation of Unique Events. *Theory and Decision* 6: 7-97.

Frankena, W. 1979. Ethics and the Environment. In *Ethics and Problems of the 21st Century*, K. Goodpaster and K. Sayre, eds., University of Notre Dame Press, Notre Dame, pp. 3-20.

Fritzell, P. 1987. The Conflicts of Ecological Conscience. In *A Companion to the Sand County Almanac*, J.B. Callicott, ed., University of Wisconsin Press, Madison.

Gilpin, M., and J. Diamond. 1984. Are Species Co-Occurrences on Islands Non-Random, and Are the Null Hypotheses Useful in Community Ecology? In *Ecological Communities*, D. Strong, et al., eds., Princeton University Press, Princeton, pp. 298-315.

Goodman, D. 1975. The Theory of Diversity-Stability Relationships in Ecology. *The Quarterly Review of Biology* 50(3): 237-266.

Heffernan, J.D. 1982. The Land Ethic: A Critical Appraisal. *Environmental Ethics* 4 (Fall): 235-247.

Hull, D. 1988. *Science as a Process*. University of Chicago Press, Chicago, pp. 102ff.

Hunt, E., and A. Bischoff. 1960. Inimical Effects on Wildlife of Periodic DDT Application to Clear Lake. *California Fish and Game* 46: 91-106.

Johnson, E. 1981. Animal Liberation Versus the Land Ethic. *Environmental Ethics* 3(3).

Jolly, A. 1980. *A World Like Our Own: Man and Nature in Madagascar*. Yale University Press, New Haven.

Kay, J. 1993. On the Nature of Ecological Integrity. In *Ecological Integrity and the Management of Ecosystems*, S. Woodley, J. Francis, J.K. Kay, eds., St. Lucie Press, Del Ray Beach, FL, pp. 201-214.

Kay, J. 1991. A Nonequilibrium Thermodynamic Framework for Discussing Ecosystem Management. *Environmental Management* 15: 483-495.

Kay, J.K., and E.D. Schneider, 1992. Thermodynamics and Measures of Ecological Integrity. In *Ecological Indicators*, D.H. McKenzie, D.E. Hyatt, and V.J. McDonald, eds., Elsevier, Ft. Lauderdale, FL, pp. 159-182.

Kiester, A. 1982. Natural Kinds, Natural History, and Ecology. In *Conceptual Issues in Ecology*, E. Saarinen, ed., D. Reidel Publishing Company, Boston, London, and Dordrecht, pp. 355ff.

Kuhn, T. 1970. *The Structure of Scientific Revolutions*. University of Chicago, Chicago.

Leopold, A. 1968. *A Sand County Almanac and Sketches Here and There*. Oxford University Press, New York.

Levin, S., and M. Harwell. 1985. Environmental Risks Associated with the Release of Genetically Engineered Organisms. *Genewatch* 2(1): 15.

Levins, R. 1966. The Structure of Model Building in Population Biology. *American Scientist* 54(4): 421-431.

Lewin, R. 1986. In Ecology, Change Brings Stability. *Science* 234(28 November): 1072.

Lewin, R. 1983. Santa Rosalia Was a Goat. *Science* 221(12 August): 636-639.

MacArthur, R. 1955. Fluctuations of Animal Populations, and a Measure of Community Stability. *Ecology* 36: 533-536.

Macfadyen, A. 1975. Some Thoughts on the Behaviour of Ecologists. *Journal of Animal Ecology* 44(2): 351.

Margules, C., A. Higgs, and R. Rafe. 1982. Modern Biogeographic Theory: Are There Any Lessons for Nature Reserve Design? *Biological Conservation* 24: 115-128.

McCoy, E.D. 1983. The Application of Island Biogeographic Theory to Patches of Habitat: How Much Land is Enough? *Biological Conservation* 25: 53-61.

McCoy, E.D. 1982. The Application of Island Biogeography to Forest Tracts: Problems in Determination of Turnover Rates. *Biological Conservation* 22: 217-227.

McIntosh, R.P. 1985. *The Background of Ecology: Concept and Theory*. Cambridge University Press, Cambridge.

Milton, J.P., and M.F. Farvar. 1972. *The Careless Technology: Ecology and International Development*. Natural History Press, New York.
Mitchell, C.G. 1986. Vampire Bat Control in Latin America. In *Ecological Knowledge and Environmental Problem Solving*, G.H. Orians, et al., eds., National Academy Press, Washington, DC, pp. 151-164.
Moriarty, F. 1983. *Ecotoxicology*. Academic Press, New York.
Murdoch, W., and J. Connell. 1971. The Ecologist's Role and the Nonsolution of Technology. In *Ecocide—and Thoughts Towards Survival*, C. Fadiman and J. White, eds., Center for the Study of Democratic Institutions, Santa Barbara, CA, pp. 47-64.
Naess, A. 1973. The Shallow and the Deep, Long-Range Ecology Movements: A Summary. *Inquiry* 16: 95-100.
Nixon, S.W. 1979. Between Coastal Marshes and Coastal Waters. In *Ecological Processes in Coastal and Marine Systems*, R.J. Livingston, ed., Plenum Press, New York, pp. 437-525.
Norton, B.G. 1987. *The Spice of Life: Why Save Natural Variety?* Princeton University Press, Princeton, NJ.
Norton, B.G. 1982. Environmental Ethics and the Rights of Nonhumans. *Environmental Ethics* 4: 17-36.
Orians, G.H., Chair, Committee on the Applications of Ecological Theory to Environmental Problems, 1986. *Ecological Knowledge and Environmental Problem Solving*. National Academy Press, Washington, DC.
Peters, R.H. 1991. *A Critique for Ecology*. Cambridge University Press, Cambridge.
Peters, R.H. 1976. Tautology in Evolution and Ecology. *The American Naturalist* 110: 1-12.
Pielou, E.C. 1981. The Usefulness of Ecological Models: A Stock-Taking. *The Quarterly Review of Biology* 56(1): 17-31.
Poole, T.W. 1977. Periodic, Pseudoperiodic, and Chaotic Population Fluctuations. *Ecology* 58: 210-213.
Regier, H. 1992a. Ecosystem Integrity in the Great Lakes Basin. *Journal of Aquatic Ecosystem Health* 25: 25-37.
Regier, H. 1992b. Indicators of Ecosystem Integrity. In *Ecological Indicators*, S. McKenzie, ed., Elsevier, Barkiep, UK, pp. 183-200.
Ricklefs, R.E. 1987. Community Diversity: Relative Roles of Local and Regional Processes. *Science* 235(9 January): 171.
Rodman, J. 1977. The Liberation of Nature. *Inquiry* 20: 91.
Rolston, H. 1988. *Environmental Ethics*. Temple University Press, PA.
Rolston, H. 1987. Duties to Ecosystems. In *Companion to "A Sand County Almanac*, J.B. Callicott, ed., University of Wisconsin Press, Madison, pp. 246-274.
Rolston, H. 1986. *Philosophy Gone Wild*. Prometheus, Buffalo.
Rolston, H. 1975. Is There an Ecological Ethic? *Ethic* 85(January): 103-109.
Rosenberg, A. 1985. *The Structure of Biological Science*. Cambridge University Press, Cambridge, UK, pp. 182-187.
Sagoff, M. 1986a. *What Ecology Can Do*. Unpublished essay.
Sagoff, M. 1986b. *On Explanation in Ecology*. Unpublished manuscript.
Sagoff, M. 1985a. Fact and Value in Ecological Science. *Environmental Ethics* 7: 99-116.
Sagoff, M. 1985b. *Environmental Science and Environmental Law*. Center for Philosophy and Public Policy, College Park, MD.
Salmon, W. 1989. Four Decades of Scientific Explanations. In *Scientific Explanation*, P. Kitcher and W. Salmon, eds., University of Minnesota Press, Minneapolis, pp. 3-219.
Salwasser, H. 1986. Conserving a Regional Spotted Owl Population. In *Ecological Knowledge and Environmental Problem Solving*, G.H. Orians, et al., eds., National Academy Press, Washington, DC, pp. 227-247.

Schneider, E.D., and J.K. Kay, 1993. Life as a Manifestation of a Second Law of Thermodynamics. In *Advances in Mathematics and Computers in Medicine*, University of Waterloo, Waterloo, Ontario.
Shrader-Frechette, K. 1991a. Ethics and the Environment. *World Health Forum* 12: 311-321.
Shrader-Frechette, K. 1991b. Pesticide Toxicity: An Ethical Perspective. In *Environmental Ethics*, K. Shrader-Frechette, ed. Boxwood Press, Pacific Grove, CA, pp. 287-324.
Shrader-Frechette, K. 1991c. A Philosophic Basis for Ecocentric Ethics. *Earth Ethics Report* 1(1): 25-31.
Shrader-Frechette, K. 1990. Interspecific Competition, Evolutionary Epistemology, and Ecology. *Evolution, Cognition, and Realism*, N. Rescher, ed., University Press of America, New York, pp. 47-61.
Shrader-Frechette, K. 1986. Organismic Biology and Ecosystems Ecology. In *Current Issues in Teleology*, N. Rescher, ed., University of Pittsburgh Center for the Philosophy of Science, Pittsburgh, pp. 77-92.
Shrader-Frechette, K., and E.D. McCoy. 1994. Applied Ecology and the Logic of Case Studies. *Philosophy of Science* 61(1): 45-70.
Shrader-Frechette, K., and E.D. McCoy. 1993. *Method in Ecology: Strategies for Conservation*. Cambridge University Press, Cambridge.
Shrader-Frechette, K., and E.D. McCoy. 1992. Statistics, Costs, and Rationality in Ecological Inference. *Trends in Ecology and Evolution* 7(3): 96-99.
Simberloff, D.S. 1983. Competition Theory, Hypothesis Testing, and Other Community Ecological Buzzwords. *American Naturalist* 122: 626-635.
Simberloff, D.S. 1981. The Sick Science of Ecology. *Eidema* 1(1).
Simberloff D., and J. Cox. 1987. Consequences and Costs of Conservation Corridors. *Conservation Biology* 1: 63-71.
Singer, P. 1979. Not for Humans Only. In *Ethics and Problems of the 21st Century*, K.E. Goodpaster and K.M. Sayre, eds., University of Notre Dame Press, Notre Dame, pp. 194-195ff.
Singer, P. 1977. *Animal Liberation*. Avon, New York.
Sokal, P., and P. Sneath. 1963. *Principles of Numerical Taxonomy*. Freeman, San Francisco.
Soulé, M., and D. Simberloff. 1986. What Do Genetics and Ecology Tell Us About the Design of Nature Reserves? *Biological Conservation* 35: 19-40.
Strong, D. 1986. Density Vagueness: Abiding the Variance in the Demography of Real Populations. In *Community Ecology*, J. Diamond and T. Case, eds., Harper and Row, New York.
Suter, G. 1981. Ecosystem Theory and NEPA Assessment. *Bulletin of the Ecological Society of America* 62(3): 186-192.
Taylor, P.W. 1986. *Respect for Nature*. Princeton University Press, Princeton, NJ.
Ulanowicz, R.E. 1985. Community Measures of Marine Food Networks and Their Possible Applications. In *Flows of Energy and Material in Marine Ecosystems*, M.J.R. Fasham, ed., Plenum, London, pp. 23-47.
Victor, P.A., J.K. Kay, and H.J. Ruitenbeek. 1991. *Economic, Ecological, and Decision Theories: Indicators of Ecologically Sustainable Development*. Canadian Environmental Advisory Council, Ottawa, Canada.
Walters, J.R. 1991. Application of Ecological Principles to the Management of Endangered Species: The Case of the Red-Cockaded Woodpecker. *Annual Review of Systematic* 22: 505-523.
Westra, L. 1994. *An Environmental Proposal for Ethics: The Principle of Integrity*. Rowman and Littlefield, Lanham, MD.
Westra, L. 1989. 'Respect', 'Dignity', and 'Integrity': An Environmental Proposal for Ethics. *Epistemologia* 12(1): 91-124.

Williamson, M. 1987. Are Communities Ever Stable? *Symposium of the British Ecological Society* 24: 353-370.

Zimmerman, B., and R. Bierregaard. 1986. Relevance of the Equilibrium Theory of Island Biogeography and Species-Area Relations to Conservation with a Case from Amazonia. *Journal of Biogeography* 13: 133-143.

Chapter 10
SCIENCE FOR THE POST NORMAL AGE

S. O. Funtowicz[1]
Jerome R. Ravetz[2]

1. Introduction

Science always evolves, responding to its leading challenges as they change through history. After centuries of triumph and optimism, science is now called on to remedy the pathologies of the global industrial system of which it forms the basis. Whereas science was previously understood as steadily advancing in the certainty of our knowledge and control of the natural world, now science is seen as coping with many uncertainties in policy issues of risks and the environment. In response, new styles of scientific activity are being developed. The reductionist, analytical world-view which divides systems into ever smaller elements, studied by ever more esoteric specialties, is being replaced by a systemic, synthetic and humanistic approach. The old dichotomies of facts and values, and of knowledge and ignorance, are being transcended. Natural systems are recognized as dynamic and complex; those involving interactions with humanity are "emergent," including properties of reflection and contradiction. The science appropriate to this new condition will be based on the assumptions of unpredictability, incomplete control, and a plurality of legitimate perspectives.

At the present time, there is no agreed description of what the future will bring; but there is a general sense that much of our intellectual inheritance now lies firmly in the past. "Post-modern" is widely used for describing contemporary cultural phenomena; it refers to an approach of unrestrained criticism of the assumptions underlying our dominant culture, and it flirts with nihilism and despair. In contrast to this, here we will introduce the term "Post-normal." This has an echo of the seminal work on modern science by Kuhn.[1] For him "normal science" referred to the unexciting, indeed anti-intellectual routine puzzle-solving by which science advances steadily between its conceptual revolutions. In this "normal" state of science, uncertainties are managed automatically, values are unspoken, and foundational problems unheard of. The post-modern phenomenon can be seen in one sense as a response to the collapse of such "normality" as the norm for science and culture. As an alternative to post-modernity, we show that a new, enriched awareness of the functions and methods of science is being developed. In this sense, the appropriate science for this epoch is "post-normal."

---

[1]CEC-Joint Research Centre, Institute for Systems Engineering and Informatics, I-21020 Ispra (Va), Italy; [2]The Research Methods Consultancy Ltd., 196 Clarence Gate Gardens, London NW1 6AU, UK.

This emerging science fosters a new methodology that helps to guide its development. In this, uncertainty is not banished but is managed, and values are not presupposed but are made explicit. The model for scientific argument is not a formalized deduction but an interactive dialogue. The paradigmatic science is no longer one in which location (in place and time) and process are irrelevant to explanations. The historical dimension, including reflection on humanity's past and future, is becoming an integral part of a scientific characterization of Nature.

Our contribution to this new methodology focuses on two aspects. One is the quality of scientific information, analyzed in terms of both the different sorts of uncertainty in knowledge and the intended functions of the information. It has hitherto been a well-kept secret that scientific "facts" can be of variable quality; and an informed awareness of this human face of science is a key to its enrichment for its future tasks. Our other contribution relates to problem-solving strategies, analyzed in terms of uncertainties in knowledge and complexities in ethics. When science is applied to policy issues, it cannot provide certainty for policy recommendations; and the conflicting values in any decision process cannot be ignored even in the problem-solving work itself. For quality of information, we have developed a transparent system of notations (NUSAP) whereby the different sorts of uncertainty that affect scientific information can be expressed. It can thereby be communicated in a concise, clear and nuanced way, among traditional and extended peer communities alike. The NUSAP approach embodies the principle that uncertainty cannot be banished from science; but that good quality of information depends on good management of its uncertainties.[2]

We use the interaction of systems uncertainties and decision stakes to provide guidance for the choice of appropriate problem-solving strategies. The heuristic tool is a set of graphical displays of three related strategies, from the most narrowly defined to the most comprehensive. Two of them are familiar from past experience of scientific or professional practice; the last, where systems uncertainties or decisions stakes are high, corresponds to the practice of the sciences of the post-normal epoch.[3] One way of distinguishing among the different sorts of research is by their goals: Applied Science is "mission-oriented;" Professional Consultancy is "client-serving;" and Post-Normal Science is "issue-driven." These three can be contrasted with Core Science, the traditional "pure" or "basic" research, which is "curiosity-motivated." In the area of Post-Normal Science the problems of quality assurance of scientific information are particularly acute, and their resolution requires new conceptions of scientific methodology.

In this new sort of science, the evaluation of scientific inputs to decision-making requires an "extended peer community."[4] This extension of legitimacy to new participants in policy dialogues has important implications for society and for science as well. With mutual respect among various perspectives and forms of knowing, there is a possibility for the development of a genuine and effective democratic element in the life of science. The new challenges for science can then become the successors of the earlier great "conquests," as of disease and then of space, in providing symbolic meaning and a renewed sense of adventure for a new generation of recruits to science in the future.

2. The Re-invasion of the Laboratory by Nature

The place of science in the industrialized world was well depicted by Bruno Latour,[5] when he imagined Pasteur as extending his laboratory to all the French countryside, and thereby conquering it for science and for himself. In this vision, Nature itself no longer needs to be approached as wild and threatening, but through the methodology of science it can be tamed and rendered useful to mankind. The miracle of modern natural science is that the laboratory experience, the study of an isolated piece of Nature that is kept unnaturally pure, stable and reproducible, can be successfully extended to the understanding and control of Nature in the raw. Our technology and medicine together have made Nature predictable and in part controllable, and they have thereby enabled very many people to enjoy a more safe, comfortable and pleasant life than was ever before imagined in our history. The obverse side of this achievement is that it may well be unsustainable, not merely in terms of equity, but even in terms of sheer survival.

The triumph of the scientific method, deploying the technically esoteric knowledge of its experts, has led to its domination over all other ways of knowing; this applies to our knowledge of Nature, and of much else besides. Common sense experience and inherited skills of making and living have lost their claim to authority; they have been displaced by the theoretically constructed objects of scientific discourse, which are necessary for dealing with invisible things such as microbes, atoms, genes and quasars. Although formally democratic (since there are now no formal barriers to the training for that expertise), science is in fact a preserve of those who can engage on a prolonged and protected course of education, and thereby of the social groups to which they belong. In a tradition stemming from the Enlightenment of the eighteenth century, the rationality of public decision-making must appear to be scientific. Hence intellectuals with a scientific style (including economists *par excellence*) have come to be seen as leading authorities, indeed the possessors and purveyors of practical wisdom. There has been a universal assumption (however superficial and laced with cynicism) that scientific expertise is the crucial component of decision-making, whether concerning Nature or society.

Now the very powers that science has created have led to a new relationship of science with the world. The extension of the laboratory has gone beyond the small-scale intervention typified by Pasteur's conquest of anthrax. We do not merely observe the familiar gross disturbances of the natural environment resulting from modern industrial and agricultural practices. The methodology for a successful coping with these novel problems cannot be the same as the one that helped to create them. Much of the success of traditional science lay in its power to abstract from uncertainty in knowledge and values; this is shown in the dominant teaching tradition in science, which created a universe of unquestionable facts, presented dogmatically for assimilation by uncritical students. Now scientific expertise has led us into policy dilemmas which it is incapable of resolving by itself. We have not merely lost control and even predictability; now we face radical uncertainty and even ignorance, as well as ethical uncertainties lying at the heart of scientific policy issues.

For understanding the new tasks and methods of science, we can fruitfully invert Latour's metaphor, and think of Nature as re-invading the lab. We see this in many ways; for example, our science-based technology, which for a while appeared to be a new man-made nature dominant over the old, is now appreciated as critically dependent on the larger eco-system in which it is embedded; and that it risks destruction of itself if that matrix becomes seriously perturbed or degraded. Similarly, the extension of modern technology to all humanity, essential if equity between peoples is to be realized under the present system, would accelerate the self-destructive tendencies of the technological system itself. Thus Nature reasserts herself on all our scientific planning, for the technical and human perspectives alike.

There have been other episodes in history when science has been transformed, when a particularly successful problem-solving activity has displaced older forms and become the paradigmatic example of science. These transformations have been identified with the names of such great scientists as Galileo, Darwin and Einstein. They have mainly affected theoretical science, because until quite recently technology and medicine were not generally influenced in the short term by the results of scientific research. The challenges to science were largely in the realm of ideas. Now, as the powers of science have given rise to threats

3. The Centrality of Uncertainty and Quality

Now that the policy issues of risks and the environment present the most urgent problems for science, uncertainty and quality are moving in from the periphery, one might say the shadows, of scientific methodology, to become the central, integrating concepts. Hitherto they have been kept at the margin of the understanding of science, for lay persons and scientists alike. A new role for scientists will involve the management of these crucial uncertainties; therein lies the task of quality assurance of the scientific information provided for policy decisions.

These new policy issues have common features that distinguish them from traditional scientific problems. They are total in their scale and long term in their impact. Data on their effects, and even data for baselines of "undisturbed" systems, are radically inadequate. The phenomena being novel, complex and variable, are themselves not well understood. Science cannot always provide well-founded theories based on experiments for explanation and prediction; but can frequently achieve at best only mathematical models and computer simulations, which are essentially untestable. On the basis of such uncertain inputs, decisions must be made, under conditions of some urgency. Therefore science cannot proceed on the basis of factual predictions, but only on policy forecasts.

Computer models are the most widely used method for producing statements about the future based on data of the past and present. For many, there is still a magical quality about computers, since they are believed to perform reasoning operations faultlessly and rapidly. But what comes out at the end of a program is not necessarily a scientific prediction; and it may not even be a particularly good policy forecast. The numerical data used for inputs may not derive from experimental or field studies; the best numbers available, as in many studies of industrial risks, may simply be guesses collected from experts. Instead of theories which give some deeper representation of the natural processes in question, there may simply be standard software packages applied with the best-fitting numerical parameters. And instead of experimental, field or historical evidence, as is normally assumed for scientific theories, there may be only the comparison of calculated outputs with those produced by other equally untestable computer models.

In spite of the enormous effort and resources that have gone into developing and applying such methods, there has been little concerted attempt to see whether they contribute significantly either to knowledge or to policy. In research related to policy for risks and the environment, which is so crucial for our well-being, there has been very little effort of quality assurance of the sort that the traditional experimental sciences take for granted in their ordinary practice. Whereas computers could in principle be used to enhance human skill and creativity by doing all the routine work swiftly and effortlessly, they have instead in many cases become substitutes for disciplined thought and scientific rigour.[6]

Even when there is empirical data for policy problems, it is not really amenable to treatment by traditional statistical techniques. As J.C. Bailar puts it:

> All the statistical algebra and all the statistical computations are of value only to the extent that they add to the process of inference. Often they do not aid in making sound inferences; indeed they may work the other way, and in my experience that is because the kinds of random variability we see in the big problems of the day tend to be small relative to other uncertainties. This is true, for example, for data on poverty or unemployment; international trade; agricultural production; and basic measures of human health and survival. Closer to

home, random variability—the stuff of p-values and confidence limits, is simply swamped by other kinds of uncertainties in assessing the health risks of chemicals exposures, or tracking the movement of an environmental contaminant, or predicting the effects of human activities on global temperature or the ozone layer.[7]

Thus, by traditional criteria of scientific method, the quality of research on these policy-related problems is dubious at best. The tasks of uncertainty management and quality assurance, managed in traditional science by individual skill and communal practice, are left in confusion in this new area. New methods must be developed for making our ignorance usable.[8] For this there must be a radical departure from the total reliance on techniques, to the exclusion of methodological, societal or ethical considerations, that has hitherto characterized traditional "normal" science.

An integrated approach to the problems of uncertainty, quality and values has been provided by the NUSAP system. In its terms different sorts of uncertainty can be expressed, and used for an evaluation of quality of scientific information. We have to distinguish among the technical, methodological and epistemological levels of uncertainty; these correspond to inexactness, unreliability and "border with ignorance," respectively.[9] Uncertainty is managed at the technical level when standard routines are adequate; these will usually be derived from statistics (which themselves are essentially symbolic manipulations) as supplemented by techniques and conventions developed for particular fields. The methodological level is involved when more complex aspects of the information, as values or reliability, are relevant. Then personal judgements depending on higher-level skills are required; and the practice in question is a Professional Consultancy, a "learned art" like medicine or engineering. Finally, the epistemological level is involved when irremediable uncertainty is at the core of the problem, as when computer modellers recognize "completeness uncertainties" which can vitiate the whole exercise, or more generally in Post-Normal Science. In NUSAP these levels of uncertainty are conveyed by the categories of spread, assessment and pedigree, respectively.

Quality assurance is as essential to science as it is to industry; and whereas in traditional research science it could be managed informally by a peer community, in the new policy issues of risks and the environment quality of science must be addressed as a matter of urgency. The inadequacy of traditional peer review has been extensively analyzed for the different areas of Core Science,[10] "mandated" science,[11] and "regulatory" science.[12] As we shall see, the evaluation of quality in this new context of science cannot be restricted to products of research; it must also include process and persons, and in the last resort purposes as well. This "p-fourth" approach to quality assurance of science necessarily involves the participation of others than the technically qualified researchers; indeed all the stakeholders in an issue form an "extended peer community" for an effective problem solving strategy for global environmental risks.

4. Problem-Solving Strategies

To characterize an issue involving risks and the environment, in what we call "Post-Normal Science," we can think of it as one where facts are uncertain, values in dispute, stakes high and decisions urgent. In such a case, the term "problem," with its connotations of an exercise where a defined methodology is likely to lead to a clear solution, is less appropriate. We would be misled if we retained the image of a process where true scientific facts simply determine the correct policy conclusions. However, the new challenges do not render traditional science irrelevant; the task is to choose the appropriate kinds of problem-solving strategies for each particular case.

Our diagram involves three distinctive features. First, (and this is an innovation for scientific methodology), it shows the interaction of the epistemic (knowledge) and axiological (values) aspects of scientific problems. These are depicted as the axes of a diagram, representing the intensity of uncertainty and of decision stakes, respectively. We notice that uncertainty and decision stakes are the opposites of attributes which had traditionally been thought to characterize science, namely its certainty and its value neutrality. (This is the second innovative feature of our analysis). Finally, the two dimensions are themselves each displayed are comprising three discrete intervals. By this means, we achieve a diagram which has three zones representing and characterizing three kinds of problem-solving strategies (Figure 1).

The term "systems uncertainties" conveys the principle that the problem is concerned not with the discovery of a particular fact, but with the comprehension or management of an inherently complex reality. By "decision stakes" we understand all the various costs, benefits, and value commitments that are involved in the issue through the various stakeholders. It is not necessary for us to attempt now to make a detailed map of these as they arise in the technical and social aspects of dialogue on any particular policy issue. It is enough for the present conceptual analysis, that it is possible in principal to identify which elements are the leading or dominant ones, and then to characterize the total systems by them.

5. Applied Science

The explanation of the diagram of problem solving strategies starts with the most familiar strategy. We call this Applied Science. This is involved when both systems uncertainties and decisions stakes are low. The systems uncertainties will be at the technical level, and will be managed by standard routines and procedures. These will include particular techniques to keep instruments operating reliably, and also statistical tools and packages for the treatment of data. The decision stakes will be simple as well as small; resources have been put into the research exercise because there is some particular straightforward external

Figure 1. Diagram of Problem-Solving Strategies.

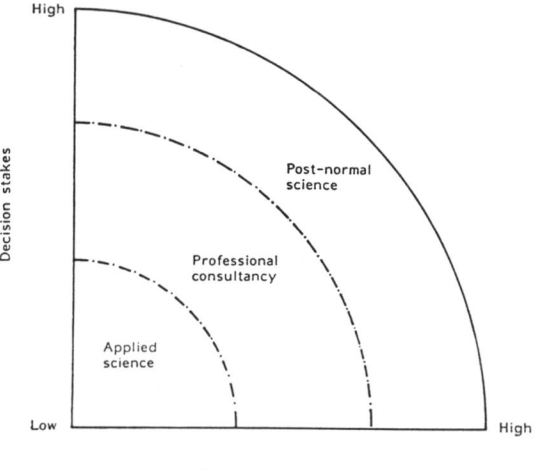

function for its results. The resulting information will be used in a larger enterprise, which is of no concern to the researcher on the job. We illustrate this in Figure 2.

In this diagram, traditional "pure," "basic," or "core" science can be considered as concentrated around the intersection of the axes. By definition, there are no external interests at stake in curiosity-motivated research, so decision stakes are low. Also, the research exercise is generally not undertaken unless there is confidence that the uncertainties are low, that is that the problem is likely to be soluble by a normal, puzzle-solving approach. Clearly, highly innovative or revolutionary research, either pure or applied, does not lie within this category, since the system uncertainties are inherently high, and for various reasons the decision stakes are also. Thus Galileo's astronomical researches involved the whole range of issues from astronomical technique to religious orthodoxy; so even though it was not directly applicable to industrial or environmental problems, it was definitely extreme both in its uncertainties and its decision stakes. The same could be said of Darwin's work in *The Origin of Species.* In this respect there is a continuity between the classic "philosophy of nature" and the Post-Normal Science that is now emerging.

We can usefully compare Core Science and Applied Science in relation to quality assurance. Where both uncertainties and external decision stakes are both low, the traditional processes of peer review of projects and refereeing of papers have worked well enough in spite of their known problems. However, when the results of the research exercise become important for some external function, the relevant peer community is extended beyond one particular research community, to include users of all sorts, and also managers. The situation in quality assessment becomes rather more like that of manufacturers and consumers, bringing different agendas and different skills to the market. For an example of how criteria of quality can differ between producers and consumers, we may consider product safety; a rare accident may be less significant to manufacturers (especially if product liability laws are lax) than for consumers. In the case of Applied Science, a result validly produced under one set of conditions may be inappropriate when applied to others; thus if measurements of a toxicant are given as an average over time, space or exposed populations, that may be

Figure 2. Applied Science.

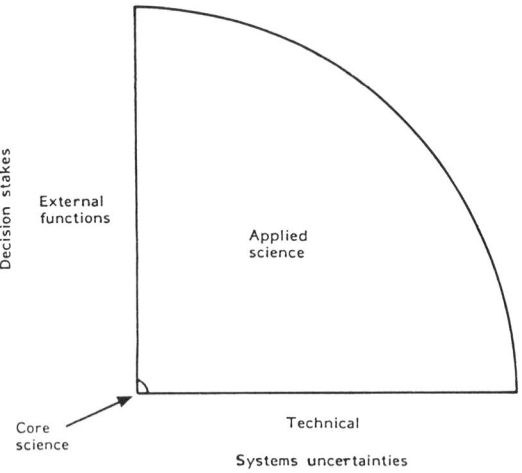

adequate for general regulatory purposes, but that set-up could ignore damaging peak concentrations or harm to susceptible groups.

It frequently happens that the results of an Applied Science project are not "public knowledge," freely available to all competent users, but rather are "corporate know-how," the "intellectual property" of the private business or State agency that sponsors the research exercise. If the information is relevant to some policy issue, the tasks of quality assurance may become controversial, involving conflicts over confidentiality; and the decision stakes may be raised over that non-scientific aspect. Then the actual problem-solving strategy is no longer Applied Science, for the issue may involve struggles over administrative and political power, and constitutional principles of "right to know" of citizens (for example, concerning environmental hazards or technological risks). The relevant peer community is thus extended beyond the direct producers, sponsors and users of the research, to include all with a stake in the product, the process, and its implications both local and global. This extension of the peer community may include investigative journalists, lawyers, and pressure groups. Thus a problem which may appear totally straightforward scientifically can become one which transcends the boundaries of Applied Science, giving rise to a more complex problem-solving strategy, such as "Post-Normal Science." When scientists with a traditionalist outlook bemoan the bad influence of "the media," it is sometimes because of their difficulty in comprehending this new feature of science when it is involved in policy.

6. Professional Consultancy

The diagram for Professional Consultancy (Figure 3) has two zones, with Applied Science nested inside. This signifies that Professional Consultancy includes Applied Science, but that it deals with problems which require a different methodology for their complete resolution. Uncertainty cannot be managed at the routine, technical level, because more complex aspects of the problem, such as reliability of theories and information, are relevant.

Figure 3. Professional Consultancy.

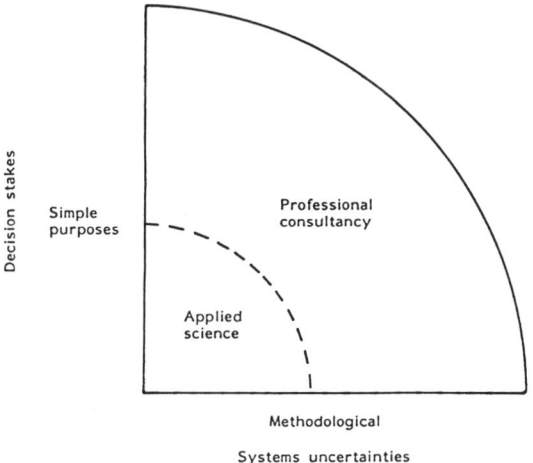

Then personal judgements, depending on higher level skills, are required, and uncertainty is at the methodological level.

The decision stakes are also more complex. Traditionally the professional task is performed for a client, whose purposes are to be served. These cannot be reduced to a clear, perfectly defined goal, for humans are not machines or bureaucracies, and are conscious of their own purposes. In the case of risks and environmental policy issues, the professionals may experience a tension between their traditional role and new demands. For the purposes relevant to the task are no longer the simple ones of clients, but will be in conflict, involving various human stakeholders and natural systems as well.

The relation between systems uncertainties and decision stakes are well illustrated by the task of incorporation of error-costs in a decision. For exercises in Applied Science, these are generally subsumed implicitly in standard statistical methods. Confidence limits, and bounds for the two types of inference-errors, are normally employed at pre-set constant values, without reflection. But in professional tasks, error-costs may so large as to endanger the continuation of a career. Hence they must be treated as risks, where some calculation may be employed but where judgement will necessarily predominate. When in a forensic situation, the professional will need to take account of the burden of proof for the particular problem, which will reflect the values of a particular society (whose harm is the more important to be prevented?). The same consideration holds for any policy issue; thus a problem of environmental pollution will be handled differently depending on whether a process is deemed safe until proved dangerous, or *vice versa*. Alternatively, we might ask whether absence of evidence of harm is interpreted as evidence of absence of harm. Although such methodological issues are quite beyond the ken of Applied Science, in Professional Consultancy they strongly condition all the work; and the simple descriptions as given here do not encompass the subtleties of burden of proof as it is used in practice.

Professional Consultancy shares many features with Applied Science, distinguishing them both from Core Science. Both operate under constraints of time and resources, with projects funded and mandated by external interests; and their products frequently lie outside the "public knowledge" domain. For much of the time professional tasks can be reduced to routine exercises, as the work becomes standardized in its technique and in the management of uncertainty. But Professional Consultancy involves the readiness to grapple with new and unexpected situations, and to bear the responsibility for their outcome. Engineering is on the border between the two, for most engineering work is done within organizations rather than for individual clients; and yet the problems cannot be completely reduced to a routine, so that "engineering judgement" is a well known aspect of the work. Of engineering we could say that most routine engineering practice is a matter of empirical craft skills using the results of Applied Science, while at its highest levels it becomes true Professional Consultancy.

A contrasting intermediate case is that of the role of the "expert." This is normally someone who advises, but whose responsibility is defined by his position as an employee; hence it is not the client's interest that defines his role but that of his employer. In that respect, his decision stakes are simpler than those of the professional consultant, and the systems uncertainties as he sees them are correspondingly reduced. It is possible for a single individual to occupy these three roles, alternately or even (to some extent) simultaneously, giving rise to confusion among his audiences or perhaps even for himself. An academic researcher may give advice on a policy-related issue; his prestige and legitimacy derive from his reputation in research, either in Core Science or Applied Science; he assumes the authority of the Professional Consultant when offering his judgements; and if his research is too closely controlled by some funding organization, then in fact he might be acting as an expert on their behalf. This is why the possibility of "conflict-of-interest" is raised when scientists make public pronouncements, without anyone impugning their personal integrity as perceived by themselves.

As a problem-solving strategy, Professional Consultancy has important differences from Applied Science. The outcomes of Applied Science exercises, like those of Core Science, have the features of reproducibility and prediction. That is, any experiment should in principle be capable of being reproduced anywhere by any competent practitioner; for they operate on isolated, controlled natural systems. Therefore the results amount to predictions of the future behaviour of natural or technical systems under similar conditions. By contrast, professional tasks deal with unique situations, however broadly similar they may be. The personal element becomes correspondingly important; thus it is legitimate to call for a second opinion without questioning the competence or integrity of a doctor in a medical case. Alternatively, who would expect two architects to produce identical designs for a single brief? In the same way, it would be unrealistic to expect two safety engineers to produce the same model (or the same conclusions) for a hazard analysis of a complex installation. The public may become confused or disillusioned at the sight of scientists disagreeing strongly on a problem apparently involving only Applied Science (and the scientists may themselves be confused!). But when it is appreciated that these policy issues involve Professional Consultancy, such disagreements should be seen as inevitable and healthy. The gain in clarity should more than compensate for the loss of mystique of scientific infallibility.

This last phenomenon reminds us of the differences in quality assurance that emerge when we extend from Applied Science to Professional Consultancy. We can envisage four components in the problem-solving task: the process, the product, the person and the purpose. This is the "p-fourth" approach to quality assurance mentioned above. In Core Science, the main focus in the task of quality assessment is on the process; the assessment is made on the basis of the research report, and it requires a community of subject-specialty peers (who can "read between the lines" of the research report) for its performance. In Applied Science, the focus of assessment extends to products, and is done by users; for it is they on whose behalf the research exercises are done. Quality assurance is then not so esoteric, since the users have less need to understand the research process; and thus there is an automatic extension of the community with a legitimate participation in evaluation. In Professional Consultancy there can be no simple, objective criteria or processes for quality assurance (beyond simple competence). The clients become an important part of the community that assesses quality of work, although they have no relevant technical expertise. Thus in these three cases, we see an expansion of the "peer community" involved in quality assurance. In this respect, the "extended peer community" of Post-Normal Science is a natural continuation of this tendency.

7. Post-Normal Science

We can now consider the third sort of problem-solving strategy, where systems uncertainties or decision stakes are high (Figure 4).

The policy issues that drive Post-Normal Science may include a large scientific component in their description, sometimes even to the point of being capable of expression in scientific language. In this sense they are analogous to the "trans-science" problems first announced by Alvin Weinberg.[13] But it seems best to distinguish the problems analyzed here from that earlier class; for Weinberg imagined problems that differed only in scale or technical feasibility from those of Applied Science. They were scarcely different from those of Professional Consultancy as we define it.[14] In the terms of our diagram, Post-Normal Science occurs when uncertainties are either of the epistemological or the ethical kind, or when decision stakes reflect conflicting purposes among stakeholders. We call it "post-normal" to indicate that the puzzle-solving exercises of normal science (in the Kuhnian sense) which were so successfully extended from the laboratory to the conquest of Nature, are no longer appropriate for the resolution of policy issues of risks and the environment. We notice that in Figures 2, 3 and 4, Applied Science appears three times and Professional Consultancy twice. Do these labels refer to the same things when they are included in a broader problem-solving strategy as when they are standing alone? In the sense of their routine practice, yes. But when they are embedded in a broader problem-solving strategy the whole activity is reinterpreted. The problems are set and the solutions evaluated by the criteria of the broader communities. Thus Post-Normal Science is indeed a sort of science, and not merely politics or public participation. However different from the varieties of problem-solving that have now become entrenched and traditional, it is a valid form of enquiry, appropriate to the needs of the present.

Examples of problems with combined high decision stakes and high systems uncertainties are familiar from the current crop of policy issues of risks and the environment. Indeed, any of the problems of major technological hazards or large scale pollution belong to this class. Post-Normal Science has the paradoxical feature that in its problem-solving activity the traditional domination of "hard facts" over "soft values" has been inverted. Because of the high level of uncertainty, approaching sheer ignorance in some cases, and the extreme decision stakes, we might even in some cases interchange the axes on our diagram, making values the horizontal, independent variable. A good example of such an inversion is provided by the actions that will need to be taken in preparation for mitigating the effects of sea-level rise consequent on global climate change. The "causal chain" here starts with the various outputs of human activity, producing changes in the biosphere, leading to changes in the climatic system, then changes in sea-level (all these interacting in complex ways with varying delay-times). Out of all this must come a set of forecasts which will provide the scientific inputs to decision processes; these will contribute to policy recommendations that must then be implemented on a broad scale. But all the causal elements are uncertain in the extreme; to wait until all the facts are in, would be another form of imprudence. At stake may be much of the built environment and the settlement patterns of people; mass migrations from low-lying districts could be required sooner or later, with the consequent economic, social and cultural upheaval.

Figure 4. Post-Normal Science.

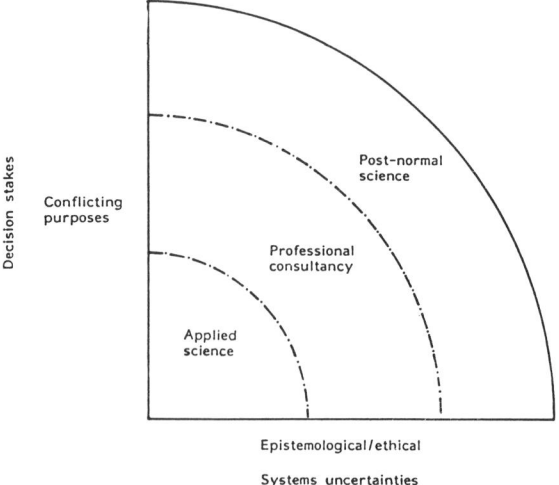

Such far-reaching societal policies will be decided on the basis of scientific information that is inherently uncertain to an extreme degree; even more so because plans for mitigation must be started with a long lead-time so that the huge rebuilding and resettlement programmes can get underway. The rise in sea level would not be like a slow tide, but more likely in the form of floods of increasing frequency and destructiveness. Unprepared harbour cities (as most of the world's political and financial centres) could be devastated. A new form of legitimation-crisis could emerge; for if the authorities try to base their appeals for sacrifice on the traditional certainties of Applied Science, as on the model of Pasteur, this will surely fail. Public agreement and participation, deriving essentially from value commitments, will be decisive for the assessment of risks and the setting of policy. Thus the traditional scientific inputs have become "soft" in the context of the "hard" value commitments that will determine the success of policies for mitigating the effects of a possible sea-level rise. In this way we see how the "systems" involved in environmental policy issues are truly "emergent," comprising dimensions of cognition and value which transcend those of the systems studied by traditional systems theory and its modelling techniques. Thus Post-Normal Science corresponds to an enriched systems theory, deriving analytical rigour from it, and providing it with experience and insights.

The traditional fact/value distinction has not merely been inverted; in Post-Normal Science the two categories cannot be realistically separated. The uncertainties go beyond those of the systems, to include ethics as well. All policy issues of risks and the environment involve new forms of equity, which had previously been considered "externalities" to the real business of the scientific-technical enterprise, that is the production and consumption of commodities. These new policy issues involve the welfare of new stakeholders, such as future generations, other species, and the planetary environment as a whole. The intimate connection between uncertainties in knowledge and in ethics is well illustrated by the problems of extinctions of species, either singly or on a global scale. It is impossible to produce a simple rationale for adjudicating between the rights of people who would benefit from some development, and those of a species of animal or plant which would be harmed. However, the ethical uncertainties should not deter us from searching for solutions; nor can decision-makers overlook the political force of those humans who have a passionate concern for those who cannot plead or vote. Only a dialogue between all sides, in which scientific expertise takes its place at the table with local and environmental concerns, can achieve creative solutions to such problems, which can then be implemented and enforced. Otherwise, either crude commercial pressures, inept bureaucratic regulations, or counterproductive protests will dominate, to the eventual detriment of all concerned.

All these complexities do not prevent the resolution of policy issues in Post-Normal Science. The diagram should not be seen statically, but rather dynamically; different aspects of the problem, located in different zones, interact and lead to its eventual solution. There is a pattern of evolution of issues, with different problem-solving strategies successively coming to prominence, which provides a means whereby dialogue can eventually contribute to their resolution. For as the debate develops from its initial confused phase, positions are clarified and new research is stimulated. Although the definition of problems is never completely free of politics, an open debate ensures that such considerations are neither one-sided nor covert. And as Applied Science exercises eventually bring in new facts, Professional Consultancy tasks become more effective. A good example of this pattern of evolution is lead in gasoline, where in spite of the absence of conclusive environmental or epidemiological information, a consensus was eventually reached that the public-health hazards were not acceptable. Such a resolution does not always come quickly or easily; some substances might be called "yo-yo risks" because of the way they go up and down in the experts' perception; dioxin seems to be one such. In those cases, effective public policy would be better based on an appreciation of the inherent uncertainties rather than on the illusion that this time Applied Science has given us the true verdict of Safe or Dangerous.

8. Extended Peer Communities

The dynamic of resolution of policy issues in Post-Normal Science involves the inclusion of an ever-growing set of legitimate participants in the process of quality assurance of the scientific inputs. As we have seen, in Applied Science and Professional Consultancy the peer communities are already extended beyond those for Core Science. In Post-Normal Science, the manifold uncertainties in both products and processes requires that the relative importance of persons becomes enhanced. Hence the establishment of the legitimacy and competence of participants will inevitably involve broader societal and cultural institutions and movements. For example, persons directly affected by an environmental problem will have a more keen awareness of its symptoms, and a more pressing concern with the quality of official reassurances, than those in any other role.[15] Thus they perform a function analogous to that of professional colleagues in the peer-review or refereeing process in traditional science, which otherwise might not occur in these new contexts.

On occasion, the legitimate work of extended peer communities can even go beyond the reactive tasks of quality assessment and policy debate. The new field of "popular epidemiology" involves concerned citizens doing the disciplined research which could, or perhaps should, have been done by established institutions but was not. In such cases they may encounter professional disapproval and hostility, being criticized either for lacking certified expertise or for being too much personally concerned about the problem. The creative conflict between popular and expert epidemiology not only leads to better control of environmental problems; it also improves scientific knowledge. A classic case is "Lyme Disease," where local citizens first identified a pattern in the vague symptoms which later characterized a previously unknown, but not uncommon tick-borne disease.

When problems lack neat solutions, when environmental and ethical aspects of the issues are prominent, when the phenomena themselves are ambiguous, and when all research techniques are open to methodological criticism, then the debates on quality are not enhanced by the exclusion of all but the specialist researchers and official experts. The extension of the peer community is then not merely an ethical or political act; it can positively enrich the processes of scientific investigation. Knowledge of local conditions may determine which data are strong and relevant, and can also help to define the policy problems. Such local, personal knowledge does not come naturally to the subject-specialty experts whose training and employment predispose them to adopt abstract, generalized conceptions of genuineness of problems and relevance of information. Those whose lives and livelihood depend on the solution of the problems will have a keen awareness of how the general principles are realized in their "back yards." They will also have "extended facts," including anecdotes, informal surveys, and official information published by unofficial means. It may be argued that they lack theoretical knowledge and are biased by self-interest; but it can equally well be argued that the experts lack practical knowledge and have their own unselfconscious forms of bias.

The new paradigm of Post-Normal Science, involving extended peer communities as essential participants, is clearly seen in the case of AIDS. Here the research scientists operate in the full glare of publicity involving sufferers, carers, journalists, ethicists, activists, and self-help groups, as well as traditional institutions for funding, regulation and commercial application. The researchers' choice of problems and evaluations of solutions are equally subjected to critical scrutiny, and their priority disputes are similarly dragged out into the public arena. There are some costs; thus it is no longer easy for scientists to exercise their benevolent dictatorship over passive test subjects in the "double-blind" procedure where some get no treatment. But unless we believe it right that the sufferers from this dread disease should depend entirely on the zeal and dedication of researchers, manufacturers and regulators, they should be included in the dialogue, however fractious it may sometimes become.

As yet, such cases are still the exception. Extended peer communities generally operate in isolation, on special policy issues in isolated localities, with no systematic means of financial support, and little training in their special skills. On many occasions, there is insufficient competence in dialogue and communication with other stakeholders.[16] Recognition of their role is very variable; in America, with its traditions of devolution of power to the local level, "intervenors" in some decision processes are provided with support; in other countries they may be ignored or actively hindered. Within such extended peer communities there will be the usual tensions between those with special-interest demands, and the outside activists with a more far-reaching agenda, along with the inevitable divisions along lines of class, ethnicity, gender and formal education. However, all such confusion is inevitable, and indeed healthy, in an embryonic movement which is fostering the transition to a new era for science. It could be that the field of health, where individual "consumer preferences" can operate more effectively on a mass scale than in environmental policy issues, the rise of Post-Normal Science will occur more smoothly. "Complementary medicine" could in many ways be considered a type-case for Post-Normal Science; and in spite of the inevitable external opposition and internal confusions, it grows steadily.

It is important to appreciate that Post-Normal Science is complementary to Applied Science and Professional Consultancy. It is not a replacement for traditional forms of science, nor does it contest the claims to reliable knowledge or certified expertise that are made on behalf of science in its legitimate contexts. The technical expertise of qualified scientists and professionals in accepted spheres of work is not being contested; what can be questioned is the quality of that work in these new contexts, especially in respect of its environmental, societal and ethical aspects. Previously the ruling assumption was that these were "externalities" to the work of science or technology; and that when such problems arose an appropriate response would somehow be invented by "society." Now the task is to see what sorts of changes in the practice of science, and in its institutions, will be entailed by the recognition of uncertainty, complexity and quality within policy-relevant research.

As in any deep transition, the present contains seeds of destruction as well as renewal. Some participants in environmental struggles come to see scientists merely as hired guns, who should provide the data that "we" need and consent to the suppression of the rest. Others will be personally impervious to any arguments and evidence that weaken their prejudged case. Are such participants legitimate members of an extended peer community? Even traditional science has always included such types, but there has been an implicit ethical commitment to integrity whereby the community as a whole has maintained the quality of its work.[17] The maintenance of quality, without which all efforts to solve policy issues of risk and the environment are doomed, is a major task for the methodology of the science of the future.

9. Conclusion

In every age, science is shaped around its leading problems, and it evolves with them. The new policy issues of risks and the environment are global not merely in their extent, but also in their complexity, pervasiveness, and novelty as a subject of scientific inquiry. Up to now, with the dominance of Applied Science, the rationality of reductionist natural-scientific research has been taken as a model for the rationality of intellectual and social activity in general. However successful it has been in the past, the recognition of the policy issues of risks and the environment shows that this ideal of rationality is no longer universally appropriate.

The activity of science now encompasses the management of irreducible uncertainties in knowledge and in ethics, and the recognition of different legitimate perspectives and ways of knowing. In this way, its practice is becoming more akin to the workings of a democratic society, characterized by extensive participation and toleration of diversity. As the political process now recognizes our obligations to future generations, to other species and indeed to the global environment, science also expands the scope of its concerns. We are living in the midst of this rapid and deep transition, so we cannot predict its outcome. But we can help to create the conditions and the intellectual tools whereby the process of change can be managed for the best benefit of the global environment and humanity.

The democratization of this aspect of science is not a matter of benevolence by the established groups, but (as in the sphere of politics) the achievement of a system which in spite of its inefficiencies is the most effective means for avoiding the disasters that result from the prolonged stifling of criticism. Recent experience has shown that such a critical presence is as important for the solution of the policy issues of risks and the environment issues as it is for society. Let us be quite clear on this; we are not arguing for the democratization of science on the basis of a generalized wish for the greatest possible extension of democracy in society. The epistemological analysis of post-normal science, rooted in the practical tasks of quality assurance, shows that such an extension of peer-communities, with the corresponding extension of facts, is necessary for the effectiveness of science in meeting the new challenges of global environmental problems.

This analysis is complementary to that of our previous paper on post-modernity.[18] Both deal with the loss of hegemony of a single world-view based on a particular vision of Science. The post-modern phenomenon is one of a deepening disillusion and a consequent fragmentation at all levels including the ideological and the societal. One reaction, as among some leading exponents of post-modernity, is despair. Another reaction is to re-assert "normality," thus some leading scientists claim that the solution of our ecological problems lies through funding their large programme of relevant basic research, in which uncertainty is never mentioned.[19] Indeed, the suppression of uncertainty in "normal" science makes it compatible with quite extreme reactions to the contemporary condition; thus it has been noticed that some religious fundamentalists find no difficulty in practicing scientific expertise of various sorts, as the two dogmatisms can, with appropriate boundary-drawing, coexist comfortably.[20] Finally, the post-normal response is to recognize the challenge, with all its dangers and promise; and then to start towards a re-integration, through the acceptance of uncertainty and the welcoming of diversity. In a later paper we will discuss these various trends.

10. References

1. T.S. Kuhn, *The Structure of Scientific Revolutions*, (Chicago, University of Chicago Press, 1962).
2. S.O. Funtowicz and J.R. Ravetz. *Uncertainty and Quality in Science for Policy* (Dordrecht, Kluwer, 1990).
3. S.O. Funtowicz and J.R. Ravetz, 'A New Scientific Methodology for Global Environmental Issues', in R.Costanza (editor) *Ecological Economics - The Science and Management of Sustainability* (New York, Columbia University Press, 1991).
4. S.O. Funtowicz, and J.R. Ravetz, 'Three Types of Risk Assessment and the Emergence of Post-Normal Science', in D. Golding and S. Krimsky, eds. *Theories of Risk*, (New York, Greenwood Press, 1991).
5. B. Latour, *The Pasteurization of France*, (Cambridge, Harvard University Press, 1988).
6. S. MacLane, 'Letters', *Science*, 241, 1988 pp 1144, and 242, 1988, pp 1623-1624.

7. J.C. Bailar, *Scientific Inferences and Environmental Problems: The Uses of Statistical Thinking*, (Chapel Hill, Institute for Environmental Studies, The University of North Carolina, 1988), pp 19.
8. J.R. Ravetz, 'Usable Knowledge, Usable Ignorance: Incomplete Science with Policy Implications', in J.R. Ravetz, *The Merger of Knowledge with Power* (London. Cassell, 1990).
9. Funtowicz and Ravetz, *op. cit.*, ref. 4.
10. J. Turney, 'End of the Peer Show?', *New Scientist*, 22 September, 1990, pp 38-42.
11. L. Salter, *Mandated Science* (Dordrecht, Kluwer, 1988)
12. Jasanoff, *The Fifth Branch*, (Cambridge, Harvard University Press, 1990).
13. A, Weinberg, 'Science and Trans-science', *Minerva*, 10, 1972, pp 209-222.
14. A. Weinberg, 'Letters', *Science*, 180, 1972, pp 1124.
15. Krimsky, 'Epistemic Considerations on the Value of Folk-Wisdom in Science and Technology', *Policy Studies Review*, 3, 1984, pp 246-262.
16. P. Brown, 'Popular Epidemiology: Community Response to Toxic Waste-Induced Disease in Woburn, Massachusetts', *Science, Technology & Human Values*, 12, 1987, pp 78-85.
17. L. Salter, *op. cit.*
18. Funtowicz and Ravetz, *op. cit.*, ref. 1.
19. J. Lubchenko et. al., The Sustainable Biosphere Initiative: An Ecological Research Agenda (A Report from the Ecological Society of America), *Ecology*, 72, 1991, pp 371-412.
20. Z. Sardar, *Explorations in Islamic Science* (London, Mansell, 1989), Chapter 2, 'Anatomy of a Confusion.'

11. Acknowledgements

Reprinted from *Futures*, 25(7): 739-755, Silvio O. Funtowicz and Jerome R. Ravetz, Science for the Post-Normal Age, Copyright 1993, with kind permission from Elsevier Science Ltd, The Boulevard, Langford Lane, Kidlington OX5 1GB, UK.

Chapter 11
THE VALUE OF INTEGRITY

Mark Sagoff[1]

1. Introduction

The concept of integrity is a normative one; it presupposes that some good is at stake. This essay does not attempt to define "integrity" in an ecological context but seeks to understand why integrity in that context is valuable and worthy of protection.

The integrity of biological communities or systems could be an instrumental good, an intrinsic good, or both. Integrity in nature is good instrumentally insofar as it enhances nature's ability to provide the goods and services people want. It is good intrinsically insofar as natural objects or systems are valuable independently of what people want, need, or enjoy. For example, we may value the objects of nature in terms of their beauty, their amazing complexity, or in relation to their histories, that is, the long, tortuous, and contingent labor of evolution that produced the creatures the world happens to contain.

In this essay, I shall argue that the integrity of nature may not be as important instrumentally as is sometimes thought. It is also difficult to explain the value of integrity in terms of the intelligible structure of ecosystems since they may not have such a structure. At any rate, it has proven difficult to define historical or mathematical qualities of ecosystems, such as complexity, diversity, resilience, stability, and so on, on which a conception of their integrity might rest. Accordingly, the value of integrity may best be explained in terms of our respect for the past and our commitment not to obliterate its remnants. It may be respect for the creative power of nature that moves us—a power to which we owe our own existence and that in many ways remains inscrutable to us.

2. Why Integrity?

The term "integrity" refers to a quality it is good to possess. But why is integrity a good thing? For human beings, the reasons are obvious: integrity is a virtue of character and makes one more likely to receive advantages such as trust and love. As a virtue, integrity is intrinsically good; it is also instrumentally valuable, moreover, insofar as it inspires the confidence of others.

Why is the quality of integrity good or important, however, insofar as it is predicated on nature? Why should we seek to maintain or restore the integrity of biological communities, for example, wetlands, rather than to "improve" those systems, let us say, by draining

---

[1] Institute for Philosophy and Public Policy, University of Maryland at College Park, College Park, MD 20742, U.S.A.

them, in favor of a farm or shopping mall, more compatible with particular economic ends? If it is some biological or hydrological service the wetland performs that concerns us, we might say it is that service, not the integrity of the system as a whole, we might wish to protect or maintain. Perhaps a pipe or culvert would do as well. We must ask why an ecosystem is itself good or what it is good for, to understand why we might value and wish to protect its integrity.

In this essay, I shall try to specify four different positions that provide reasons to value the integrity of natural communities or systems. "There is no conflict, in principle," Laura Westra has rightly said, "in recognizing a number of possible ethical (and prudential) positions, supporting the *same* action and the *same* value (integrity). As long as they are clearly acknowledged and specified, this does not present a problem." [1]

The four positions I shall describe can be defined in relation to two distinctions. The first, which Westra mentions, distinguishes between prudential and ethical reasons for protecting the natural world; it distinguishes, that is to say, between instrumental and intrinsic values. The instrumental value of nature, which lies in its usefulness to us, involves our need for sources of raw materials and sinks for wastes. Those who value nature on intrinsic grounds believe that its systems or objects possess characteristics that make them worthy of admiration, appreciation, and respect quite apart from the uses we may make of them. From this point of view, whales, for example, did not lose their essential worth when their oil was no longer used as a fuel or as a lubricant. On the contrary, these magnificent mammals gained value—they became "sacred"—when people appreciated their qualities as important in themselves.

The second distinction divides those who find in nature balance and harmony, viewing it as essentially orderly, from those who see natural communities as random, contingent, or even chaotic—"path-dependent," to use the jargon of economics. By combining these distinctions, we can generate four positions which, as I think, constitute the principal perspectives in environmental ethics on the value of the integrity of nature.

Since there is not space to consider all four positions in detail, I shall discuss primarily the two that value nature on moral or intrinsic grounds but differ in whether they view nature as orderly or contingent. These two positions each provide a rich understanding of the reasons we may value the integrity of the natural world. I shall first touch briefly, however, on prudential or instrumental views of the significance of ecological "integrity," if only to suggest that the uses we make of nature are constantly changing. The importance of integrity in the context of instrumental or prudential value, as I shall propose, is largely a function of the state of technology. Insofar as the intrinsic value of natural objects depends on their history and not on their utility, however, that kind of value is not as likely to be affected by advances in technology.

3. Cowboy, Spaceman, and Bull-in-the-China-Shop

In an excellent essay, Herman Daly has distinguished among three different attitudes to the natural environment that are consistent with an essentially prudential or instrumental focus.[2] He identifies these perspectives as those of the "cowboy," the "spaceman," and the "bull in the china shop." Although Daly distinguishes among these attitudes for a somewhat different purpose, they admirably capture the major views of ecosystem "integrity" that may arise within an instrumental context.

The "cowboy" mentality, as Daly explains, views nature as an unlimited frontier for exploitation. On this view, there is no need to worry about the "integrity" of nature—no need to recycle anything—because technological progress makes resources essentially infinite. In fact, the "cowboy" is likely to regard nature not only as an unlimited frontier to exploit but also as an implacable foe to conquer. Insofar as nature has a good of its own—and in that sense an "integrity"—it seems to the "cowboy" far more often to oppose than promote the good of human beings.

Those who regard nature as essentially unpredictable, disorderly, or even hostile believe that the environment as we find it provides very little that we need or use. On this approach, we employ technology to reduce nature it to its elements. In the place of wild or pristine ecosystems, we build from these elements a "second" nature more tractable to our desires. On this view, the "integrity" of ecosystems would have little instrumental value for which technology cannot provide a substitute. For example, the "integrity" of coastal fisheries will matter more to a hunter-gatherer society than to an economy that has developed "large scale, vertically integrated, high technology, centrally controlled, aquabusiness food production systems."[3]

A distrustful, indeed, fearful attitude toward nature accompanied the early European settlers into the American wilderness, which they described as a "howling waste" and as a "devil's den." A reading of William Bradford's *On Plymouth Plantation* tells us that about half of the Pilgrims died miserably the first winter from cold, hunger, and disease in the desolate wilderness. If nature is basically hostile to human purposes, as Pilgrims and pioneers thought, it must be (as it has been) dammed, plowed, blasted, cut, drained, dredged, poisoned, fenced, hunted, exterminated, genetically reengineered, and, in general, controlled, if it is to do us more good than harm. Nature in all its integrity is full of dangers. Some would argue, therefore, that the destruction of nature in favor of industrial development was on balance good not bad for human beings.

Those who analogize the natural environment to a "spaceship" or a "lifeboat" believe, on the contrary, that nature surrounds us with life support systems minutely calibrated to our needs. Any change in their functioning could upset delicate natural balances and destroy us all. They foresee an ecological crisis and believe the global economy exceeds the limits nature sets. "The spaceman in a small capsule," Daly writes, "lives off tight material cycles and immediate feedbacks, all under total control subservient to his needs." For the "cowboy," the integrity of ecosystems has either negligible or negative value. For the spaceman, it is all-important.

"Between the cowboy and the spaceman economies is a whole range of larger and smaller 'bull-in-the-china-shop economies,'" Daly writes, which, as he notes, better approximate where we are. In these circumstances, while it remains true that everything connects with everything else (as in the "spaceman" economy), some connections are deemed more important than others. So the question of "integrity," at least from a prudential point of view, may come down to the problem of identifying the crucial connections in nature—the systemic relationships—on which human activity relies and is most likely to upset. The bull-in-the-china shop approach does not prohibit every change we make in the natural world. It seeks to distinguish those changes that prudence allows from those it may forbid.

The "integrity" of the proportional mix of various chemicals in the atmosphere would count on these criteria as crucial because human beings depend upon but are also altering that mix. Other systems (gravitation, for example), while essential, are not threatened, while still others (the life cycles of certain noxious pests and parasites, for example), if they are threatened, may have no beneficial effects.

The "bull in the china shop" image of humanity's place in the natural environment suggests an approach to the concept of integrity which is largely empirical. What is needed to unpack the concept of "integrity," in other words, is a description of the particularly crucial systems the "bull" threatens. Of course, breaking a lot of china might not bother the bull at all; one might suggest, indeed, that a bull, since it has no sense of the intrinsic or aesthetic value of the china, might as well destroy a good deal of it. The question of the value of integrity would turn on the question of which environmental assets ("china") human beings have instrumental or economic reasons to protect and which are either unnecessary or inimical to our purposes.

The ozone layer provides an obvious example of an asset we need. It protects humanity from the actinic rays of the sun. Its "integrity," for all intents and purposes, consists simply in its thickness, since that is the quality upon which we depend. The relevant concept of integrity refers precisely to properties or functionings of biospheric and ecological systems that provide the goods or services we need. Aside from examples widely mentioned—the ozone layer is one—it is often unclear what these crucial properties or functions are.

The state of technology generally determines the goods and services we require of nature—for example, whether we hunt meat in the wild or produce livestock on farms. One must approach the question of the instrumental value of nature, then, on an empirical or case-by-case basis. It is easy to refer in general to the "life support" functions of ecosystems, but in the policy trade-offs that confront society, we need to know precisely what these "life support" systems are, and to what extent technology can substitute for them.

4. Integrity as an Instrumental Value

If "integrity" is a normative concept and the context of valuation is exclusively instrumental, it relates to those aspects of an ecosystem that make it useful for us. These aspects of nature will continually change with changes in technology. For example, the savanna ecosystem where Chicago and its suburbs stand presumably provided life-support systems for the Amerindians who inhabited the area centuries ago; one might say the same thing about the relationship between buffalo ecosystems and the plains Indians. For 20th century Americans, however, ecological restoration is not always advantageous. Indeed, if Chicago and its suburbs were removed and the millions who live there were left to the mercies of the original ecosystem, rich with integrity, they would starve or, in the winter, freeze to death.

The "integrity" of agroecological systems—wheat and corn farms, for example—has become more important than buffalo herds to our economy. As agriculture continues to industrialize, other changes loom. Agricultural economists foresee the day when enormous fermenting vats located just outside urban markets will turn cheap biomass into orange juice, tomato soup, apple sauce, or anything else we make from the cells of plants. If these cells will replicate *in vitro*, farmers do not need to grow them *in situ*. It will be cheaper to slurry biomass in mammoth fermenting cauldrons where plant cells will replicate to provide "just in time" commodities to fill day-to-day consumer demand.[4]

There is no general, abstract, or universal approach to analyzing the concept of "integrity" in an instrumental context, since the crucial qualities of ecosystems will be the ones, in the particular circumstances, that benefit human beings. These circumstances change and will continue to vary with changes in human needs, interests, and technology.

This point can also be made in relation to the concept of "carrying capacity." No one "really knows scientifically how large the carrying capacity of the earth is now or could be in the twenty-first century" Stephen Schneider of the National Center for Atmospheric Research has written. It depends on "social, economic, industrial, and agricultural practices."[5] It is these practices that determine, in large part, the characteristics of the environment that we value, at any time, on prudential or instrumental grounds.

In view of the relation between technology and carrying capacity, those who regard the integrity of nature as fundamentally an instrumental or prudential good must identify which ecosystem services humans depend upon and at the same time threaten. Where this has been done, for example, with respect to the ozone layer, we know both what "integrity" of the system consists in and why it is important. All too often, discussions of the "integrity" of ecological systems are abstract and general, leaving us to intuit the instrumental or prudential good in question. Instrumental goods, however, tend to be concrete not abstract qualities.

5. Integrity as a Moral Concept

If we turn from prudential to moral considerations—or from instrumental to intrinsic valuation—the concept of integrity will take on a very different meaning and use. We shall now be concerned not with our own well-being or flourishing but that of the biotic community itself. The familiar words of Aldo Leopold may guide us in defining the value of ecosystems in terms of *their* well-being. "A thing is right when it tends to preserve the integrity, stability, and beauty of the biotic community. It is wrong when it tends otherwise."[6]

On this criterion, of course, the conversion of the American continent from its prelapsarian pristine integrity to its present fallen condition represents an enormous loss of the integrity, stability, and beauty of biotic communities and systems. This conversion of the landscape to our economic purposes may be wrong, as Also Leopold suggests. It is wrong, however, not for instrumental or prudential reasons—these are the very purposes this conversion serves—but because it sacrifices the good of nature to the presumed good of human beings. The pioneers and settlers who survived in the wilderness did so precisely by sacrificing the beauty, stability, and integrity of the natural world to build an industrial and urban economy in its place. This was the wrong thing to do on Leopold's principle, but on instrumental and prudential grounds, it did make sense.

What are we to understand by the "integrity, stability, and beauty of the biotic community" in an ethical as distinct from instrumental context? Answers to this question, as we have said, depend on whether we view nature as orderly or as chaotic, that is, as organized around properties of homeostasis, balance, and harmony or as structured by random, contingent, and unpredictable events.

I shall now argue that if we take a Platonic view of nature—in other words, if we assume that it has an intelligible structure and is organized in terms of general law-like principles—we will base conceptions of integrity on that essential order, telos, or design. Modern systems ecology, building on a framework inherited from Neoplatonic conceptions of an orderly Nature, provides the scientific approach most suitable to efforts to understand integrity as a property of the structure or design of ecosystems.

To ground a notion of integrity, systems ecology offer such concepts as stability, diversity, complexity, homeostasis, optimization, resilience, and many other system-level concepts. One must assume that nature has an intelligible structure to which concepts such as these can be predicated. The concept of "integrity" might not differ logically, then, from any other quality that depends on the intelligible systematic structure of the ecological world.

If one assumes, on the contrary, that there is no intelligible order to nature—in other words, that a platonizing science is an empty pretense—then one is thrown back to the Darwinian setting that makes species the principal record of the contingent path that evolution happened to take. It is then the sheer chance and contingency of nature, not its necessary order, that the concept of integrity addresses. Integrity as a moral concept would honor and derive its force from the history that made the world and us what we are, rather than from an *eidos* of a formal design or a final cause underlying ecological systems.

6. The Happy Beast in Pre-Darwinian Thought

The idea that nature possesses intelligible order, structure, or balance has always been an appealing one. The obvious strength, skill, and ingenuity of animals, for example, has always astounded humanity. From the early Greeks till our own time, human observers have been amazed at the complexity of the spider's web, the architecture of an eagle's nest, the lyric of a thrush's song, and, in all, the incredible ways by which nature designs creatures to survive each other and their surroundings. Admiration for nature unites all cultures; in the West, it finds early classic statements in writing stretching from Plutarch's essay "On the Industry of the Beasts," through Montaigne's essays on the wonders of the natural world. The point of this literature, as George Boas has argued, was not merely to demonstrate the skillfulness and intelligence of animals but to show that they are "as noble and virtuous, as worthy of respect as man."[7]

The most sentimental of the successors to this ancient and medieval literature were the "anthropomorphizers" whom John Burroughs and others castigated as "Nature Fakers."[8] Today, this extremely benign view of life persists in the images of nature associated with Bambi, Pogo, and Barney. The term "integrity" applies to these creatures and the worlds they inhabit in exactly the same sense in which it applies to human beings and our societies, since these animals are basically human, indeed, many can speak. Interestingly, Pogo, an opossum, has proven so lovable and enduring a character because he provides a standard of honesty and integrity which human beings themselves seldom meet.

Natural historians at least prior to Darwin, who studied actual animals and biological communities, did not include Barney in their bestiaries, but they described animals and biological communities in terms that made them morally attractive, indeed, exemplary. Gilbert White (1720-1793), the most prominent nature writer of his time, waxed eloquently on the "wonderful spirit of sociality in the brute creation," observing, for example, the love a tortoise showed to a woman who fed it—proof that "the most abject reptile and torpid of all beings distinguishes the hand that feeds it, and is touched with feelings of gratitude."[9]

Eighteenth century poets such as William Cowper and Thomas Gray similarly presented an Arcadian or bucolic image of nature—consistent with the art of the time—and advocated a return to a lost harmony between human beings and the natural world. The overwhelming impression the 18th century Arcadian tradition creates, Donald Worster has written, "is of a man eager to accept all nature into his parish sympathies. That desire is what the rediscovery of pagan literature in the eighteenth century was primarily about: a longing to establish an inner sense of harmony between man and nature through an outer physical reconciliation."[10]

The concept of "integrity" makes perfect sense when applied to this Arcadian or Edenic vision of nature. The bucolic world is seen to have an order, hierarchy, and balance arranged by its inner logic of harmony and natural sympathy. When human beings impose on nature an alien technology—the illicit fruit of the Edenic tree—they separate themselves from the other animals, who are their natural comrades and companions. According to this view, what Montaigne condemned as "our natural and original malady," that is, a Faustian commitment to technology, prevents us from achieving a harmonious or, as we might now say, a "sustainable" relation with the natural world.[11]

The Arcadian imagery that made up the stock-in-trade of eighteenth century nature writing combined with the subjectivism of nineteenth century romanticism to produce the highly personal accounts of nature-as-experienced found in Emerson and Thoreau, who saw the appreciation of the natural world less as a scientific enterprise or economic necessity than as a moral test. These writers discovered and discussed the aesthetic and ethical far more than the prudential or economic relation between humanity and nature. They considered the meaning of natural objects as symbols not their uses as resources. "God did not make this world in jest; no, nor in indifference," Thoreau wrote. "These migrating sparrows all bear messages that concern my life."[12]

Leo Marx, in the *Machine in the Garden,* details the growing tension between the oneness-with-nature pastoralism of the Transcendentalists and the myth of economic progress that characterized the American Dream.[13] According to Marx, the conflict between the moral harmony of nature's "garden" and the ascendance of a dog-eat-dog market-driven industrial civilization produced a conflict that defines much of American thought. Since 1960, this tension has led to a "wholly new conception of the precariousness of our relations with nature," Marx observes, which "is bound to bring forth new versions of the pastoral." Current attempts by philosophers, ecologists, and others to predicate notions of "integrity" and "health" to ecosystems may be understood as an effort of just this kind, that is, new versions of the pastoral relevant to contemporary conditions.

In the past few decades, moreover, theoretical ecologists, especially in research centering on stability, complexity, diversity, homeostasis, balance, hierarchy, resilience, and other structural properties of ecosystems, have sought a scientific basis on which one may predicate concepts of order and, therefore, "integrity" and "health" to the natural world. This work in theoretical ecology, which has been described, for example, by Robert McIntosh[14] and Donald Worster[15] among others, suggests that nature's course "is not an aimless wandering to and fro but a steady flow toward stability that can easily be plotted by the scientist."[16]

7. Great Chain of Being Cosmology and Ecological Science

The tradition of mathematical or theoretical ecology, which seeks to discover the intelligible principles that govern population biology and ecological communities and systems, traces its intellectual roots back to the image of the Great Chain of Being, which epitomizes the moral and religious attention people within the Western tradition have long paid to the diversity of life. From Plato's theory of perfect Forms to the quest of many recent ecologists to find order and balance in nature, philosophers, poets, painters, and scientists have attempted to describe the living world in ways that answer to religious and moral expectations. Ecologists in this century—like theologians and poets in previous centuries—have argued that the diversity of living things results not from mere contingency or chaos but serves larger purposes, instantiates universal principles and ideas, follows law-like general principles, or expresses an intelligible order or plan.

In the 11th Century, the French theologian Abelard, following Plato's *Timaeus* (30c), defined one aspect of the Chain-of-Being theme, namely that a sufficient reason explains the existence of every kind of organism. "Whatever is generated is generated by some necessary cause, for nothing comes into being except there be some due cause and reason for it."[17] Along with the idea of sufficient reason, the principles of plenitude, continuity, and gradation determined the order of creatures from the least to the greatest in a vast Chain of Being.

These principles have analogies in the ecological theory of recent decades. Plenitude—the principle that the richness and diversity of creation is so great because it expresses the fullness of God's perfection—is found in various versions of the diversity-stability hypothesis, for example, in G. E. Hutchinson's speculation that there are so many species "at least partly because a complex trophic organization of a community is more stable than a simple one."[18] The themes of gradual continuity and gradation likewise echo in hierarchy theory, in theories of trophic levels, food chains and webs, and in the concept of orderly succession in forest and other ecosystems that characterized ecology earlier this century.

Fundamental to the idea of the Great Chain of Being was a belief that God creates nothing in vain. Accordingly, we are obliged to care as much for the least creature in nature as for the greatest. Today, the medieval principle of sufficient reason echoes in the popular analogy that likens species to rivets in the wing of an airplane. The rivet-popping analogy captures in one image the thought of this well-known passage in Alexander Pope's *Essay on Man*:

> Vast chain of being! which from God began,
> Natures aethereal, human, angel, man,
> Beast, bird, fish, insect, what no eye can see...

Where, one step broken, the great scale destroyed,
From Nature's chain whatever link you strike,
Tenth, or ten thousandth, breaks the chain alike.

Commenting upon the centrality of the Chain of Being metaphor, historian A. O. Lovejoy observed that according to this tradition, the diversity of nature corresponds to law-like principles that establish its order; the "universe was at least not a many-ringed circus."[19] Lovejoy notes, however, that in the eighteenth century, a controversy arose pitting philosophers like Spinoza and Leibniz, who believed that the principle of sufficient reason necessitated such a hierarchical order in the variety of nature, against those who followed the British philosopher Samuel Clarke in arguing that only God's essence implied existence, and that contingency pervaded the created world. In 1712, a British poet put that thesis as follows:

Might not other animals arise
Of different figure and of diff'rent size?
In the wide womb of possibility
Lie many things which ne'er actual may be:
And more productions of a various kind
Will cause no contradiction in the mind . . .
These shifting scenes, these quick rotations show
Things from necessity could never flow,
But must to mind and choice precarious beings owe.[20]

A controversy that rages between those who believe that nature must exhibit a "balance" or "order" and those who argue that it is all chaos and contingency—a many-ringed circus—characterizes ecological debates today as it did cosmological debates in medieval times. Earlier this century, ecologists such as Paul Sears and Frederic Clements, for example, remaining firmly within Great Chain of Being tradition, approached ecology as the study of harmony, continuity, gradation, and equilibrium. Following Clements, Gaian theorists recast the Great Chain of Being in modern terms, representing the earth as a vast superorganism, possessing as much internal order as the organisms that make up its functioning parts.

Eugene P. Odum, an ecologist who seems among those most indebted to Great Chain of Being cosmology, has restated the 18th Century principle of plenitude as "the strategy of ecosystem development" which is "directed toward achieving as large and diverse an organic structure as is possible within the limits set by the available energy input and the prevailing conditions of existence." This "strategy" is supposed to lead ecosystems in law-like ways through a series of orderly successive changes to species composition to achieve a state of mature homeostasis in which the stability and diversity of the system are the greatest it can achieve under given conditions. In Odum's version, this happens, for example, when weedy generalists (r-selected species) are replaced by a greater variety of specialized organisms (K-selected species) who exploit all the niches available to them. In such a system, as in the Chain of Being, every possible creature finds its niche.

This tradition in ecology emphasizes the interconnectivity within ecosystems, the interdependency of their parts, and their progress toward increased stability and diversity, all of which support conceptions of ecological integrity. Yet this traditional way of regarding nature, although helpful in grounding conceptions of ecological health and integrity, has been criticized as understating the random or contingent elements in nature. Indeed, it is customary today to question the ecosystem concept and the kind of integrated order it presupposes.

As Frank Golley has written, the ecosystem research community has been unable or unwilling to respond to persistent criticisms regarding the existence of the units of analysis it studies and about the testability of their hypotheses. "Rather, ecosystem ecologists tended to restate the ideas underlying the field and further define their concepts."[21] Great Chain of Being Cosmology, especially when scientized to meet the needs of a secular age, does not die easily. "Ecologists were not questioning the cultural paradigms," Golley observes, "they were working within them."[22]

8. Criticisms of Great-Chain-of-Being Ecology

Among the many criticisms of the ecosystem concept and the ecological theory that employs it, a number stand out that are pertinent to the development of the concept of ecological integrity. First, the central concepts of systems ecology have not been defined— or been defined in so many different ways that no consensus emerges concerning the meaning of major principles or assumptions. As Stuart Pimm has argued in *The Balance of Nature?*, concepts of ecological "resilience," "complexity," "equilibrium," "stability," and "diversity" are used in so many senses that it is very difficult to determine exactly what claims about them mean.[23]

Pimm distinguishes, for example, at least five definitions of "stability" and observes that "complexity" has been "taken variously to mean the number of species in a system, the degree of connectance of the food web, and the relative abundance of a species in the community." Pimm believes that ecologists rarely look at the same question; for example, they may interpret data in relation to "the combinations of five definitions of stability, three definitions of complexity, and three levels of organization—a total of forty-five possible questions about the relationships between community complexity and stability."

In spite of notorious difficulties in establishing definitions of key terms such as "stability" and "diversity" and in establishing clear links between them, the tradition of systems ecology that runs from Sears and Clements to Odum captures in secular terms the concepts of balance, order, harmony, plenitude, and sufficient reason previously associated with Chain of Being cosmology. The problem is that this school of ecology, by secularizing a religious vision of nature—by clothing it as science rather than as theology—has demystified it. This led to two kinds of difficulties. First, the central theories that linked stability and diversity, that called for an orderly succession of communities, and that arranged creatures in trophic levels and webs, opened themselves to empirical and theoretical refutations. In a kind of war between the generations, the students of Odum and other founders of community ecology set out to test and in the process debunked the theory of forest succession,[24] the "stability-diversity"[25] hypothesis, the "outwelling hypothesis,"[26] and other tenets basic to the discipline.

As a result of this kind of self-criticism, many ecologists would agree with Van Valen and Pitelka that "ecology has no known regularities."[27] Kristin Schrader-Frechette and E.D. McCoy have summarized the findings of many reviewers that such "laws" as are proposed in ecology "are frequently not generalizable and are indistinguishable from mere principles. Moreover, ecologists do not agree on what the basic principles or laws are."[28] General laws in ecology, when they are introduced, as Shrader-Frechette and McCoy add, are often "trivial, tautological, or not testable."[29] The others, as a rule, are unintelligible.

Second and more relevant to our purposes, biologists who emphasized ecosystem-level properties and processes, such as productivity, energy flow, respiration, trophic webs, nutrient cycling, and efficiency, have shown less and less interest in the minute particularities of individual organisms. These ecologists remained committed to a philosophy of science that insisted on abstract and general mathematical theories.[30]

Thus, the science began to pursue models that looked like those found in physics but had little if anything to do with the concrete properties of organisms. According to one historian, community ecologists came to emulate "the language of systems scientists" and began to work on models at the intersection of biology and engineering.[31] As a result, the minute particulars of natural history were lost in theoretical abstractions often borrowed from economics, engineering, information theory, cybernetics, and other mathematical sciences.

In the context of these developments in ecology, especially the branch that became known as "systems ecology," both scientists and policy makers have found it easier to think of living creatures as resources to be manipulated than as—in John Muir's expression—"conductors of divinity."[32] To be sure, both community and systems ecology retained faith with the central thesis of the Great Chain of Being that nature exemplifies a timeless and intelligible order rather than sheer historical contingency. By secularizing this religious intuition, however, ecosystem science replaced a priesthood of theologians with one of engineers and mathematical modelers.

9. Systems Ecology as an Engineering Science

It is not surprising that many environmental engineers and other planners found in ecological theories that secularized Great Chain of Being themes grounds not to venerate but to manipulate nature or to manage the earth for improved efficiency. As Donald Worster notes, "'Governing' of all nature was the dream of these ecosystem technocrats."[33] Indeed, one such ecologist, writing in the *Bulletin of the Ecological Society of America*, called upon his colleagues to embrace the "biotechnologist" credo "to engineer and produce plants, animals, and microbes that better suit the presumed needs and aspirations of the human population." He continued:

> Ecologists are the people most fit to develop the conceptual directions of biotechnology. We are the ones who should have the best idea as to what successful plants and animals should look like and how they should behave....Armed with such expertise, are we going to continue investing nearly all our talents in Natural History?...Or should we take the forefront in biotechnology, and provide the rationale for choosing species, traits, and processes to be engineered? I suspect this latter approach will be more profitable for the world at large as well as for ourselves.[34]

To summarize: Poets and theologians who described the Great Chain of Being understood the principles of sufficient reason, plenitude, continuity, and gradation in terms of religious beliefs and moral values. Ecological concepts developed earlier this century and in popular vogue today adhered closely to these same principles—optimization is the current version of sufficient reason; diversity of plenitude; succession of continuity; and hierarchy of gradation. Many analysts look to concepts such as these to explain a conception of ecological "integrity," as they might, insofar as they retain the normative elements they once possessed within the original cosmology.

Yet, when these concepts occur in scientific theories, they may be shorn of their religious and moral significance. They may then be open to empirical and theoretical refutation. They may also support arguments that back efforts not necessarily to protect nature for its own sake but to manipulate it to meet our economic needs and demands. A systems approach in ecology, in other words, can as easily lead to a "cowboy" as to a "spaceship" view of our place in nature; it can put human beings at the controls rather than in the back seat of our biospheric ark.

Great Chain of Being cosmology provided the initial legitimacy or credibility for the systems attack in ecology, which insisted, against the grain of Darwinian thought and piecemeal natural history, that nature contains a telos, balance, structure, or intelligible order that would meet the expectations of a Platonic sense of the Universe. A Neoplatonic science of nature, insofar as it is captured in terms of the interrelationships of stability, resilience, complexity, hierarchy, and so on—would answer to our need to find an objective basis on which to predicate concepts of "integrity" and "health" to the natural environment.

On the other hand, if these concepts lose their Neoplatonic connotations and become merely principles for environmental engineering, we may wonder how much we wish the ecological and biological sciences—including genetic engineering—to succeed. Knowledge is power—but it can be the power to control or the power to protect—the power to bend nature to our purposes or to appreciate nature for its own sake.

Many analysts have attempted to found a notion of ecological integrity on conceptions of resilience, stability, complexity, diversity, and so on, found in systems ecology. It may be possible, however, to work in the other direction. We may first seek to define ecological "integrity" and "health," in fundamentally normative terms and then infer from those definitions corollary conceptions of resilience, complexity, stability, and other ecosystem-level properties. Another possibility might be to concede that nature does not have a "soul"—that is, an intelligible structure, *telos*, or order—but a history. This idea, consistent with Darwinian insights, offers a different approach to the concept of "integrity" in the natural world.

10. Threading the Needle of Evolution

In 1977, the British ecologist John Harper wrote that every discovery in ecology must take its meaning from evolutionary phenomena. "Evolutionary thinking," he wrote, "concentrates attention on the behavior of the individual and his descendants If nothing in biology has meaning except in the light of evolution and if evolution is about individuals and their descendants—i.e. fitness—we should not expect to reach any depth of understanding from studies that are based at the level of the superindividual."[35]

Henry Gleason, who opposed the a priori attribution of balance, equilibrium, succession, and other "systems" properties to nature, argued early this century that nature is more like a Heraclitean flux than like a Chain of Being. In an article titled "The Individualistic Concept of the Plant Association," he wrote that each species of plant "is a law unto itself."[36] Many ecologists—Robert Colwell is an example[37]—have emphasized the same point, namely, that ecological events depend on the minute particulars of organisms and not on mathematical or theoretical properties of abstract systems or communities. Insofar as ecosystem, community, and population ecology abstract from these particulars, they play *Hamlet* without the Prince of Denmark.

During the almost 70 years since Gleason wrote, ecologists have emphasized the search for universal theories, mathematical principles, and general properties over the historical study of individual organisms. This may be the reason that ecosystem modelers and theory-builders outnumber trained taxonomists today by a ratio perhaps of 1,000 to 1. Nevertheless, some ecologists are now turning away from system-level analogies with engineering and other mathematical sciences toward "rich descriptions" of individual organisms in their habitats.[38]

"Whenever we seek to find consistency" in nature, an ecologist has recently written, "we discover change"[39] —thus echoing Gleason's remark that each species is a law unto itself. This biologist compares nature not to a three-ring circus, but to several musical compositions played in the same hall at once, each intruding on the pace and rhythms of the others. Appreciation then comes down to the intense and patient observation of details not speculation about overarching organizing principles. This kind of patient observation and rich or "thick" interpretative description characterizes the study of natural history in contrast to theoretical ecology.

The empirical work of natural history, including taxonomy, has been ignored, even ridiculed, paleontologist S. J. Gould has written, because it does not indulge in the a-priori mathematical modeling thought to characterize "hard" science. Yet our knowledge of species depends entirely on "the historical sciences, treating immensely complex and non-repeatable events (and therefore eschewing prediction while seeking explanation for what has happened) and using the methods of observation and comparison."[40]

The essence of historical explanation, Gould writes, is contingency. He explains:
> A historical explanation does not rest on direct deductions from laws of nature, but on an unpredictable sequence of antecedent states, where any major change in any step of the sequence would have altered the final result. This final result is therefore dependent, or contingent, upon everything that came before—the unerasable and determining signature of history.[41]

Gould observes that historical narratives that explain the minute particulars of organisms at specific times and places "are endlessly fascinating in themselves, in many ways more intriguing to the human psyche than the inexorable consequences of nature's laws."[42] Biologist E. O. Wilson elegantly takes up this theme in arguing that every creature, large and small—the flower in the crannied wall—"is a miracle," but one that makes sense—is explicable—in the context of a rich historical narrative. "Every kind of organism has reached this moment in time by threading one needle after another, throwing up brilliant artifices to survive and reproduce against nearly impossible odds."[43]

It is the enormous and timeless labor of evolution that invests its products—the plants and animals we encounter—with a dignity and meaning. Their legitimacy is based not in any purpose they may serve—ours or that of some superorganism that contains them—but in the circumstances of their coming hither. They survive to tell the story of random mutation and natural selection, of chance and matter, which turns out to be more magnificent and harrowing than anything any one of us could imagine. As E. O. Wilson has remarked, "The more we know of other forms of life, the more we enjoy and respect ourselves. Humanity is exalted not because we are so far above other living creatures, but because knowing them will elevate the very concept of life."[44]

11. Conclusion

The value of the integrity of natural things depends on the value of those objects. If they are valuable instrumentally, their integrity, I have argued, will be important only in specific and discrete instances, as, for example, the "integrity" of the ozone layer. As technology advances, natural objects, communities, and systems become epiphenomenal to economic activity or may be viewed as an obstruction to it. Instrumental or prudential concerns, then, would provide at best a poor and ephemeral basis for the value of integrity of the natural world.

If natural objects and communities have intrinsic value, it is either because of their ahistorical or their historical qualities. Systems and other ecologists have not been able to obtain clarity in their attempts to define and to model ecosystem-level qualities, such as stability, resilience, complexity, diversity, homeostasis, and so on. This is not to deny that many deep and mathematically sophisticated treatises have been written on these and related concepts. It is only to say that these attempts have become a Tower of Babel in which each ecologist speaks his or her own language without much prospect of clarity or consensus.

We have the greatest chance of understanding the value of integrity—and therefore the concept itself—in relation to the historical qualities of the natural world. This explains why integrity applies to organisms and communities that are authentic in nature as distinct from those that we might create through bioengineering. "What we believe important," as Ronald

Dworkin has written, "is not that there be any particular number of species but that a species that now exists not be extinguished by us. We consider it a kind of cosmic shame when a species that nature has developed ceases, through human actions, to exist."[45]

12. References

1. Laura Westra, *An Environmental Proposal for Ethics: The Principle of Integrity*, (Lanham, MD. Rowman and Littlefield, 1994), p. 30.
2. Herman Daly, "Elements of Environmental Macroeconomics," in Robert Costanza, ed., *Ecological Economics: The Science and Management of Sustainability* (New York: Columbia University Press, 1991), pp. 33-46; especially, pp. 38-40.
3. Harold Webber, "Aquabusiness," in R. Colwell, A. Sinskey, and E. Pariser, eds., *Biotechnology in the Marine Sciences*, (New York: Wiley, 1984), pp. 115-116.
4. The literature offers two broad ways of responding to this argument. Many writers take the view that no matter how the technology of producing food and fiber progresses—no matter how far agricultural production becomes industrialized and thus becomes independent of farming—nevertheless prudence demands that we protect the soil for utilitarian reasons. For arguments of that kind, see R. Neil Sampson, *Farmland or Wasteland: A Time to Choose* (Emmaus, PA: Rodale Press, 1981) and John M. Harlin and Gigi M. Berardi, eds., *Agricultural Soil Loss: Processes, Policies, and Prospects* (Boulder: Westview Press, 1987).

    Many other writers take the view that even if science and technology can create abundant and cheap substitutes for soil, water, or whatever, these go only to the uses we make of the environment not to the value farms and other place shave for us. For an excellent presentation of this position, see Donald Worster, "A Sense of Soil: Agricultural Conservation and American Culture," *Agriculture and Human Values* 2 (Fall 1985): 28-35, arguing that the soil has a "powerful meaning" (p. 33) for farmers and others and from this point of view, "The soil is a resource for which there is no substitute" (p. 34).

    For an instructive exchange of views, see Martin H. Rogoff and Stephen L. Rawlins, "Food Security: A Technological Alternative," *BioScience* 37(11) (December 1987): 800-807, arguing that biotechnology can convert cheap and readily available biomass into a stable food supply, and David Orr, "Food Alchemy and Sustainable Agriculture," *BioScience* 38(11) (December 1988): 801-802, arguing that even if technology is capable of making almost anything edible, we "should not confuse the results with a genuinely sustainable agriculture or culture, which rest on different premises" (p. 802). Orr observes: "Good agriculture is concern for the long-term health of the land and rural communities" (p. 802). See also the exchange of letters in *Bioscience* 39(6) (June 1989): 356; and Orr's reply two issues later (p. 588).
5. Stephen H. Schneider, "Climate and Food: Signs of Hope, Despair, and Opportunity," in Paul Erlich and John Holdren, eds., *The Cassandra Conference* (College Station: Texas A&M Press, 1985): 17-51; quotation at p. 42.
6. Aldo Leopold, *A Sand County Almanac* (New York: Oxford University Press, 1966), p. 231.
7. George Boas, *The Happy Beast: In French Thought of the Seventeenth Century* (Baltimore: John Hopkins Press, 1933), p. 6.
8. See Ralph H. Lutts, *The Nature Fakers: Wildlife, Science, and Sentiment* (Denver: Fulcrum, 1990) and Frank Stewart, *A Natural History of Nature Writing* (Washington, DC: Island Press, 1995), esp. Chapter 5.

9. Gilbert White, *The Natural History and Antiquities of Selbourne* (London: Bensley, 1789), pp. 193, 149; cited and quoted in Lawrence Buell, *The Environmental Imagination: Thoreau, Nature Writing, and the Formation of American Culture* (Cambridge, MA: Harvard University Press, 1995), p. 186.
10. Donald Worster, *Nature's Economy: A History of Ecological Ideas* (New York: Cambridge University Press, 1977), p. 10.
11. See Montaigne's Essays II, xii, 2, 164.
12. Henry David Thoreau, *Journal*, vol. III, March 31, 1852.
13. Leo Marx, *The Machine in the Garden: Technology and the Pastoral Idea in America* (New York: Oxford University Press, 1964, 1977).
14. Robert P. McIntosh, *The Background of Ecology: Concept and Theory* (New York: Cambridge University Press, 1995).
15. See, for example, Donald Worster, "The Ecology of Order and Chaos," *Environmental History and Review* 14 (Spring-Summer, 1990): 1-18.
16. Donald Worster, *Nature's Economy*, p. 210.
17. Quoted and cited in A. O. Lovejoy, *The Great Chain of Being* (New York: Harper Torchbooks, 1936, pbk. 1960), p. 71.
18. G. E. Hutchinson, "Homage to Santa Rosalia or Why are There So Many Kinds of Animals," *American Naturalist* 93 (1959): 145-159.
19. Op. Cit., p. 143.
20. Sir Richard Blackmore, *Creation*, Bk. 12, 1712; quoted by Lovejoy, op. cit., p. 195.
21. Frank B. Golley, *A History of the Ecosystem Concept in Ecology*, (New Haven: Yale University Press, 1993), p. 107.
22. Golley, p. 108.
23. Stuart L. Pimm, *The Balance of Nature?* : Ecological Issues in the Conservation of Species and Communities (Chicago: University of Chicago Press, 1991), p. 15.
24. See, for example, William H. Drury and Ian Nisbet, "Succession," *Journal of the Arnold Arboretum* 54 (July 1973) and Joseph H. Connell and Ralph Slayter, "Mechanisms of Succession in Natural Communities and Their Role in Community Stability and Organization," *The American Naturalist* 111 (Nov.-Dec. 1977): 1119-44.
25. For a review, see Daniel Goodman, "The Theory of Diversity-Stability Relationships in Ecology," *Quarterly Review of Biology* 50 (1975): 237-261.
26. For a review, see Scott Nixon, "Between Coastal Marshes and Coastal Waters—A Review of Twenty Years of Speculation and Research on the Role of Salt Marshes in Estuarine Productivity and Water Chemistry," in P. Hamilton and K. MacDonald, eds., *Estuarine and Wetland Processes* (New York: Plenum Press,1980).
27. L. Van Valen and F. Pitelka, "Intellectual Censorship in Ecology," *Ecology* 55 (1974): 925-926.
28. Kristin Shrader-Frechette and E. D. McCoy, *Method in Ecology: Strategies for Conservation* (New York: Cambridge University Press, 1993), p. 81; cf. K. Shrader-Frechette and E. D. McCoy, "Theory Reduction and Explanation in Ecology," *Oikos* 58(1990): 109-114.
29. Shrader-Frechette and McCoy, *Method in Ecology*, p. 81.
30. For a review, see M. Sagoff, "Ethics, Ecology, and the Environment: Integrating Science and Law," *Tennessee Law Review* 56 (Fall 1988): 77-229.
31. See "Preface" in Bernard Patten, ed., *Systems Analysis and Simulation in Ecology* vol. I (1971), p. xiv.
32. Linnie Marsh Wolfe, ed., *John of the Mountains: The Unpublished Journals of John Muir* (Madison, WI: University of Wisconsin Press, 1979), p. 118.

33. Donald Worster, "The Ecology of Order and Chaos," in Worster, *The Wealth of Nature* (New York: Oxford University Press, 1993), p. 161. Worster cites as an example the work of Odum's brother, H. T. Odum, who developed engineering models of ecosystems. See, e.g., Peter Taylor, "Technological Optimism, H.T. Odum, and the Partial Transformation of the Ecological Metaphor after World War II," *Journal of the History of Biology* 21 (Summer 1988): 213-244.
34. Frank Forcella, "Commentary: Ecological Biotechnology," *Bulletin of the Ecological Society of America* 65 (1984), p. 434.
35. J. L. Harper, "The Contributions of Terrestrial Plant Studies to the Development of the Theory of Ecology," in C.E. Goulden, ed., *The Changing Scenes in the Natural Sciences: 1776-1976* (Philadelphia: Academy of Natural Sciences, 1977), pp. 139-57; quotation at p. 148.
36. H. A. Gleason, "The Individualistic Concept of the Plant Association," *Bulletin of the Torrey Botanical Club* 53 (1926), p. 25.
37. See Robert Colwell, "What's New? Community Ecology Discovers Biology," in P. Price, C. Slobodchikoff, and W. Gaud, eds., *A New Ecology: Novel Approaches to Interactive Systems* (New York: Wiley, 1984).
38. See L. B. Slobodkin, D. B. Botkin, B. Maguire, B. Moore, III, and H. Morowitz, "On the Epistemology of Ecosystem Analysis," in V. Kennedy, ed., *Estuarine Perspectives* (New York: Academic Press,1980), esp. p. 506.
39. Daniel Botkin, *Discordant Harmonies: A New Ecology for the Twenty-first Century* (New York: Oxford University Press, 1990), p. 62.
40. S. J. Gould, "Balzan Prize to Ernst Mayr," *Science* 222 (1984), p. 255.
41. S. J. Gould, *Wonderful Life: The Burgess Shale and the Nature of History* (New York: WW. Norton, 1989), p. 283.
42. Ibid., p. 284
43. E. O. Wilson, *The Diversity of Life* (Cambridge, MA: Harvard University Press, 1992), p. 345.
44. E. O. Wilson, *Biophilia: The Human Bond With Other Species* (Cambridge, MA: Harvard University Press, 1984), p. 115.
45. Ronald Dworkin, *Life's Dominion* (New York: Knopf, 1993), p. 75.

Chapter 12
ECOLOGICAL INTEGRITY AND NATIONAL PARKS

John Lemons[1]

1. Introduction

Recently, concepts of ecological integrity have been proposed to facilitate enhanced protection of biological and ecological resources against the threat of human activities because ecosystems that encompass facets of integrity would be protected better against activities that cause ecological change or impairment (Johnson 1993, Westra 1994). Angermeier and Karr (1994) propose that ecological integrity refers most appropriately to ecosystems whose operations and evolution have been minimally influenced by human interventions. Consequently, they propose that the natural integrity of national parks, nature preserves, and other similar areas be protected. However, few case studies have been conducted on the theoretical and practical implications that might arise from the use of ecological integrity as the basis for management decisions about natural resources.

Most definitions of ecological integrity focus on the ability of ecosystems to cope with stress and maintain their self-organizational capacities (Kay 1992, Schneider and Kay in press); however, the concept is not defined precisely. Concepts of ecological integrity have been derived from studies of ecosystems based upon complex systems theories. Such systems are described as nonlinear whose properties or behavior cannot be explained or predicted by knowledge of lower levels of hierarchical organization within them. In addition, complex systems have multiple organizational states and processes based upon nonequilibrium paradigms that include the following notions: (1)ecosystems are open, (2)processes rather than end points are emphasized, (3)a variety of temporal and spatial scales are emphasized, and (4)episodic disturbances are recognized.

Some of the nation's most important ecological and scenic resources are protected in national parks, whose fundamental purpose is threefold: (1)to conserve scenery, natural and historic objects, and wildlife; (2)to promote the enjoyment of scenery, natural and historic objects, and wildlife; and (3)to provide for public enjoyment of these areas so that the scenery, natural and historical objects, and wildlife are unimpaired for future generations (16 United States Code [U.S.C.] 1). The National Park Service (NPS) manages over 63 natural area units ranging between 30,000 to over 3 million acres in size (NPS 1980). Included in this category are parks such as Yosemite, Yellowstone, Glacier, Grand Canyon, Great Smoky Mountains, and Everglades. Twelve of the units are Biosphere Reserve Parks, which are dedicated to long-range ecosystem monitoring under the UNESCO Man and Biosphere program.

---

[1]Department of Life Sciences, University of New England, Biddeford, ME 04005; U.S.A.

Importantly, the U.S. National Park Biosphere Reserves are considered to be ecological standards for the network of International Biosphere Reserves.

The management of national parks has been controversial because of the difficulties of protecting the parks' resources from the impacts of high levels of visitor use. Controversies have focused on questions regarding: (1)what constitutes appropriate protection of parks' resources, (2)what constitutes appropriate use of parks' resources and levels of use, (3)whether visitor facilities should be limited to those deemed necessary as opposed to merely convenient, and (4)whether and how to restore parks' biological resources or ecosystems perturbed by human activities (McNeely and Miller 1984, Lemons 1987b). National parks can be used as a case study to explore theoretical and practical implications of basing management decisions upon concepts of ecological integrity because parks encompass areas where resources already must be protected while at the same time allowances must be made for their use and enjoyment.

Several implications of using concepts of ecological integrity as a basis for national park management need to be explored in order to determine whether or to what extent they might be used for such a purpose. First, to fulfill the legislative mandates to protect parks' resources, scientific information must be used to provide a basis for management decisions. However, there is a potential conflict between the prescriptive legislative mandates to protect parks' resources consistent with concepts of ecological integrity if the capabilities of science to ascertain the consequences of human activities are limited because they are able to explain better some of the complexities of nature but are less able to provide reasonably certain information to decisionmakers for predictive purposes. Consequently, an assessment of the role of science in decisions about parks' resources must be made. Second, as will be shown, although ecosystems may respond to environmental change in one of several qualitative ways, none of the ways indicate whether a loss of integrity has occurred. While science can provide information about the responses of ecosystems to environmental change, it cannot provide a basis for deciding whether one type of change is more acceptable than another. In other words, decisions regarding the acceptability of ecosystem changes must rely on value-laden human judgment. Accordingly, the role of values in national park management decisions must be assessed. Third, an assessment of whether existing national park legislation is likely to allow management policies based upon concepts of ecological integrity must be made. The reason for this assessment is because national park legislation traditionally has been interpreted in a manner that allows for wide oscillations of policy regarding how to balance protection and use of parks' resources. A new management policy based upon concepts of ecological integrity would most likely result in limitations of levels or types of visitor uses in parks because policies could not be based simply on maintaining present ecosystems that may be deemed to be desirable or even "healthy" if they lacked the ability to cope with environmental changes or to continue their self-organization. In this chapter, I assess the practical and theoretical implications of using ecological integrity as a basis for national park management.

2. The Role of Science in Decisions About Parks' Resources

It is a common perception that parks' resources are impaired by threats originating both internally and externally to parks' boundaries (NPS 1980, Lemons 1986b and 1987b, Keiter and Boyce 1991, Wright 1992). Over 70 different kinds of threats in the categories of aesthetics, air pollution, physical removal of resources, encroachment by exotic species, visitor physical impacts, water quality pollution and water quantity changes, and park operations and planning of facilities have been identified, and over 2,000 internal threats and over 2,400 external threats have been identified for biological, physical, aesthetic, cultural, and park operations resources. An overwhelming consensus exists that there is a lack of adequate understanding and documentation of most threats to parks' resources. Lack of understanding of the threats to parks' resources is a barrier to more effective protection of resources under existing policies as well as to management policies that might be based upon concepts of ecological integrity.

Ideally, national park decisionmakers would like reasonably sound scientific information on which to base decisions. Politically speaking, decisions initiated to protect parks' resources before science can describe with reasonable certainty the relationships between a proposed action and its effects on resources often are perceived to be irrational because they are said to be without a scientific basis that compels or supports the decisions. Because decisionmakers are sensitive to the preferences of park users regarding questions of resource use as well as to the economic and developmental consequences of decisions to protect parks' resources, they often are reluctant to propose or approve protection measures that conflict with the preferences or that might slow or conflict with economic concerns in situations where scientific uncertainty exists. Even more troublesome is the fact that public policy and legal questions about protection of parks' resources often are shaped in public debate and legal proceedings through the placement of the burden of proof (Brown 1990). Generally speaking, NPS legislation authorizing governmental regulation places the burden of proof for showing environmental impact or risk on the agency, since it is responsible for rule-making. One consequence of the placement of the burden of proof is that those advocating the enhanced protection of parks' resources must demonstrate with reasonable certainty that there is a scientific need for protection and that recommended solutions have a reasonable chance of success. Placement of this burden of proof often imposes a requirement for reasonably certain scientific information that is not possible to meet even under existing legislative mandates and that would be more difficult to meet if management decisions were to be based upon concepts of ecological integrity.

The problems posed by a lack of adequate data constrain NPS efforts to mitigate internal threats associated with high levels of visitor use (Lemons 1987b). Any policy of limiting visitor access to parks' resources has political consequences and raises questions of equity because proposals for limited access have been said to imply cultural elitism strongly and because it is difficult to determine how to accomplish limited access equitably. Because government agencies attempt to take a middle-of-the-road approach, it is not surprising that the NPS avoids limiting access whenever possible (wilderness designations in parks are an exception).

Policies to enhance protection of parks' resources might entail the removal of concessioner facilities from certain park locations. Many of these facilities were built in the prime scenic and ecologically important areas of parks. These facilities contribute to overcrowding and damage of soils, vegetation, wildlife, water quantity and quality, air quality, archeological and historical sites, geological and ecological processes, and visitors' enjoyment of natural resources. Park concessioners influence park policy as it affects their interests and typically oppose curtailment or relocation of their facilities (Lemons and Stout 1984, Lemons 1987a). Further, 16 U.S.C. 20(b) guarantees park concessioners a property right in the form of a possessory interest meant to protect the concessioner's investment and the chance to realize controlled profits (see below). As a result, if the NPS wishes to terminate or alter its contract with a concessioner, or require relocation of facilities, the concessioner must be reimbursed for full construction costs and possibly for loss of profits; these costs amount to many millions of dollars. A decision to enhance protection of parks' resources that would necessitate reimbursement to a park concessioner under 16 U.S.C. 20(b) likely would not be made in the absence of compelling scientific information that important parks' resources were threatened.

Absence of factual knowledge will also seriously constrain mitigation efforts directed at external threats as well as efforts to base management of parks on concepts of ecological integrity if changes in the use of surrounding nonpark lands were required. Congress has given the NPS little explicit authority to regulate activities on adjacent park lands that pose external threats to park resources. To the extent that the NPS lacks authority for dealing with problems of incompatible uses on adjacent nonpark lands, the agency must rely on cooperation with private landowners and representatives from local, state, and other federal agencies to resolve conflicts. Such cooperation will be based on the NPS's ability to convince those responsible for activities on nonpark lands that their actions are needed to protect park resources. Such actions will not be forthcoming unless arguments are predicated upon compelling factual knowledge because of conflicting goals and interests of private landowners and other governmental agencies. Keiter and Boyce (1991) have provided an excellent account of scientific and political efforts to provide greater protection to Yellowstone National Park and its surrounding ecosystem.

Specifically, problems of fulfilling the burden of proof would be exacerbated if park management were to be based upon concepts of ecological integrity because: (1)the definition of "ecological integrity" is ambiguous, (2)it is not known what ecological attributes for measuring and monitoring ecological stress and ecological integrity might be most appropriate, and (3)the value of most ecological theory and information is heuristic as opposed to predictive.

2.1. THE DEFINITIONS AND MEANINGS OF INTEGRITY

Presumably, the meaning of ecological integrity must be understood and accepted by park decisionmakers if it is to be used as a basis for decisionmaking. Kay (1992) has proposed that ecological integrity encompasses three facets of ecosystems: (1)the ability to maintain optimum operations under normal conditions; (2)the ability to cope with changes in environmental conditions (i.e., stress); and (3)the ability to continue the process of self-organization on an ongoing basis, i.e., the ability to continue to evolve, develop, and proceed with the birth, death, and renewal cycle. By optimum operations, Kay means the situation where the external environmental fluctuations that tend to disorganize ecosystems, i.e., make them less effective at dissipating solar energy, and the organizing thermodynamic forces that make ecosystems more effective at dissipating solar energy are balanced. It is not clear what Kay means by "normal environmental conditions" and, hence, to what extent ecological integrity can or should refer to human interventions in ecosystems. Westra (1994) proposes a definition slightly different from Kay's, wherein she says that ecosystems can be said to have ecological integrity when they have the ability to maintain operations under conditions as free as possible from human intervention, the ability to withstand anthropocentric changes in environmental conditions (i.e., stress), and the ability to continue the process of self-organization on an ongoing basis. She argues that concepts inherent in ecological integrity emerge from continuing scientific, legal, and ethical analysis and that while they correspond in her mind to more or less "pristine nature," they cannot be described or predicted precisely because ecosystems are constantly changing and evolving.

Regier (1993) provides an abstract definition stating that ecological integrity exists when an ecosystem is perceived to be in a state of well-being. In part, a more precise definition of ecological integrity is dependent on peoples' perspectives of what constitutes complete ecosystems. In addition to reflecting the concerns and values of scientists, definitions of ecological integrity also must reflect various social and ethical values relevant for public policy decisions regarding protection of ecosystems. One reason for inclusion of these various values is because there is no a priori scientific definition of ecological integrity, and therefore the concept encompasses perspectives or ways of viewing the world that inevitably reflect value-laden judgments. The ambiguity of ecological integrity is a recognition that its definition, like many ecological concepts, is determined, in part, on the basis of value-laden judgments and not solely on so-called value-free or precisely defined scientific criteria.

According to Schneider and Kay (in press), ecosystems can respond to environmental changes in five qualitatively different ways: (1) after undergoing some initial structural/functional changes, they can operate in the same manner prior to the changes; (2) they can operate with an increase or decrease in the same structures they had prior to the changes; (3) they can operate with the emergence of new structures that replace or augment existing structures; (4) different ecosystems with significantly different structures can emerge; and (5) they can collapse with little or no regeneration. Although ecosystems can respond to environmental changes in one of these five ways, there is no inherent or predetermined state to which they will return. Further, none of these ways indicates a priori whether a loss of integrity has occurred.

As Kay points out, any change that permanently alters the normal operations of an ecosystem could be said to affect its integrity. Accordingly, the last four types of ecosystem responses would constitute a loss of integrity. One problem with this view is that it seems to reflect a commitment to preserve ecosystem attributes as they exist and does not recognize sufficiently the fact that ecosystems are dynamic and evolve; attempts to maintain them in existing states often require intensive management interventions that can have adverse ecological consequences. Further, there is no scientific reason why a changed ecosystem necessarily has less ability to maintain optimum operations under normal environmental conditions, cope with changes in environmental conditions less effectively, or be limited in its ability to continue the process of self-organization on an ongoing basis. Alternatively, it could be said that any ecosystem that can maintain itself without collapsing has integrity. Accordingly, the first four types of ecosystem responses would constitute integrity. One problem with this view is that it would not be helpful to park decisionmakers because it accepts all ecosystem responses to change as constituting integrity, with the exception of total collapse, which occurs rarely and is clearly undesirable. Consequently, Kay concludes that between these two extremes is the option that some ecosystem changes might represent a loss of integrity, while others might not. However, while in theory science can inform park decisionmakers about the responses of ecosystems to environmental change, it cannot provide a scientific or so-called objective basis for deciding whether one change is more desirable than another. In other words, the selection of criteria to use in such a decision must be based on human judgment regarding the acceptability of a particular change.

If concepts of ecological integrity are to be used as a basis for enhanced protection of parks' resources, judgments about whether particular ecological responses to environmental change are consistent with the concepts will have to be made and will have to be reconciled with other value-laden judgments that have to be made about the meaning of prescriptive but ambiguous normative words and phrases contained in the NPS legislation, as well as with the implications of the fact that the methods of science that are used to provide information for decisionmakers are themselves embedded with value-laden judgments that may influence assessments of ecological integrity and other normative language in park legislation.

2.2. VALUE-LADEN LANGUAGE OF LEGISLATION

National park legislation contains undefined words such as "natural" and "unimpaired" as used in 16 U.S.C. 1 & 1(a)-1 and "park values" as used in 16 U.S.C. 20. These terms are prescriptive because the legislation mandates that parks be managed to preserve resources in their natural and unimpaired state and that park values be preserved, and they are normative because they provide criteria upon which to base management decisions. Despite both the prescriptive and normative nature of these terms, the legislation does not provide standards that can be used to indicate whether decisions are consistent with the norms. Nevertheless, if decisions about parks' resources were to be based upon concepts of ecological integrity, they would have to be deemed compatible with the legislative mandates of parks.

The term "natural" often is used in a sense that implies freedom from human influence, especially that of modern humans (Devall and Sessions 1984, Smith and Theberge 1986). Used in this manner, the concept of "natural" would appear to be consistent with Karr's (1992) and Westra's (1994) concepts of ecological integrity, since the latter imply minimal human intervention in ecosystems. However, another meaning of "natural" may differ from Westra's and Karr's insofar as some people argue that human use and influence should not necessarily be excluded from protected areas if such use is compatible and harmonious with the ecosystem (Sax 1980, Parks Canada 1982, Callicott 1991). This latter meaning of "natural" might be more consistent with Kay's (1992) concept of ecological integrity, since he does not explicitly specify the role of humans in his concept.

Ambiguity about the meaning of "natural" has permitted wide oscillations in park policy regarding the acceptability of phenomena such as nonhuman-caused fires, floods, and fluctuations of animal populations as well as of human intrusions in park ecosystems (NPS 1980, Lemons 1987b). For instance, the primary goal for many of the Alaskan parks, Isle Royal National Park, and the large wilderness areas of other parks is to manage them in a condition relatively free from human impact. On the other hand, management of some parks (e.g., Acadia National Park) and some areas within parks (e.g., Yosemite Valley) permits significant human alteration of the landscape. Confusion also exists with respect to understanding of the word "natural" even when it is being used in a sense that implies freedom from human influence, especially that of modern humans. For example, as a guide to management decisions for parks' ecosystems, Parsons et al. (1986) define "natural" as "…the unimpeded interaction of native ecosystem processes and structural elements." Alternatively, Bonnicksen and Stone (1985) define a natural ecosystem as one that "…portrays, to the extent feasible, either the same scene that was observed by the first European visitor to the area or the scene that would have existed today, or at some time in the future, if European settlers had not interfered with natural processes." Consequently, the concept of "natural" can lead to differences of opinion regarding whether a management emphasis should be placed upon specific structural or functional attributes of ecosystems. Park policies designed to maintain ecosystems in particular structural states would conflict with concepts of ecological integrity that emphasize ecosystems' ability to continue the process of self-organization on an ongoing basis.

Despite the legislative prescription to preserve parks' resources in their natural state, the ability of the ecological sciences to specify the meaning of "natural" is highly questionable (Lemons 1987b). "Natural" often is defined as a condition existing prior to human perturbation of ecosystems. Jorling (1976) argues that this definition ignores the fact that humans are a part of nature and therefore need be included in the definition of "natural." Given the historical and current global impacts of humans on ecosystems, there are probably few or no ecosystems in existence that have not been or are not affected by humans. Finally, it is difficult or impossible to know with reasonable certainty ecological conditions existing in ecosystems prior to human influence. Consequently, ecology cannot provide an unambiguous or noncontroversial definition of "natural." Ferré (1993) has argued that the interactions between humans and the environment result in ecosystems possessing different degrees of similarity to pristine ecosystems and that the task of those concerned with management policies based upon concepts of ecological integrity or "natural" need to ascertain what types of ecosystems count as being acceptable for management purposes. Rolston (1991) has argued that humans have a responsibility not to deny ecological processes relatively undisturbed by human intervention but to affirm them. One of the practical difficulties of using concepts of ecological integrity and "natural" in the management of national parks is that although we might be predisposed to consider parks' ecosystems to be relatively pristine, many are cultural artifacts because of the consequences of historical and present uses and interventions (NPS 1980). Consequently, not only have humans influenced present ecosystem structure and function but also their future evolutionary states. Accordingly, the ecology park managers would be affirming in Rolston's sense would be one emergent from both past and present human interventions in parks' ecosystems.

Likewise, although park legislation states that resources should be maintained in an "unimpaired" condition, no standards are provided to allow an assessment of what constitutes "unimpaired." Presumably, "unimpaired" means ecosystems relatively free from human activities or those that possess either structural or functional naturalistic attributes deemed desirable by humans. As will be discussed, various attributes potentially useful for measuring and monitoring ecosystem responses to environmental changes have been proposed, but ecologists disagree about their utility as indicators of environmental impacts or ecological integrity. Further, there is no a priori choice of what criteria should be used to assess whether parks' biological or ecological resources are impaired. Consequently, the selection of one criteria over another may reflect management goals to monitor, assess, or protect particular resources, or it may reflect the public's or a manager's values concerning the importance of respective biological or ecological attributes (Lemons and Stout 1984).

Despite the fact that legislation mandates the protection of park values, nowhere does it define the values or indicate how decisions regarding conflicts between them are to be resolved. Such definitions and decisions are left up to NPS decisionmakers. Rolston (1985) has provided the most comprehensive list and analysis of values often attributed to national parks resources: (1)market or commercial values, (2)life support values, (3)recreational values, (4)scientific values, (5)genetic and biodiversity values, (6)aesthetic values, (7)cultural symbolization values, (8)historical values, (9)character-building values, (10)therapeutic values, (11)sacramental values, (12)aspirational or option values, and (13)intrinsic values. Generally speaking, park management based upon concepts of ecological integrity would most likely conflict with market or commercial values, some recreational values, and possibly some aesthetic, cultural symbolization, or historical values if they were based upon particular structural attributes of ecosystems. Rolston (1985) and Lemons (1987b) have provided an extensive analysis of approaches to use in making decisions about conflicting park values in order to enhance the protection of parks' resources.

2.3. THE STATUS OF ECOLOGY AS A BASIS FOR MANAGEMENT

The management of national parks based upon concepts of ecological integrity requires that decisionmakers have adequate knowledge of the impacts of human activities upon biological and ecosystem attributes for use in the decisionmaking process regarding protection of parks' resources.

A number of researchers have critically analyzed the extent to which the methods and techniques of science are capable of yielding reasonably certain information appropriate to serve as a basis for management and public policy decisions (see, e.g., Sagoff 1985, Shrader-Frechette and McCoy 1993, Lemons 1995). The conclusion of these studies is that the capabilities of the ecological sciences primarily are heuristic as opposed to being able to predict the consequences of human activities with reasonable certainty.

Sagoff argues that the role of ecology should be to identify ecological indicators that might allow scientists to diagnose perturbations in species or ecosystems early enough so that mitigation measures could be implemented. This type of diagnosis does not depend upon knowing generalizable laws and basing predictions upon them; rather, it involves the integration of diverse information to make a general argument for one rather than another interpretation of the causes or consequences of ecological impacts. Recently, Bella et al. (1994) have argued that the role of the ecological sciences in problems of environmental change ought to be in the identification of indicators of ecological change rather than in the prediction of the consequences of human activities with reasonable certainty. Lemons (1986a) has analyzed the different meanings of "stress" as applied to organisms, populations, species, and ecosystems and has concluded that the theoretical differences between the different meanings are so great that, when combined with informational uncertainty concerning the assessment and evaluation of the causes of stress and their effects, little basis for reasonably certain predictions exists. Lubchenco et al. (1991) have identified numerous scientific uncertainties regarding protection of biodiversity and have proposed a research agenda to obtain more information about it. In their report, they acknowledge the limited role scientists can play in making reasonably accurate predictions about the effects of human interventions in ecological systems. Cairns and McCormick (1992) note that in theory both structural and functional attributes of ecosystems can be used as a basis for ecological predictions but that practically speaking there is a significant lack of knowledge about them. Finally, the Committee for the National Institute of the Environment has developed a comprehensive proposal to reform environmental research in the United States in order to provide more suitable information for decisionmaking (CNIE 1994).

Despite the difficulties of identifying appropriate indicators of ecological change, a number of researchers have proposed various ecosystem attributes that might serve as a basis for measuring and monitoring. Mack (1983) analyzed long-term monitoring efforts devoted to vegetation, macroclimate, disturbances caused by exotic species, chemical and physical parameters in aquatic systems, anthropogenic disturbances, natural disturbances, and biological parameters in aquatic systems in Biosphere Reserve Parks in the United States, and Wright (1992) surveyed research goals and techniques for wildlife management in national parks.

Schaeffer et al. (1988) proposed the following attributes for measuring and monitoring ecosystem changes: (1)habitat for desired diversity and reproduction of desired organisms, (2)phenotypic and genotypic diversity among organisms, (3)relationships between food chains and desired biota, (4)relationships between nutrient pools and cycling and desired organisms, (5)relationships between energy flux and maintenance of trophic structure, (6)homeostatic mechanisms and damping of undesirable oscillations, and (7)the capacity to decompose, transfer, chelate, or bind anthropogenic chemicals and radionuclides until they are no longer toxic in the system. Parameters necessary to study to assess these ecosystem characteristics include: (1)individual measures such as fitness, disease, mutation, reproduction, physiology, acclimation, and individual behavior; (2)measures for populations such as intraspecific behaviors, epidemiology, genotypic variation, phenotypic variation, reproduction, physiology, and adaption; (3)system-level factors such as interspecific behavior, decomposition, production, recovery, resilience, resistance, connectivity, indicator species, keystone species, successional patterns, guild theory, species diversity, and vegetative diversity; and (4)abiotic elements such as nutrient and mineral retention, leaching, physiography, structural diversity, and chemical composition. Based upon thermodynamic considerations, Schneider and Kay (in press) have provided the following indicators of ecosystem responses to stress: (1)energy flow, (2)use of energy, (3)flow of energy through the system, (4)cycling of energy and materials, (5)trophic level structure, (6)articulation of food webs, (7)respiration, (8)transpiration rates, (9)biomass, and (10)species diversity.

Although these types of attributes might be of potential use in measuring and monitoring the responses of species or ecosystems to changes, their use is limited by the fact that there often is no a priori way to determine which attribute(s) should be measured or monitored for a particular type of ecosystem or ecosystem state or for a particular type of environmental change, or what threshold values and spatial or temporal scales should be used as indicators of undesirable or significant changes. There also is a lack of information regarding the linkages or synergistic effects between changes in one attribute and another and how to separate normal variations of attributes' measurements from those signifying long-term changes or changes due to human activities. Finally, the attributes are of limited use in the assessment and prediction of the ability of ecosystems to cope with environmental changes or to continue the process of self-organization because the nonlinear emergent behavior of systems cannot be explained fully by knowledge of its component parts.

Based upon a review of the role of science in environmental assessment, Lemons (1994) concluded that scientific information should be used in the planning process in order to help decisionmakers make more environmentally sound decisions but that it is not adequate to serve as a basis for firm predictions. Shrader-Frechette and McCoy (1993) maintain that science may have some role in informing public policy decisions in matters of biodiversity, but that its role should be considered to be descriptive and explanatory. More specifically, they argue that site-specific case studies may yield information useful for decisionmaking but that such information should not be used to make more generalizable laws for predictive purposes. Accordingly, it appears that both theoretical and informational uncertainty is so pervasive that the science of ecology should be considered primarily as having heuristic value and relatively fewer predictive capabilities suitable as a firm basis for decisions about management of national parks' resources. Consequently, decisionmakers should adapt their decisionmaking processes and policies to reflect the capabilities of science to provide information as a basis for decisonmaking.

The scientific uncertainties inherent in concepts of ecological integrity and attributes to measure and monitor ecological changes and conditions in parks' resources are exacerbated by the fact that the concepts and methods of ecology to study and apply them are embedded with value-laden judgments (Mayo and Hollander 1991, Shrader-Frechette 1991, Cairns et al. 1992, Cranor 1993, Shrader-Frechette and McCoy 1993). For example, scientists often have to make judgments about which species or ecosystem attributes to study without having a firm scientific knowledge base to inform their choice. Often, ecologists make value-laden judgments when they use simplified models with many built-in assumptions that cannot be validated or verified, such as those used in global environmental change or complex systems analysis. Assumptions have to be made regarding what spatial and temporal scales to use for studies and management policies when there are no firm scientific principles that can be used to determine the appropriate choice of scale. Further, many studies are by necessity limited to small spatial and temporal scales, yet scientists often make long-term predictions extrapolated from them even though such predictions cannot be verified or validated. In many ecological studies, assumptions have to be made, often without direct empirical evidence, whether ecosystem parameters should be considered independently or synergistically and whether threshold values for environmental impacts exist, and if so, what such values should be. Assumptions also have to be made concerning whether and to what extent results from laboratory studies can be extrapolated to field conditions. In addition, a lack of empirical data cannot be separated entirely from practical limitations imposed on environmental scientists. Decisionmakers require information in a relatively short time period and at reasonable cost. These factors constrain the focus of most ecological studies to lower levels of hierarchical organization, the short-term, small spatial areas, and measurement of relatively small numbers of parameters. Accordingly, adequate knowledge is difficult to obtain for practical as well as scientific reasons. Scientists also have to make judgments about whether to minimize type I or type II errors in their evaluation of acceptance or rejection of testable hypotheses. In other words, they must decide whether it is better to have a higher probability of accepting false positive or false negative results.

Carpenter (1990), Carter et al. (1994), and Schneider and Kay (in press) also note how ecological, economic, and cultural parameters must be selected and synthesized in scientific research in order to promote the goals of resource protection and ecological integrity. The selection of such parameters and the manner and extent in which they are synthesized are based on human judgment as opposed to objective criteria. Recently, science-based conservation policies have been criticized as being predicated upon elitist and Western cultural attitudes and traditions. Gómez-Pompa and Kaus (1992) maintain that conservation policies are predicated upon Western beliefs about nature and that they ignore perspectives of indigenous rural peoples. They argue that rural peoples have long maintained a relationship with nature and that their views and practices in terms of both utilizing the land and caring for it must be taken into account in conservation plans. According to this view, conservation must reflect the values and practices of rural indigenous people who depend on the land for their physical and cultural subsistence. In the United States, the inclusion of the values and practices of indigenous peoples in conservation policies would have implications for the management of Alaskan parks.

The fact that conclusions based upon scientific methods are embedded with value-laden judgments often is unrecognized by many scientists and environmental decisionmakers. The failure to recognize the existence of the value-laden methodologies of science, the value-laden dimensions of park legislation, and the value-laden aspects of concepts of ecological integrity can cast serious doubts about the best and most thorough of so-called scientific and technical studies used to inform decisions about protection of parks' resources because the decisions will appear to be compelled by value-neutral scientific reasoning when, in fact, they will be based, in part, on often controversial and conflicting values of scientists and decisionmakers. This situation implies that appropriate scientific and decisionmaking processes be adopted that would be reflective of populist, ethical, and interdisciplinary public policy considerations.

2.4. UNCERTAINTY AND COST-BENEFIT ANALYSIS

Many important decisions regarding protection of parks' resources require resolving conflicts between protection of the resources and economic concerns. Traditionally, methods of cost-benefit analysis have been used as a means for assessing the values of species, ecosystems, and economic commodities that would be affected by the decisions. In particular, such methods are used in the assessment of whether concessioner facilities in parks should be relocated to enhance the protection of critical park habitat and in the assessment of options to mitigate external threats to parks' resources. A variety of methods are used to assess the values of species and ecosystems, and it is beyond the scope of this chapter to discuss such methods in detail. Extensive discussions of the use of cost-benefit analysis in natural resources problems can be found in Norton (1987).

Proponents of cost-benefit analysis assume that its techniques can provide quantitative and objective information to decisionmakers on the present and future values of species or ecosystems. However, the use of cost-benefit analysis in decisions about parks' resources may conflict with management policies based upon concepts of ecological integrity for several reasons: (1)it may systematically bias decisions by ignoring, discounting, or miscalculating values; (2)different methods of assigning dollar or other quantitative values to species or ecosystems can lead to different approximations of benefits; (3)pervasive scientific uncertainties imply that it is not possible to compute accurately present or future values for most species or ecosystems; (4)it ignores distributional problems such as when the future worth of species or ecosystems is discounted and therefore greater benefits are distributed to present generations and greater potential costs or harm are distributed to future generations; (5)a decision not to proceed with a development project based upon cost-benefit analysis is reversible, whereas a decision not to enhance protection of parks' resources might not be; (6)the use of cost-benefit analysis ignores so-called intrinsic values which are independent of market or instrumental values; and (7)from the standpoint of ecological integrity, what matters is a species' role in maintaining or contributing to the ability of an ecosystem to cope with stress and to maintain its self-organizational abilities and not the instrumental values that cost-benefit analysts seek to determine.

In addition, the treatment of species and ecosystem attributes as traditional commodities has been identified as a significant cause of ecological degradation for several reasons (Goodland and El Serafy 1993, Cairns and Meganck 1994): (1) those who receive the benefits of exploiting biological and ecological resources usually do not pay the full costs of the exploitation; (2) the benefits of utilizing biotic resources are easier to quantify than the benefits of preserving them; (3) many biological resources are publicly owned and treated as free or inexpensive commodities; and (4) discount rates in cost-benefit analysis often are set too high compared with biological growth rates, thereby enabling more efficient depletion of biotic resources.

2.5. RECOMMENDATIONS FOR BETTER SCIENCE

Numerous studies focusing on the role of science in protecting national parks resources have provided specific examples and analyses of the contributions of scientific knowledge to understanding of park resources as well as in identifying areas where more scientific knowledge is needed for informed management decisions (see, e.g., NPS 1980, McNeely and Miller 1984, Lemons 1986b, Wright 1992). In these studies, few examples of specific resources problems were provided where scientific information was conclusive. Consequently, the studies recommended that scientific research in national parks be improved. Generally speaking, the studies have analyzed the types and extent of problems requiring scientific information for their solution, levels of financial and administrative support for scientific research, organizational structures to support scientific research, the quality and quantity of scientific researchers, and agency attitudes and philosophies regarding levels of support for scientific research. The underlying assumption of all of these studies was that by making improvements in these areas, scientific capabilities to fulfill prescriptive legislative mandates and burden of proof requirements would be enhanced.

Murphy (1990) and Drew (1994) have criticized the use of science in decisionmaking about resources such as those that occur in national parks. Their criticisms are that most researchers have engaged in descriptive or inductive studies and that they have not conducted more useful experimental studies based upon hypothesis-deduction or other scientific methods that would yield more robust and less speculative scientific knowledge. In other words, they propose that science used as a basis for management about natural resources conform to classical methodologies that utilize rigorous scientific methods to form hypotheses and apply the results to specific management problems. Peters (1991) claims that the primary way to improve the utility of the ecological sciences in decisionmaking and cost-benefit analysis is to judge every ecological theory on the basis of its ability to predict. I am unaware of any studies that focused on the application of concepts of ecological integrity to national parks' resources in any detail.

While the recommendations to improve scientific research in national parks' resources are worthwhile, they seem to imply that problems of protecting parks' resources are primarily or significantly a function of a lack of scientific knowledge and that as such knowledge increases, species and ecosystems can then be managed more effectively. Further, these recommendations seem to imply that increases in knowledge can be obtained rapidly enough to be of real assistance in redirecting the management of parks' resources in the time required to protect them against increasing threats. If this interpretation is correct, then the recommendations do not explicitly address the fact that ecological methods, concepts of ecological integrity, and park legislation are embedded with pervasive theoretical and informational uncertainty and value-laden dimensions that are not amenable to decisionmakers' use of classical methodologies of science as a basis for reasonably certain predictions.

Other scientific approaches ("post-normal" science) might be more amenable for use in protection of parks' resources (Miller 1993, Shrader-Frechette and McCoy 1993). The post-normal science approach emphasizes: (1)adequate formulation of problems so that data will contribute to public policy goals; (2)that most results from scientific studies will not yield reasonably certain predictions about future consequences of human activities and that many problems of protecting parks' resources therefore should be considered to be "trans-science" problems requiring research directed toward useful indicators of change rather than precise predictions (Bella et al. 1994); and (3)the need to evaluate and interpret the logical assumptions underlying the empirical beliefs of scientists with a view toward ascertaining more fully the validity of scientific claims and their implications. While post-normal science is not easy to characterize, it seeks a broad and integrated view of problems and places more emphasis on professional judgment and intuition and is less bound by analytically derived empirical facts. Accordingly, it allows for consideration of mixed questions of science and values in public policy decisions. Proponents of post-normal approaches acknowledge that they are based on retroduction and conceptual analysis and that by necessity they emphasize explanation or heuristic understanding of the complexities of nature rather than predictions based upon deductive thinking.

Post-normal scientific approaches rarely produce results that are acceptable according to classical scientific norms and hence are of limited utility as long as stringent burden-of-proof requirements exist. Post-normal approaches recognize the pervasive uncertainty surrounding problems of protecting parks' resources and also recognize that the problems include social, political, economic, and ethical dimensions not amenable to classical scientific analysis. In this sense, post-normal approaches would appear to be more suitable for national park management based upon concepts of ecological integrity and park legislation, both of which are embedded with a mix of scientific and value-laden issues. Post-normal approaches might focus on improving the understanding of responses of ecosystems to human activities and on the monitoring and analysis of responses so that reliable indicators of ecological change can be identified. In this sense, post-normal approaches would focus less on obtaining reasonably certain predictions in the classical scientific sense. This distinction between the classical and post-normal scientific approaches is subtle but important. According to the former, we might predict that certain ecosystems are more sensitive to environmental change than others, although specific outcomes within these ecosystems cannot be predicted. The focus of the classical approach is to accomplish the latter prediction. Typically, proponents of post-normal approaches advocate a minimization of type II errors that lead to acceptance of false negative results. In other words, they would favor a higher chance of accepting false positive results and a lesser chance of accepting false negative results that would imply no environmental harm. However, the post-normal approach has been criticized by some scientists and public policy experts as being too speculative and riddled with bias (Murphy 1990, Drew 1994). Neither scientific approach would ensure that complex problems will be understood before serious environmental consequences arise.

3. National Park Legislation

The traditional interpretation of the NPS's legislative mandate is that parks are to be managed for two conflicting purposes: (1)maintenance of park resources in an unimpaired condition and (2)provision for their use and enjoyment by the public (Lemons 1987b). The NPS has not been entirely explicit or consistent about how it interprets the conflicting purposes. At times it has stated that park resources should be managed: (1)according to a strict policy of preservation (NPS 1980), (2)to balance equally the goals of preservation and use (NPS 1978), and (3)to maximize use and enjoyment of parks (Shabecoff 1981).

Unfortunately, few judicial interpretations of the legislative mandates indicate the extent to which parks should be managed for preservation versus visitor use and enjoyment. National park scholars have argued that the traditional interpretation of NPS legislation has led to oscillating policies and degradation of parks' resources (see, e.g., Sax 1980, Lemons and Stout 1984).

Scholars of national park policy traditionally have believed that an examination of the history and meaning of park legislation was of limited value because Congress has never resolved the difficult questions of competing uses or the dilemma of preservation versus development. For example, Sax (1980) maintains that the statutes are of little help in balancing preservation and use. Mantell (1979) and Runte (1979) indicate that because the legislation does not adequately define the purpose of the parks, it cannot be used to resolve conflicts. However, if management of national parks is to be based upon concepts of ecological integrity that have the potential to restrict some uses of parks' resources in order to enhance their protection, a determination must be made that such a policy would be consistent with legislative mandates. In the following section, I argue that a reasonable reinterpretation of NPS legislation allows policies that would protect parks' resources better than the traditional interpretation that seeks to balance protection and use.

3.1. THE PURPOSE AND POLICY OF THE ORGANIC ACT

In 1916, Congress passed the National Park Service Organic Act (Title 16 U.S.C.) to place all national parks and monuments and certain reservations under the protection, management, and administration of the NPS. The fundamental purpose expressed in Section 1 of the act is threefold: (1) to conserve scenery, natural and historic objects, and wildlife; (2) to promote the enjoyment of scenery, natural and historic objects, and wildlife; and (3) to provide for public enjoyment of these areas so that the scenery, natural and historical objects, and wildlife are unimpaired for future generations.

The purpose of the Organic Act must be analyzed to determine the intent of Congress. Although congressional debate centered primarily on grazing issues and the inclusion of certain monuments within forest reserves, the House of Representatives in its report on the Organic Act expressed the great value of parks and compared them with national forests. "The segregation of national-park areas necessarily involves the question of preservation of nature as it exists," whereas national forest areas are "specifically created for the conservation of the natural resources of timber and other national assets" (U.S. Congress 1916b). The major distinction between national parks and forests is, respectively, the preservation of nature as it exists and the conservation of national assets. By comparing national parks and forests, the House expressed its belief that a higher standard of environmental protection was required in national parks.

Congress clearly intended, however, to allow the public to use and enjoy parks. People require shelter, water, food and supplies, facilities for disposal of waste products, and transportation. While the NPS was created in part to manage parks more efficiently and to deal with increases in park visitation, Congress also specifically provided for preservation of natural park resources as a primary management objective. In urging Congress to approve the establishment of the NPS, President Taft stated: "I earnestly recommend the establishment of a bureau of national parks. Such legislation is essential to the proper management of those wondrous manifestations of nature, so startling and so beautiful that everyone recognizes the obligations of the Government to preserve them for the edification and recreation of the people" (Taft 1911). The secretary of the interior recognized in his 1915 administrative report that national parks "alone have the seclusion and other conditions essential for the protection and propagation of wild animal life" (U.S. Congress 1916a).

Other statements in the report demonstrated that Congress recognized the parks as natural laboratories and classrooms.

Congressional intent thus seemed to permit the accommodation of visitors, but with two significant conditions: (1) enjoyment meant enjoyment of the parks' scenery, natural and historic objects, and wildlife; and (2) visitation and accommodation would not impair the preservation of park resources. Spectacular scenery, natural and historic objects, and undisturbed wildlife constitute a national park, and Congress sought to preserve these assets unimpaired for future generations. To despoil parks and fail to conserve these assets would violate the fundamental purpose of 16 U.S.C. 1.

Section 1 traditionally has been interpreted to promote both preservation and use. This interpretation allowed park resources to be compromised in response to increasing demand and changing visitor expectations. The plain meaning of 16 U.S.C. 1 is that use is legitimate, but only limited use that does not cause impairment. This interpretation is recognized in the management of park backcountry where unlimited use is not permitted. Instead, use of the backcountry is limited by ecological carrying capacities. Although the NPS is attempting to mitigate the impact of visitors and development in, for example, Yosemite Valley, the remaining numbers of visitors and levels and types of development still significantly harm park resources. According to the plain language of 16 U.S.C. 1, resource impairment would violate the fundamental purpose of parks. Section 1 implies that the carrying capacity concept should also be applied to managing all park areas and that nonconforming uses and facilities should be prohibited.

Congress passed 16 U.S.C. 3 at the same time it passed 16 U.S.C. 1. Section 3 creates penal sanctions for violation of rules and regulations promulgated by the secretary of the interior governing the use and management of parks, and expressly authorizes the NPS to act in ways that, if not expressly authorized, might be beyond the authority of the NPS under 16 U.S.C. 1. Express authorizations include: (1) disposing of timber when necessary to control attacks of insects or diseases, (2) disposing of wildlife and plant life where it is detrimental to parks, and (3) granting restricted leases and privileges to accommodate the public for terms not exceeding 30 years.

At first reading, 16 U.S.C. 1 & 3 may appear to establish the conflict of preservation versus use because the first and simplest intrusion of human objects into parks may impede the fundamental purpose of those sections. However, 16 U.S.C. 3 does not conflict with 16 U.S.C. 1, because some intrusions must be permitted if parks are to be enjoyed. Therefore, 16 U.S.C. 3 cannot reasonably support an argument that Congress intended to promote development of parks at the expense of preservation. Section 3, which provides criminal punishments for violation of park rules and regulations, is consistent with and strengthens the purpose expressed in 16 U.S.C. 1. Under 16 U.S.C. 1, the NPS must establish rules and regulations to protect parks. Section 3 complements 16 U.S.C. 1 by establishing penal sanctions to deter violations of those rules. Nor does the authorization to sell or dispose of timber conflict with the fundamental purpose of parks. The secretary of the interior is only authorized to sell or dispose of timber where, in his or her judgment, its removal is required to control the attacks of insects or diseases or to conserve the scenery or the natural or historic objects in parks. This provision not only is consistent with 16 U.S.C. 1, but also it adopts much of Section 1's language. The destruction of timber by the secretary under any other conditions would exceed the authority vested in him or her by 16 U.S.C. 1. Thus, this provision in 16 U.S.C. 3 is intended to preserve as much healthy timber as possible.

Section 3 also allows the secretary to destroy animals and plant life that may be detrimental to the parks. Because the purpose of natural parks is to preserve pristine ecological processes, this section only authorizes destruction or disposition of nonnative flora and fauna. Finally, 16 U.S.C. 3 provides that the secretary may grant privileges, leases, and permits for the use of land to accommodate visitors in the parks. Leasing park lands, in and of itself, could be construed as a violation of 16 U.S.C. 1. However, 16 U.S.C. 3 clearly authorizes leasing only insofar as it is necessary to accommodate visitors and is conducted in a manner that would not violate the fundamental purpose of 16 U.S.C. 1.

The lease provision of 16 U.S.C. 3 was not designed to authorize the secretary of the interior to turn national parks into major development areas. Section 3, as amended in 1958, expressly states that such privileges, leases, and permits for the use of land to accommodate visitors shall not exceed 30 years and shall not interfere with free public access to any natural curiosities, wonders, or objects of interest. Again, the secretary has broad discretion to grant privileges, leases, and permits, but only to the extent of the authority set forth in 16 U.S.C. 1. The specific mention of a time limitation and a prohibition against impairing public access to natural curiosities and objects of interest are additional safeguards Congress included to ensure that the preservation of the national parks conformed with the stated fundamental purpose.

Between 1916 and 1965, the conflict between preservation and use became more apparent and deepened as a result of a larger national population, greater mobility, changing visitor expectations, and a longstanding NPS policy to encourage increased visitation to the parks. Consequently, Congress supplemented the Organic Act's statement of national park purpose with the Concessions Policy Act. Section 20 of the Act (16 U.S.C. 20) is a statement of congressional findings and purpose regarding park concessions, accommodations, facilities, and services, which appears to resolve conflicts of preservation versus use. It provides:

In furtherance...of this title,...which directs the Secretary of the Interior to administer national park system areas in accordance with the fundamental purpose of conserving their scenery, wildlife, natural and historic objects, and providing for their enjoyment in a manner that will leave them unimpaired for the enjoyment of future generations, the Congress hereby finds that the preservation of park values requires that such public accommodations, facilities, and services as have to be provided within these areas should be provided only under carefully controlled safeguards against unregulated and indiscriminate use, so that the heavy visitation will not unduly impair these values and so that development of such facilities can be limited to locations where the least damage to park values will be caused. It is the policy of the Congress that such development shall be limited to those that are necessary and appropriate for public use and enjoyment of the national park area in which they are located and that are consistent to the highest practicable degree with the preservation and conservation of the areas.

In enacting 16 U.S.C. 20, Congress intended to establish a financial policy for concessioners and to provide the proper atmosphere for private investment to meet the demands of increased park use (Aspinall 1965). Section 20(a) states that the interest of the NPS is to promote private as opposed to public concessioners. Section 20(b) guarantees the concessioner a property right in the form of a possessory interest and a chance to realize controlled profits. Sections 20(c) and 20(d) grant monopoly rights and preferential rights for renewal of contracts to existing concessioners. Sections 20(b) and 20(d) subordinate concessioner franchise fees to the objectives of preserving park resources.

Park concessioners argue that these sections were intended to encourage development and use of concession facilities (U.S. Congress 1976). Their rationale is that the operative sections of law, such as 16 U.S.C. 20(c) & 20(d), are needed only to protect large investments, which implies that substantial commercial developments are permitted. Concessioners also argue that the guarantee of a possessory interest mandates continuation of existing facilities in parks, since substantial (and perhaps unavailable) compensatory funds would be required to terminate a concession contract.

The legislative history of 16 U.S.C. 20 reveals that the primary purpose of the Concessions Policy Act was to codify policies that, with certain exceptions, had been followed by the NPS in administering concessions within national parks. Such policies had been in force before 1950 and were favored by park concessioners (U.S. Congress 1965).

During legislative hearings, opponents of 16 U.S.C. 20 argued that it favored concessioners over the government and the public and that it virtually granted concessioners a permanent right to conduct business (U.S. Congress 1964). The potential conflict between preservation and use was, however, expressly acknowledged by Congress in the Concessions Policy Act. By repeating in 16 U.S.C. 20 the intent of 16 U.S.C. 1 to preserve parks, Congress further dispelled arguments that promotion and development of parks at the expense of preservation was permitted. Section 20 stresses that accommodations and facilities shall be regulated carefully and limited to locations that will minimize the damage to park resources and that are "consistent to the highest practicable degree with preservation and conservation of the areas."

A reasonable interpretation of the Concessions Policy Act is, therefore, that it reaffirms the fundamental policy of preservation. Only those facilities necessary for the enjoyment of natural park features and located where such features will not be impaired should be permitted. The erection of facilities not needed for such enjoyment is specifically forbidden. Such an interpretation ensures that concessioner facilities will be "consistent to the highest practicable degree with the preservation and conservation of the areas." Although 16 U.S.C. 20 creates a relationship between the NPS and concessioner that permits a fair financial return and protection of investment, that relationship must conform to the fundamental purpose of preservation set forth in 16 U.S.C. 1 & 20. Notwithstanding the language of 16 U.S.C. 20, most major development centers are within or very near the prime scenic attractions of parks. Such development detracts from park scenery and causes significant ecological change. Many development centers are holdovers from an era when travel was much slower and when former park administrators thought that development centers needed to be near scenic attractions (Runte 1979).

Another congressional statement provides increased support for preservation. Originally enacted in 1970 and amended in 1978, 16 U.S.C. 1(a)-1 reaffirms the findings and purpose of Congress concerning national parks. Section 1(a)-1 states that those areas known as national parks derive their increased national dignity and recognition from their superb environmental quality. Section 1(a)-1 provides that:

Congress further reaffirms, declares, and directs that the promotion and regulation of the various areas of the National Park System...shall be consistent with and founded in the purpose established by Section 1 of this title, to the common benefit of all the people of the United States....The authorization of activities shall be construed and the protection, management, and administration of these areas shall be conducted in light of the high public value and integrity of the National Park System, and shall not be exercised in derogation of the values and purposes for which these various areas have been established, except as may have been or shall be directly and specifically provided by Congress.

3.2.   SPECIFIC STATUTES: SEQUOIA, YELLOWSTONE, AND YOSEMITE

In addition to legislation already discussed, specific statutes further indicate that the primary objective of Congress is to promote preservation of park resources. Each national park is created by its own enabling legislation that generally sets forth its purpose, park boundaries, and monetary allotment. Specific sections deal with the character of a particular park. Sections 1 through 20, in contrast, are general administrative sections and apply to all parks to the extent that they do not conflict with more specific legislation applicable to particular parks. Thus, the secretary of the interior's discretion may be restricted because specific law controls when that law conflicts with the general law of 16 U.S.C. 1-20.

Sections 22, 43, 47-1, 47(a)-(f), 51, 55, and 61 direct the secretary to make and publish rules and regulations for Sequoia, Yellowstone, and Yosemite National Parks "for the preservation from injury of all timber, mineral deposits, natural curiosities or wonders...and their retention in their natural condition." Some provisions in these sections permit construction of hotels and roads. Nonetheless, these sections must be read within the context of the purposes for which the parks were created (16 U.S.C. 41-79) and the general provisions of 16 U.S.C. 1-20.

Sections 1 through 20 and 41 through 79 together establish that although accommodation of visitors is one objective, preservation of park resources is necessary to achieve the meaning and spirit of the Organic Act. If the resources are not preserved, the park system will lose the very characteristics for which it was established. Congress did not intend to allow indiscriminate or unrestrained development, but rather to balance preservation and use so that use is consistent with preservation. While Congress did not specifically limit development, it is implicit that the secretary must establish a carrying capacity that promotes the primary objective of preservation of unimpaired park resources and the enjoyment of the objects of preservation. This reasoning is supported by 16 U.S.C. 47-1, which authorizes the NPS to obtain 1,200 acres of land just outside of an already overcrowded Yosemite Valley to create "an administrative site...in order that utilities, facilities, and services required in the operation and administration of Yosemite National Park may be located on such site outside the park." The purpose of the grant of authority is "to preserve the extraordinary natural qualities of Yosemite National Park."

Other legislation shows that Congress intended to restrict the secretary's discretion further and to promote preservation. For example, the secretary is authorized to lease lands to park concessioners in Yosemite under 16 U.S.C. 3 and 20. Such leasing is, however, restricted by 16 U.S.C. 55 in the following manner:

The Secretary of the Interior is authorized and empowered to grant leases, for periods of not exceeding twenty years, at annual rentals, and under terms and conditions to be determined by him, to any person, corporation, or company he may authorize to transact business in Yosemite National Park, for separate tracts of land, not exceeding twenty acres each, at such places, not to exceed ten in number, to any person, corporation or company in said park, as the comfort and convenience of visitors may require, for the construction and maintenance of substantial hotel buildings and buildings for the protection of motor cars, stages, stock and equipment and so forth.

Development of facilities under 16 U.S.C. 55 is further limited by 16 U.S.C. 1, 1(a)-1, and 20 to structures that are necessary for the enjoyment of park resources but do not impair such resources (Lemons 1987a). Section 55 limits leases to 20 years (despite this section, most concessioner contracts have been for 30 years). Section 55 limits each concessioner to 10 units; currently, the park concessioner operates over 25 concessioner units in Yosemite. Section 55 limits each unit to 20 acres; currently, 5 concessioner units substantially exceed 20 acres. Finally, the types of buildings in Yosemite are limited to one hotel and various garages; currently, the park concessioner operates several hotels and numerous buildings such as gas stations, lodges, a downhill ski area, a bank, restaurants, swimming pools, golf courses, souvenir shops, dress shops, bars, grocery stores, and coin-operated laundry facilities.

3.3. SUBSEQUENT LEGISLATION

In 1964, Congress passed the Wilderness Act (16 U.S.C. 1131) to protect and preserve lands for their wilderness characteristics. The Wilderness Act prohibits vehicles, motorboats, permanent roads, temporary roads (except for emergency uses), and structures or dwellings in such areas. Congress may designate wilderness areas within national parks, thereby additionally protecting the areas' natural characteristics.

It can, perhaps, be argued that because Congress did not designate all national park areas as wilderness, then some standard other than preservation is applicable. However, this argument is not substantiated by the intent of 16 U.S.C. 1-20 and the specific legislation for individual parks. The Organic Act and the enabling legislation for Sequoia, Yellowstone, and Yosemite mandate the preservation of the scenery, natural and historic objects, and wildlife and provide for facilities necessary to enjoy these resources in an unimpaired state. The Organic Act prohibits construction of accommodations and concessioners' leases on lands protected by a wilderness classification under the Wilderness Act. Thus, the Wilderness Act permits additional preservation without detracting from the purposes or objectives of the Organic Act or the specific legislation for particular parks.

The Wild and Scenic River Act of 1968 (16 U.S.C. 1271-1287) also significantly protects designated rivers. Congress intended that designated rivers within national parks receive greater protection of their tributaries and watersheds. This protection thus strengthens the preservation objective that the NPS must use to govern its decisions to allow an activity or to pass a regulation within a park.

3.4. JUDICIAL INTERPRETATION

Although few judicial decisions examine the Organic Act, specific statutory directives, or enabling legislation for individual parks, existing cases support a strong preservation mandate for the NPS. The authority vested in the secretary of the interior admittedly is broad. The secretary has the discretion to promulgate regulations and make policy decisions and is vested with the responsibility to administer the NPS. He or she must, however, exercise discretion consistently with 16 U.S.C. 1 and the other sections previously discussed.

In *National Parks & Conservation Association* v. *Kleppe* (1976), the courts established the general rule that the NPS, under 16 U.S.C. 20, has extensive regulatory authority over all facets of park operations and is charged with implementing the congressional policy of preserving park resources for enjoyment of future generations. In *Udall* v. *Washington, Virginia & Maryland Coach Co.* (1969), the court enunciated the federal standard of review to be applied in challenging NPS decisions. Because Congress delegated the task of weighing competing uses of federal property to the secretary of the interior, the balance that the secretary established should not be upset unless it was "arbitrary or beyond his authority....Where administrative control has been congressionally authorized, the judicial function is exhausted once there is found some 'rational basis' for the action taken." This standard of review was also used in *Sierra Club* v. *Andrus* (1981), in which the court considered whether the secretary's decision not to assert federal reserved water rights had been arbitrary, capricious, and an abuse of discretion. The court's standard presumes that agency action is valid, forbids a court from substituting its judgment for that of the agency, and requires affirmation if the agency has a rational basis for its decision.

The secretary's discretion in discharging his or her duty of managing and protecting park resources is not unlimited. When Congress provided in 16 U.S.C. 1(a)-1 that the protection, management, and administration of NPS resources "shall not be exercised in derogation of the values and purposes for which these various areas have been established," it curtailed the secretary's discretion. This reasoning was applied in *County of Trinity* v. *Andrus* (1977), in which the California federal district court held that provisos in statues such as 16 U.S.C. 1(a)-1 would be meaningless if they did not limit the secretary's discretion to administer resources in accord with statutory standards. Although the secretary does have broad discretion to promulgate specific regulations, he or she may now, in light of his or her statutory and general trust duties, not fail to act to fulfill the fundamental purposes of preserving park resources.

The Ninth Circuit Court of Appeals in *Rockbridge* v. *Lincoln* (1971), considering whether the secretary could be judicially directed to adopt and enforce regulations governing traders on the Navajo Indian Reservation as he or she was empowered to do by 25 U.S.C. 261 & 262, reversed a district court dismissal of the action. Because of the trust relationship of the secretary toward the Indians and the purpose of the statutes to prevent abuses within the unregulated trading post system, the secretary did not have unbridled discretion to refuse to regulate but only discretion to decide which specific regulations to promulgate. No valid distinction existed between a public official who had a statutory duty to act and failed to do so and one who had a duty not to act and acted. Similarly, in *Citizens to Preserve Overton Park, Inc.* v. *Volpe* (1971), the United States Supreme Court, considering the propriety of the secretary of transportation's authorization of highway construction through a public park, reversed a district court dismissal. The Court held that the secretary's action was subject to judicial review, that any discretion was not to be wide-ranging, and that the legislative history indicated that protection of the parkland was to be given paramount importance.

In *Sierra Club* v. *Department of the Interior* (1975), the plaintiff sued the secretary for failure to enter into agreements to protect Redwood National Park lands from the effects of logging. The evidence demonstrated that logging on adjoining lands caused destruction of park resources. Studies conducted by the plaintiff suggested ways to alleviate further damages, including: (1)the establishment of buffer zones around the perimeter of the park; (2)the adoption of a landscape or master plan for the park; (3)the purchase of additional lands or less-than-fee management easements; and (4)the negotiation of cooperative agreements with lumber companies, so that logging practices most detrimental to park resources could be mitigated. The court held that the department's actions, which consisted of three attempts to negotiate cooperative agreements with the timber companies, were inadequate. The court further ruled that the secretary had failed to perform duties imposed both by the Organic Act and by 16 U.S.C. 79(c), the enabling legislation for Redwood National Park.

In the same case, the court found that the secretary owed general fiduciary obligations to Redwood National Park land under 16 U.S.C. 1. Moreover, the court held that the specific statutory directives contained in 16 U.S.C. 1-18(f) and the enabling legislation for Redwood National Park established that the secretary's duty to protect scenery, natural and historic objects, and wildlife was a duty as law and not a matter of discretion. As a result of this holding, any discretion vested in the secretary concerning time, place, and the specifics of exercising such powers was subordinate to the paramount duty to protect the park. The court ordered the NPS, under 16 U.S.C. 79(c), to use all its powers to protect the lands from adjacent logging, to attempt to negotiate contracts with private loggers, and to consider acquiring private lands. In addition, the court ordered the NPS to lobby Congress for funds to acquire private landholdings and, retaining jurisdiction in the case, ordered the NPS to report back periodically. Although it later was found that the secretary acted within his legal power and that further legislation or an executive order was needed to authorize the protection required, this fact does not affect the reasoning that held the secretary to a legal obligation to protect and preserve the park (*Sierra Club* v. *Department of Interior* 1976).

The concept of a public trust also may facilitate the legal argument for preservation. The U.S. Supreme Court described the secretary's powers and duties in *Knight* v. *United States Land Association* (1891). The Court held that the secretary of the interior was the guardian of the people of the United States over the public lands and that he or she was obligated to see that none of the public domain was wasted or disposed of to a party not entitled to it.

The language of 16 U.S.C. 1 is similar to the language used to create a public trust, imposing strict obligations on the NPS by referring to the "high public value and integrity of the National Park System." This interpretation is supported by the holding in *Sierra Club* v.

Department of the Interior (1975), in which the court ruled that duties to preserve Redwood National Park were imposed upon the secretary by 16 U.S.C. 1, by the enabling legislation for Redwood National Park, and by public trust obligations of the secretary over the public lands.

The Redwood National Park litigation fits the description of public law litigation to which the public trust doctrine is appropriately applied. According to Chayes (1976) and the decision in *Natural Resources Defense Council, Inc.* v. *Morton* (1976), the characteristics of public law litigation include: (1)an amorphous party structure rather than a bilateral one; (2)inquiry into legislative rather than adjudicative facts; (3)relief that is forward-looking and fashioned ad hoc on flexible and broadly remedial lines; (4)a court decree that requires continued judicial participation in the administration of the remedy rather than termination of judicial involvement; and (5)a subject matter that concerns the operation of public policy rather than a dispute between private individuals about private rights. The litigation over Redwood National Park fits this description of the public law litigation model.

Unfortunately, there is no definitive resolution of the relevance of public trust doctrine to park management. A public trust argument was expressly discarded in *Sierra Club* v. *Andrus* (1981). In that case, the plaintiff asserted that in addition to the statutory duties, the defendants held the NPS resources in trust for the public and, therefore, were charged with the duties and obligations of a trustee. The court concluded there were no duties as a trustee distinguishable from the statutory duties of 16 U.S.C. 1-20. The defendant admitted, however, that the legislative history indicated that 16 U.S.C. 1(a)-1 imposed "an absolute duty which is not to be compromised, to fulfill the mandate of the 1916 Organic Act to take whatever actions and seek whatever relief as will safeguard the units of the National Park System." Although the court found that in the event of a threat to the scenic, natural, historic, or biotic resources of the parks, the secretary must take appropriate action, 16 U.S.C. 1 or 1(a)-1 did not indicate how park resources were to be protected. The secretary had broad discretion in determining what actions were best calculated to protect park resources, but that discretion was limited by the language of 16 U.S.C. 1 & 1(a)-1 so that the discretion "shall not be exercised in derogation of the values and purposes for which these areas have been established."

Although *Sierra Club* v. *Andrus* held that there were no distinguishable public trust obligations above those statutory duties with which the secretary was charged, the argument that the language of the Organic Act was intended to impose trust-like obligations upon the secretary remains valid. When the statute provides for trust-like duties, there are trust-like obligations; such an interpretation would not permit overuse by the beneficiaries of the trust. While these cases involve trust doctrine, the statutory language and the language in *Sierra Club* v. *Andrus* may support a cause of action against the secretary's discretion to act or failure to act to fulfill obligations under 16 U.S.C. 1, 1(a)-1, & 20. No distinguishable public trust obligations exist, because they were subsumed by the statutory enactment.

In this analysis, I have argued that the traditional interpretation of park legislation ignores the express language of 16 U.S.C. 1, and makes no attempt to clarify legislative intent by placing 16 U.S.C. 1 in the context of other relevant sections. A reasonable interpretation of 16 U.S.C. 1 indicates that Congress intended to create certain legal duties of the secretary of the interior. These duties have three major objectives: (1)to conserve the scenic, natural, historic, and biotic values of parks; (2)to promote enjoyment of park resources by the public; and (3)to ensure the enjoyment of park resources so that they are left unimpaired for future generations. Although the actual balance between preservation and use of park resources depends on the individual park, the most basic fiduciary duties of the NPS are to reduce development and promote preservation of resources. Legislation subsequent to 16 U.S.C. 1 consistently reaffirms its stated fundamental purpose; judicial interpretation, although limited, holds that preservation is an absolute duty of the secretary.

Because national park policy literature fails to clarify the meaning of legislative enactments, and because judicial review of park legislation is limited, this interpretation is necessarily based upon reasoned argument and is not intended to be definitive. However, it might provide a basis to support national park policies based upon concepts of ecological integrity, because it would support the idea of enhanced protection of parks' resources. Even if national park legislation was to be reinterpreted, decisions to protect parks' resources have to be based upon reasonably certain information. Consequently, the meaning of ecological integrity would have to be clarified and methods to measure and monitor it would have to be established.

4. Conclusion

Concepts of ecological integrity have been proposed as a basis to enhance protection of ecological resources. Using national parks as a case study, an analysis of practical and theoretical prospects and implications of using ecological integrity as a basis for decisionmaking to protect resources suggests several issues that would have to be addressed prior to adoption of such a proposal.

First, national park legislation traditionally has been interpreted to mandate a balance between protection of parks' resources and their use. National park policies based upon concepts of ecological integrity might require restrictions on levels or types of visitor use in order to protect parks' resources. Prior to implementing such policies, it would have to be determined that they were consistent with the legislative mandates of national parks. My analysis suggests that although the secretary of interior has wide discretion in his or her interpretation of the legislation, a reasonable argument can be made that his or her primary fiduciary duty is to protect parks' resources and only to allow levels and types of use consistent with their protection. Theoretically, this reinterpretation of park legislation would allow for management based upon concepts of ecological integrity, because it is more preservation-oriented. However, because park legislation is prescriptive, proposals to base protection of parks' resources upon concepts of ecological integrity must demonstrate that they would be consistent with the mandates and normative language of the legislation. Because my analysis of the meaning of park legislation cannot be construed as definitive, additional analysis needs to be conducted to determine whether national park legislation can be reinterpreted to support more restrictive policies that might be required if management were to be based upon concepts of ecological integrity.

Second, the essential conflict between legislative prescriptions and scientific uncertainty must be recognized by those who promulgate laws affecting national parks and the decisionmakers who carry out the mandates of those laws. In other words, it must be recognized that while the legislation has tended to be prescriptive, the science used in protection of national parks resources generally is descriptive or explanatory regarding the complexities of nature, as opposed to having the ability to make predictions with reasonable certainty. Consequently, if the burden of proof for demonstrating harm to parks' resources is placed on rule-makers or those calling for enhanced protection of parks' resources, it is likely that the burden will not be met, because people contesting decisions for enhanced protection will be able to demonstrate the scientific uncertainties inherent in information used in decisionmaking. The problem of fulfilling stringent burden-of-proof requirements for protection of parks' resources would be exacerbated if park policies were to be based upon concepts of ecological integrity because of the additional conceptual ambiguity and scientific uncertainty surrounding the concepts. Given these problems, consideration should be given either to changing the legislative mandates governing the protection of parks resources so that they are less prescriptive and therefore more consistent with the descriptive nature of the ecological sciences or to shifting the burden of proof to those who argue in a decisionmaking process against enhanced protection of parks' resources.

Third, the language of park legislation, the concepts of ecological integrity, and the methods of science are value-laden. Consequently, approaches to the protection of national parks' resources and the implementation of management policies based upon concepts of ecological integrity should strive to integrate science with other disciplinary knowledge and values in the formulation of park policy. An appropriate role for science in the protection of parks' resources might be to increase its capacity to assist in a more adequate formulation of public policies and goals by directing research toward useful indicators of change rather than precise predictions. In this manner science would contribute to a more broad and integrated view of park problems and would place more emphasis on professional judgment and intuition, and be less bound by analytically derived empirical facts; it would seek to assist in the identification of goals for human interactions with ecosystems rather than attempt to manage them toward predictable and predetermined ends. Accordingly, scientific information should be viewed as one element under consideration in the decisionmaking process, but its review and evaluation would not be left to scientific experts entirely; the general public and decisionmakers therefore would need to become literate in the basic epistemological issues of evidence, uncertainty, and hypothesis. Scientists also would have to learn to make concise and articulate defenses of why the best available evidence supports one conclusion instead of another, and both scientists and decisionmakers would have to understand the value-laden dimensions and implications of scientific methods and scientific uncertainty. In addition, all people interested in enhanced protection of parks' resources and in ecological integrity would need to grapple with the value-laden dimensions and meanings of the concepts and the normative language of park legislation in order to minimize conflicts between them.

Finally, one other problem needs to be addressed in order for management policies to be based upon concepts of ecological integrity. This has to do with the fact that the concepts seem to imply that large enough areas of land with appropriate ecosystem boundaries would have to be set aside to ensure that they were sufficiently protected against direct and indirect human intrusions that might compromise their integrity. Such land areas might consist of central core areas where only minimal human intrusions might be permitted surrounded by buffer zones where more extensive human activities would be permitted as long as they did not adversely affect the core areas (Westra 1994). However, scientists cannot yet say how large such areas might have to be or how their boundaries should be drawn.

The best prospects for implementing concepts of ecological integrity as a basis for management policies of natural resources might be national parks, since they already are protected areas governed by preservation-oriented legislative mandates. As this analysis has shown, numerous issues would have to be addressed to achieve such policies. The prospects for implementing ecological integrity as a basis for management policies for less protected areas would be more difficult.

5. References

Angermeier, P.L. and J.R. Karr. 1994. Biological Integrity versus Biological Diversity as Policy Directives. *BioScience* 44: 690-697.

Aspinall, L. 1965. 111 *Congressional Record* 23,636.

Bella, D.A., R. Jacobs, and L. Hiram. 1994. Ecological Indicators of Global Climate Change: A Research Framework. *Environmental Management* 18: 489-500.

Bonnicksen, T.M., and E.C. Stone. 1985. Restoring Naturalness to National Parks. *Environmental Management* 9: 479-486.

Brown, D.A. 1990. Integrating Environmental Ethics with Science and Law. *The Environmental Professional* 12: 344-350.
Cairns, J., Jr., and P.V. McCormick. 1992. Developing an Ecosystem-Based Capability for Ecological Risk Assessment. *The Environmental Professional* 14: 186-196.
Cairns, J., Jr., B.R. Niederlehner, and D.R. Orvos. 1992. *Predicting Ecosystem Risk.* Princeton Scientific Publishing Co., Inc., Princeton, NJ.
Cairns, M.A., and R.A. Meganck. 1994. Carbon Sequestration, Biological Diversity, and Sustainable Development: Integrated Forest Management. *Environmental Management* 18: 13-22.
Callicott, J.B. 1991. The Wilderness Idea Revisited: The Sustainable Development Alternative. *The Environmental Professional* 13: 235-248.
Carpenter, R.A. 1990. Biophysical Measurements of Sustainable Development. *The Environmental Professional* 12: 356-359.
Carter, J., J. Gibson, A. Carr III, and J. Azueta. 1994. Creation of the Hol Chan Marine Reserve in Belize: A Grass-Roots Approach to Barrier Reef Conservation. *The Environmental Professional* 16: 220-231.
Chayes, D. 1976. The Role of the Judge in Public Law Litigation. *Harvard Law Review* 89: 1281, 1288-1304.
Citizens to Preserve Overton Park, Inc. v. Volpe. 1971. 401 U.S. 402 (1971).
(CNIE) Committee for the National Institute of the Environment. 1994. A Proposal to Create a National Institue for the Environment (NIE). *The Environmental Professional* 16: 93-191.
County of Trinity v. Andrus. 1977. 438 F. Supp. 1368, 1377-78 (E.D. Cal. 1977).
Cranor, C.F. 1993. *Regulating Toxic Substances.* Oxford University Press, Oxford, UK.
Devall, B., and G. Sessions. 1984. The Development of Natural Resources and the Integrity of Nature. *Environmental Ethics* 6: 293-322.
Drew, G.S. 1994. The Scientific Method Revisited. *Conservation Biology* 8: 596-597.
Ferré, F. 1993. Persons in Nature: Toward an Applicable and Unified Environmental Ethic. *Zygon* 28: 441-453.
Gómez-Pompa, A., and A. Kaus. 1992. Taming the Wilderness Myth. *BioScience* 42: 271-279.
Goodland, R.J.A., and S. El Serafy. 1993. The Urgent Need for Rapid Transition to Global Environmental Sustainability. *Environmental Conservation* 20: 297-309.
Johnson, S.P. (ed.). 1993. *The Earth Summit. The United Nations Conference on Environment and Development (UNCED).* Graham & Trotman/Martinus Nijhoff, London.
Jorling, T.C. 1976. Incorporating Ecological Principles into Public Policy. *Environmental Policy and Law* 2: 140-146.
Karr, J.R. 1992. Ecological Integrity: Protecting Earth's Life Support Systems. In *Ecosystem Health*, R. Costanza, B.G. Norton, and B.D. Haskell, eds. Island Press, Washington, DC, pp. 223-238.
Kay, J. 1992. A Non-Equilibrium Thermodynamics Framework for Discussing Ecosystem Integrity. *Environmental Management* 15: 483-495.
Keiter, R.B., and M.S. Boyce (eds.). 1991. *The Greater Yellowstone Ecosystem.* Yale University Press, New Haven, CT.
Knight v. United States Land Association. 1891. 142 U.S. 161, 181 (1891).
Lemons, J. 1986a. Ecological Stress Phenomena and Holistic Environmental Ethics-A Viewpoint. *International Journal of Environmental Studies* 27: 9-30.
Lemons, J. 1986b. Research in the National Parks. *The Environmental Professional* 8: 127-137.

Lemons, J. 1987a. Title 16 United States Code §55 and Its Implications for Management of Concession Facilities in Yosemite National Park. *Environmental Management* 11: 461-472.
Lemons, J. 1987b. United States' National Park Management: Values, Policy, and Possible Hints for Others. *Environmental Conservation* 14: 329-340, 328.
Lemons, J. 1994. The Use of Science in Environmental Impact Assessment. *International Journal of Ecology and Environmental Science* 20: 303-315.
Lemons, J. (ed.). 1995. *Scientific Uncertainty and Environmental Problem-Solving.* Blackwell Science Inc., Cambridge, MA.
Lemons, J., and D. Stout. 1984. A Reinterpretation of National Park Legislation. *Environmental Law* 15: 41-65.
Lubchenco, J., A.M. Olson, L.B. Brubaker, S.R. Carpenter, M.M. Holland, S.P. Hubbell, S.A. Levin, J.A. MacMahon, P.A. Matson, J.M. Melillo, H.A. Mooney, C.H. Peterson, H.R. Pulliam, L.A. Real, P.J. Regal, and P.G. Risser. 1991. The Sustainable Biosphere Initiative: An Ecological Research Agenda. *Ecology* 72: 371-412.
Mack, A. 1983. A Survey of Ecological Inventory, Monitoring, and Research in the U.S. National Park Service Biosphere Reserves. *Biological Conservation* 26: 33-45.
Mantell, M. 1979. Preservation and Use: Concessions in National Parks. *Ecology Law Quarterly* 8: 1-54.
Mayo, D.G., and R.D. Hollander (eds.). 1991. *Acceptable Evidence.* Oxford University Press, Oxford, UK.
McNeely, J.A., and K.R. Miller (eds.). 1984. National Parks, Conservation, and Development. Smithsonian Institution Press, Washington, DC.
Miller, A. 1993. The Role of Analytical Science in Natural Resource Decision Making. *Environmental Management* 17: 563-574.
Murphy, D. 1990. Conservation Biology and Scientific Method. *Conservation Biology* 4: 203-204.
National Parks & Conservation Association v. Kleppe. 1976. 547 F. 2d 673 (D.C. Cir. 1976).
Natural Resources Defense Council, Inc. v. Morton. 1976. 388 F. Supp. 829 (D.D.C. 1974), aff'd, 527 F. 2d 1386 (D.C. Cir.), cert. denied, 427 U.S. 913 (1976).
Norton, B.G. 1987. *Why Preserve Natural Variety.* Princeton University Press, Princeton, NJ.
(NPS) National Park Service. 1978. *Draft Environmental Statement, General Management Plan, Yosemite National Park/California.* US Department of the Interior, Washington, DC.
(NPS) National Park Service. 1980. *State of the Parks-1980, A Report to Congress.* Office of Science and Technology, US Department of the Interior, Washington, DC.
Parks Canada. 1982. *Parks Canada Policy.* Department of the Environment, Ottawa, Canada.
Parsons, D.J., D.M. Graber, J.K. Agee, and J.W.V. Wagtendonk. 1986. Natural Fire Management in National Parks. *Environmental Management* 10: 21-40.
Peters, R.H. 1991. *A Critique for Ecology.* Cambridge University Press, Cambridge, UK.
Regier, H.A. 1993. The Notion of Natural and Cultural Integrity. In *Ecological Integrity and the Management of Ecosystems*, S. Woodley, J. Francis, and J. Kay, eds. St. Lucie Press, Delray Beach, FL, pp. 3-18.
Rockbridge v. Lincoln. 1971. 449 F. 2d 567 (9th Cir. 1971).
Rolston, H., III. 1985. Valuing Wildlands. *Environmental Ethics* 7: 23-48.
Rolston, H., III. 1991. The Wilderness Idea Reaffirmed. *The Environmental Professional* 13: 370-377.
Runte, A. 1979. *National Parks: The American Experience.* University of Nebraska Press, Lincoln, NB.

Sagoff, M. 1985. Fact and Value in Environmental Ethics. *Environmental Ethics* 7: 99-116.
Sax, J. 1980. *Mountains Without Handrails: Reflections on the National Parks.* University of Michigan Press, Ann Arbor, MI.
Schaeffer, D.J., E.E. Herricks, and H.W. Kerster. 1988. Ecosystem Health: I. Measuring Ecosystem Health. *Environmental Management* 12: 445-455.
Schneider, E.D., and J.J. Kay. In press. Order from Disorder: The Thermodynamics of Complexity in Biology. In *What Is Life: The Next Fifty Years. Reflections on the Future of Biology*, M.P. Murphy, A. Luke, and J. O'Neill, eds. Cambridge University Press, Cambridge, UK.
Shabecoff, D. 1981. *Administration Seeks Greater Role for Entrepreneurs at Federal Parks*, New York Times, March 29, §1, p. 1, col. 1.
Shrader-Frechette, K.S. 1991. *Risk and Rationality.* University of California Press, Berkeley, CA.
Shrader-Frechette, K.S., and E.D. McCoy. 1993. *Method in Ecology.* Cambridge University Press, Cambridge, UK. 328 pp.
Sierra Club v. Andrus. 1981. 487 F. Supp. 443, 450 (D.D.C. 1980), aff'd, 659 F. 2d 203 (D.C. Cir. 1981).
Sierra Club v. Department of the Interior. 1975. 398 F. Supp. 284 (N.D. Cal. 1975).
Sierra Club v. Department of Interior. 1976. 424 F. Supp. 172 (N.D. Cal. 1976).
Smith, P.G.R., and J.B. Theberge. 1986. A Review of Criteria for Evaluating Natural Areas. *Environmental Management* 10: 715-734.
Taft, W. 1911. President Taft's Message to Congress (Feb. 3, 1911), reprinted in House of Representatives Document No. 515, 64th Congress, 1st Sess. 9 (1916).
Udall v. Washington, Virginia & Maryland Coach Co. 1969. 398 F. 2d 765 (D.C. Cir. 1968), cert. denied, 393 U.S. 1017 (1969).
U.S. Congress. 1916a. H.R. Document No. 90, 64th Congress, 1st Sess. 122-123.
U.S. Congress. 1916b. H.R. Report No. 700, 64th Congress, 1st Sess. 3.
U.S. Congress. 1964. Park Concession Policy: Hearings on H.R. 5872, H.R. 5887, H.R. 5873, H.R. 5796, and H.R. 5886 Before the Subcommittee on National Parks of the House Committee on Interior and Insular Affairs, 88th Congress, 2d Sess. 34-49.
U.S. Congress. 1965. S. Report. No. 765, 89th Congress, 1st Sess. 1.
U.S. Congress. 1976. H.R. Report No. 869, 94th Congress, 2nd Sess.
Westra, L. 1994. *An Environmental Proposal for Ethics: The Principle of Integrity.* Rowman and Littlefield, Lanham, MD.
Wright, R.G. 1992. *Wildlife Research and Management in the National Parks.* University of Illinois Press, Chicago, IL.

Chapter 13
THE IMPORTANCE OF LANDSCAPE IN ECOSYSTEM INTEGRITY: THE EXAMPLE OF EVERGLADES RESTORATION EFFORTS

D. Martin Fleming[1]
D.L. DeAngelis[2]
W.F. Wolff[3]

1. Introduction

Ecosystem integrity consists of two parts: functional integrity and structural integrity (Westra 1994). An ecosystem's functional integrity is the maintenance of characteristic ecosystem processes, such as primary production, decomposition, energy flows and nutrient cycling. Structural integrity encompasses the persistence of specific organisms and biotic communities in the ecosystem. It also includes the spatial extent, heterogeneity, configuration (or spatial arrangement) and connectivity of landscape or terrain patterns representative of the ecosystem, upon which the organisms depend. Functional integrity might be sustained even if an ecosystem affected by human impact has lower species diversity compared with the natural ecosystem, but the affected system would be lacking in structural integrity.

We value the Florida Everglades, among other things, for its diverse array of wading birds. Recent declines across all feeding guilds in wading bird populations and reproductive success (Fleming et al. in press a, b) can be considered a decline in structural integrity. In the present paper we discuss these declines in relation to changes in the Everglades landscape. We argue that restoration of spatial extent, heterogeneity, configuration, and connectivity of the wetland types that once comprised the Everglades landscape is essential to the integrity of the ecosystem.

In recent years there has been increasing recognition of the need to understand populations in the context of the landscapes they occupy. The traditional approach of ecosystem theory, which stresses energetics in the absence of spatially explicit landscapes, is inadequate for answering questions concerning the persistence of populations. There are several reasons for this. (1) Spatial configuration of food resources is important. Movement costs energy and thus the relative spatial locations of food sources can determine whether consumers will utilize the resources. (2) Timing of the occurrence of resources has a spatial aspect. Resources may not always be available at a given spatial location, but their absence at a given time at one place may be compensated for by their availability elsewhere. (3) Other factors, such as the presence of predators or the absence of cover, may limit the areas of a landscape that a population can use.

---

[1]National Biological Service, Everglades National Park Field Station, 40001 State Road 9336, Homestead, FL 33034-6733, U.S.A.; [2]National Biological Service, South Florida Field Laboratory, University of Miami, Coral Gables, FL 33124-0421; [3]Forschungszentrum Juelich, Postfach 1913, Juelich, Germany.

Numerous other theoretical studies have confirmed the importance of an explicit consideration of the landscape. For example, the spatial arrangement and density of plants influences the success of herbivores in finding food (Risch 1981, Stanton 1982, Kareiva 1983, Cain 1985). Pearson et al. (in press) showed that "when suitable habitat is less abundant, patterning and resource utilization scales become increasingly important." Fahrig and Paloheimo (1988), simulating dispersing animals in a patchy landscape, found that when the animals could only disperse short distances, the resource patches that had close neighbors received many more dispersers than those that did not. They concluded that the spatial relationship and connectivity among patches are important determinants of local population size.

The long-term persistence of populations appears to be dependent on landscape extent, heterogeneity, spatial configuration, and connectivity. Den Boer (1981) introduced the concept of "spreading of risk," which means that on a sufficiently large and heterogeneous landscape, whatever temporal variations there are in environmental factors will probably vary from place to place. He explained the regional persistence of populations of carabid beetles on this basis. Local extinctions caused by short-term unfavorable conditions were compensated for by immigration from areas where conditions are more favorable. Fahrig and Merriam (1985) showed that a similar idea applied to a white-footed mouse population in a patchy environment.

A large number of models, reviewed by DeAngelis and Waterhouse (1987), have shown that model communities subjected to local extinctions are able to persist for long periods on landscapes with sufficient extent, heterogeneity, configuration, and connectivity. For example, Hilborn (1975) modeled Huffaker's system of herbivorous and predatory mites on a spatially heterogenous array of oranges, among which they were able to disperse. A one-cell or two-cell system was unstable, but the addition of further cells stabilized the model system, as it did Huffaker's laboratory system. Within protected natural areas, native flora and faunal communities are also influenced by regional processes - disturbances on various spatio-temporal scales and dispersal and recolonization by members of a population. Some of these natural processes are necessarily dependent on the size and spatial arrangement of important elements (land areas or terrain patterns) in the ecosystem.

Land ecosystems also vary in scale and are comprised of terrain patterns with distinct biological and physical characteristics. Discernible boundaries (ecotones) exist between these terrain patterns or land areas. The theory of ecological land classification (Rowe and Sheard 1981) also recognizes that interactive characteristics of landscapes or terrain patterns include the ecological diurnal and seasonal flux of energy, material, and species across macro- (biomes), meso- (physiographic subregions and ecotones), and micro- (local ecosystems) scales. Each of these terrain patterns are characterized by energy regimes whose expressions in the landscape are most easily recognized by biotic and abiotic components. Such terrain patterns, with discernible biotic and abiotic components and related energy regimes, occur as patterns or areal components of the earth's surface that represent ecosystems, both large and small, nested within one another in a hierarchy of spatial sizes.

Large-scale landscapes are thus interacting matrices in which populations are distributed. Such populations depend on the spatial extent, heterogeneity, configuration, and connectivity of terrain patterns comprising a landscape to create a mosaic of habitats. Across such natural or undisturbed mosaics, there will exist some habitats that meet the various needs of individuals in a population in their different life stages, despite environmental fluctuations that temporarily reduce the usefulness of other local habitats.

2. Spatially Explicit Landscape Simulation Study

These theoretical ideas concerning the landscape formed the background of our simulation study (Fleming et al. 1994a) to investigate causes of the wading bird decline. We adopted both a landscape view and an individual-oriented approach in modeling.

The model (Wolff 1994) used in our simulation study explicitly takes short time-step variability and small spatial scale heterogeneity into account. This was done by simulating wading bird activities at 15-minute time steps, daily changes in water levels and related changes in prey availability, and by using fine scale resolution (1/4 km x 1/4 km pixel size) to represent landscape or habitat heterogeneity. The feasibility of a specific hypothesis was tested concerning observed declines in Everglades wading birds and in their reproductive success. This hypothesis linked the declines to changes in landscape characteristics of the Everglades (Table 1). Two of these changes hypothesized to have the greatest adverse effects on wading bird reproduction were: (1)the decline of landscape heterogeneity through the loss of high-elevation wetlands peripheral to the central sloughs; i.e, areas having short periods of inundation, or hydroperiods. Such shallow water areas provide high prey availability early in the breeding season when interior wetlands and the headwater regions of downstream estuaries are still deeply inundated; and (2)alteration in the hydrologic periodicity or frequency of major drydowns in interior wetlands and sloughs. Such drydowns can affect prey availability late in the breeding season and have adverse interannular effects on prey abundance in subsequent breeding seasons (Loftus and Eklund 1994). General loss of habitat (removal of both short- and long-hydroperiod wetlands) was also tested as a cause of declines in wading bird reproductive success. These simulations showed that for the same loss of total area, the specific habitat removal and occurrence of drydowns in interior wetlands caused a significantly greater reduction in wading bird reproduction at these traditional colony sites than did general habitat loss.

Prior to the building of drainage canals and dikes in the eastern parts of the Everglades during the early 1950s, extensive high elevation wetlands along the eastern Everglades provided broad shallow water, foraging habitat during the late wet and early dry season, the prenesting period, when water depths within interior wetlands and the headwater regions of downstream estuaries are normally too high for efficient foraging by wading birds. To sustain

Table 1. A Comparison of Pre- versus Post-Drainage Everglades Landscape Characteristics.

Landscape Characteristics	Pre-drainage	Post-drainage
Wet season sheet flow	Uninterrupted sheet flow	Sheet flow fragmented by levees into impoundments and/or overdrained marshes
Short-hydroperiod marshes (peripheral wetlands)	Extensive	Lost to development or degraded by drainage
Seasonal changes in water depths	Attenuated changes	Pronounced fluctuation
Major drydown in sloughs	Rare	Frequent
Dry season freshwater flows into the estuaries	Greater flows	Reduced flows

the level of prey capture necessary to raise offspring, wading birds require shallow, receding water levels. Without the presence of these short-hydroperiod wetlands, nesting initiation at traditional colony sites (located in the headwater regions of downstream, coastal estuaries) is delayed until seasonal drying of the interior marshes and headwater regions of estuaries begins. There has been a disproportionate reduction (85%) in the functional area of this specific habitat, primarily in the eastern Everglades, due to loss from development and/or degradation (overdrainage) since the early 1950s (Fleming et al. in press a, b).

This landscape-scale change was proposed as a major cause of late colony formation of wading birds at traditional colony sites (Fleming et al. in press a, b). In the pre-drainage landscape, two broad zones of predictable shallow water areas existed during the wet and early dry season. Such shallow water areas occurred in the tidal coastal zone of estuaries along the Gulf west coastline and in the high elevation wetlands of the eastern Everglades. Large rookeries were located between these zones in or near the headwaters of the upper reaches of the estuaries. Wading birds that staged at these nesting sites during the early dry season (prior to nesting) were within foraging distances that would allow efficient utilization of these shallow water landscapes. However, with the drainage of the eastern high elevation wetlands, predictable early dry season shallow water areas near these traditional colony sites now occur only in limited areas of higher elevation, short-hydroperiod wetlands west of the central sloughs and in the narrow, tidal coastal zones of the estuaries (Figure 1) (Fleming et al. in press b).

A simulation approach was selected to test the feasibility of this peripheral wetland loss hypothesis (Fleming et al. 1994a) based on a model of a wood stork breeding colony (Wolff 1994) located at a traditional wading bird colony site in the estuary headwater region of the southern Everglades. In the simulations, each individual wood stork, parent and nestling, was modeled. A bioenergetics model was attached to each model wading bird to keep track of its energetic status on 15-minute time steps. The allocation of time between foraging, resting, and parental care of the adult birds was also followed on 15-minute time steps, using a set of decision rules derived from a thorough study of the literature on wood storks (Wolff 1994). Foraging in the model was performed on a 40 km x 40 km landscape divided into 25,600 cells, each 1/4 km x 1/4 km. The water depth and, therefore, the availability of prey in each cell followed a hydrologic scenario typical for either the pre- or post-drainage Everglades during a nesting cycle (primarily during winter and spring). In the pre-drainage scenario, the landscape included the historical short- to long-hydroperiod wetlands of the southern Everglades in their original spatial configuration. In the post-drainage scenario, water levels in these high elevation wetlands were lowered to simulate the present extent of overdrainage of these wetlands. This overdrainage results in the loss of functional integrity of these wetlands (e.g. inefficient foraging habitat for wading birds).

For purposes of comparison, a "baseline" simulation that approximates the pre-drainage landscape was performed first. Then alternative scenarios were examined using the model. The alternative scenarios in our study (Fleming et al. 1994a) included conditions of 10%, 20%, 30%, 40%, 50%, 60%, 70% and 80% loss of functional, high elevation wetlands. Compared to the baseline situation, the last of these alternative scenarios approximates the present, post-drainage conditions.

The pattern of nesting performance predicted by the colony model for both the pre-drainage and post-drainage hydrologic scenarios agrees quite closely with the observed dynamics of nesting colonies that was documented for the traditional sites in the southern Everglades (Fleming et al. in press b). Under the present, post-drainage conditions, nesting at these sites is delayed during most moderate to higher rainfall years and related water levels. Due to late nesting, insufficient time remains for most nesting pairs to rear their young before the advent of normal spring rains and subsequent water level increases, causing the dispersal of prey resources. Subsequent reproductive success is low.

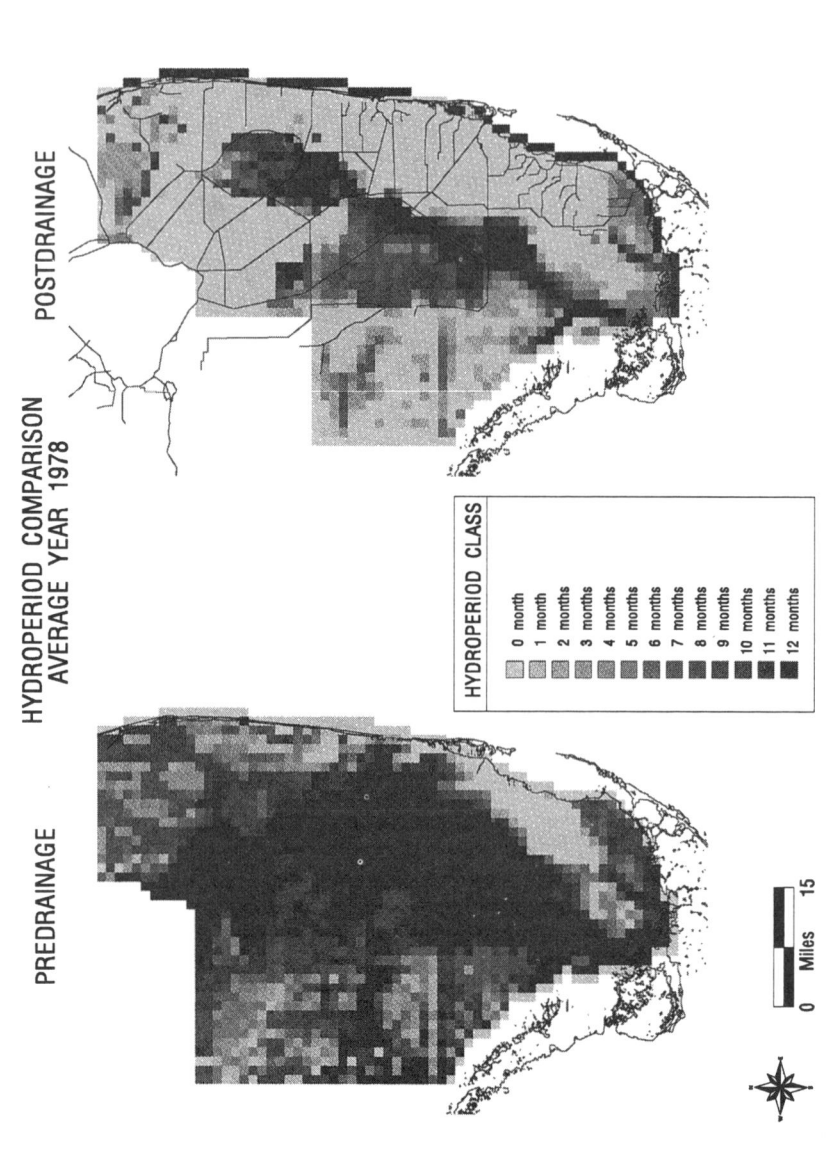

Figure 1. Hydroperiod Comparison of Pre- and Post-drainage Everglades Landscapes. Comparison based on hydroperiods estimated from natural system (approximates pre-drainage) and managed system versions of a hydrologic simulation model (Fennema et al. 1994) of south Florida using data from an average rainfall year (1978).

This pattern of both late nesting and low reproductive success was predicted by the model in the various scenarios with substantial loss of high elevation wetlands. These scenarios most closely approximate present environmental conditions. The unique characteristics of these wetlands are discussed below. These characteristics and their close proximity to traditional colony sites were important aspects of the spatial heterogeneity and configuration of the pre-drainage landscape. For wading birds staging at traditional colony sites, such proximity allowed efficient foraging by potential nesting females to build energy reserves required for egg formation. Compared to post-drainage conditions, the close proximity of these terrains also reduces time and energy budget constraints of nesting pairs at these sites during the annual nesting cycle (Fleming et al. 1994a).

Our study showed that a sufficiently large and heterogeneous Everglades landscape provides sufficient conditions for successful wading bird reproduction under natural conditions, i.e., the ability of nesting pairs to reproduce successfully in moderate to high rainfall years, as predicted by the model if sufficient short-hydroperiod wetland area is inundated, as was the case in the pre-drainage landscape. Under such conditions, successful nesting is dependent upon both adequate landscape spatial heterogeneity (ratio of short- versus long-hydroperiod wetlands) and configuration (proximity of short- and long-hydroperiod wetlands to colony sites), as well as temporal heterogeneity or hydrologic periodicity, i.e., long intervals between major drydown events.

Thus, the interval between any two successive wet seasons represents a "window of opportunity" for successful nesting. In most years, nesting must be initiated early enough in the dry season to be completed before the advent of normal spring rains. However, this "window of opportunity" may be wide or narrow depending upon the existing areal extent and heterogeneity of the landscape (ratio of short- versus long-hydroperiod wetlands) and the spatial configuration of these terrains. This result is important because it suggests that any attempt to restore and manage the Everglades system must include an appreciation of the interaction of the key biota with the landscape. We showed that certain elements of the landscape, the short-hydroperiod wetlands of the southern Everglades, may be key to successful wood stork reproduction at traditional colony sites. Restoring historical water flows in the absence of an appropriately diverse landscape that includes these areas may be counterproductive. In view of the above comments, plans to manipulate water flows to optimize wading bird success (Walters et al. 1992) that do not address the issue of the need to restore landscape or terrain patterns to sufficient spatial extent, heterogeneity, and configuration for the system may not be successful.

Wading bird colony responses to hydrologic conditions at traditional colony sites located in the southern Everglades demonstrate this principle. Nesting success at these traditional colony sites remained low in response to high regulatory flows that occurred there from 1976-1985. Such flows did not restore sheet flow to the short-hydroperiod, eastern peripheral wetlands, but occurred primarily through the central slough and shorter hydroperiod, interior wetlands to the west of this central flowway. The shorter hydroperiod wetlands west of the flowway, however, lack the highly eroded limestone bedrock that characterize the eastern, short-hydroperiod wetlands. Loftus et al. (1992) demonstrated that solution holes in this highly karstic landscape provide access to groundwater refuges important in maintaining fish populations in wetlands subjected to frequent, prolonged drydowns. Therefore, these unique wetlands cannot be functionally replaced by the substitution of existing overdrained, long-hydroperiod marsh, or by restoration of sheet flow to short-hydroperiod marshes that lack this karstic landscape.

On the basis of our simulation study and others (Fleming et al. in press a,b), therefore, we hold that effective restoration of the Everglades must consider the fundamental spatial and temporal characteristics of the system. In preserving or restoring the integrity of a system such as the Everglades, the integrity of the landscape must be the fundamental issue in these tasks. Spatial extent and heterogeneity (including spatial configuration) are two aspects of the landscape that need particular attention in restoration efforts for the southern Everglades. Heterogeneity was recognized in our study (Fleming et al. 1994a) as essential in wading bird reproductive success because both short- and long-hydroperiod wetlands, in proximity to colony sites, were needed to provide a constant supply of prey throughout a typical nesting cycle.

The need for a certain minimal size of a biotic reserve is also important because size and heterogeneity are related. An area of larger size is more likely to have greater habitat diversity. Larger reserve size also means that the size of protected species population will be larger and thus safer from stochastic, demographic or environmental fluctuations (e.g., DeAngelis and Waterhouse 1987, Soulé 1987).

Reliance on manipulation of the driving forces to maintain a system, without an emphasis on the landscape as a whole, can lead to the reduced valuation of the areal extent, heterogeneity, and configuration of terrains comprising the system. There is a danger that the reliance on manipulation could lull managers into allowing the areal extent of the system to be reduced to a size at which it could not function under natural conditions as viable habitat for native animal assemblages. Populations in such landscapes would be reduced in size, would lack the buffer that landscape extent, heterogeneity, and a functional configuration of habitats provide, and would be dependent on finely tuned manipulation. Inevitably, since manipulation can never be perfect, populations will be subject to demographic and/or environmental stochasticity that could result in their extinction.

For protected natural areas that occur in human-dominated landscapes, therefore, maintaining the natural processes upon which these assemblages depend will require a mosaic of habitats of adequate size, heterogeneity, spatial configuration, and connectivity. Adequate landscape size also helps to ensure diversity of habitat through time, which increases the persistence of both individual species and overall community diversity. However, attempts to artificially manage a system of smaller areal extent than that required for a functional landscape mosaic may either be impossible or run the risk of management errors that could be irreparable.

The definition of a functional landscape mosaic must, therefore, include the threshold of spatial scales at which dependent biotic scale processes may persist. Dependence of organisms upon such spatial scales increases from lower to higher trophic levels in a system. Higher trophic level organisms have larger spatial area needs, because the food resources upon which they depend occur at lower densities than those of lower trophic level organisms. Therefore, higher trophic level organisms or top-level carnivores, in general, have evolved more complex behavioral patterns or decision-making processes as obtaining sufficient food requires greater effort. This greater effort also contributes to energy transfer efficiency losses from lower to higher trophic levels within a system (Colinvaux 1978). For most ecosystems, therefore, the threshold scale at which dependent biotic processes may persist is frequently related to the spatial heterogeneity requirements of the top-level carnivores. In open systems as the Everglades, where dominant top-level carnivores (such as wading birds) may be migratory, preservation of top-level carnivores will also require coordinated management efforts across regional scales, i.e., the identification and protection of adequate core zones at the extremities of migratory animal ranges (with interconnecting corridors) to preserve the trophic community of a targeted area.

3. Evaluation of Other Landscape Concepts

With respect to the preceding discussion of the landscape perspective, we need to re-evaluate past approaches for preserving biodiversity in protected natural areas.

3.1. SPECIES-AREA CURVE CONCEPT

Species-area curves are often proposed as a guideline for designing functional landscapes (Diamond 1975, Wilson and Willis 1975, Diamond and May 1976). Such relationships do not, however, directly assess the landscape heterogeneity or spatial configuration requirements of a species or species within a landscape, but only provide a relationship between the number of species found within a given area to increases or decreases in the size of an area. Changes in the size of an area, however, may not be an accurate measure of concurrent changes in the area's spatial heterogeneity and configuration of terrain patterns. Such heterogeneity and configuration will vary in relation to the design of a selected area, i.e., its geometric shape and orientation to landscape topographical contours or terrain patterns. Designs based on species-area relationships may, therefore, fail to incorporate sufficient spatial (and temporal) heterogeneity or the spatial configuration of terrain patterns necessary to provide the mosaic of habitats required to sustain many top-level carnivore populations.

3.2. CORE-AREA CONCEPT

Most conservation strategies seek to preserve the biodiversity of a natural landscape by maintaining its natural dynamics and preventing anthropogenic deterioration (Noss 1983, 1987). Recent landscape-level conservation strategies include the concept of preserving a core zone or area of an ecosystem. The core zone or preserve concept appears to be derived from the "minimum dynamic area" concept defined by Pickett and Thompson (1978). As described elsewhere in the scientific literature (Harris 1984, Noss 1987), and applied in biosphere reserve management programs, there is a clear need to better define and identify core zones that actually preserve critical attributes of landscape heterogeneity and configuration important to a system's structural and functional integrity. Failure in doing so has historically caused a loss of higher-order carnivores, resulting in a depauperate trophic structure and the incomplete preservation of native animal assemblages in many protected natural areas.

3.3. CORE-CORRIDOR CONCEPT

Noss and Harris (1986) pointed out that the problem with the minimum dynamic area concept is that "few if any reserves, anywhere in the world, are large enough to constitute minimum dynamic areas." They proposed instead "a system of natural areas, interconnected with each other and integrated with the land use of the surrounding landscape." These preserve network designs were proposed "to provide some of the functions of a minimum dynamic area..." by connecting natural areas by corridors or zones of suitable habitat. Within the context of regional land-use planning, a central core zone or preserve would receive the highest intensity of protection, decreasing outward by a gradation of multiple-use buffer zones.

Such a core-corridor design, in which a central zone only receives the highest intensity of protection, appears to be more appropriate and effective for terrestrial than wetland systems in which connectivity on small spatial scales is more critical to the persistence of biotic processes. For example, maintaining dispersal and recolonization processes of contiguous marsh areas by micro/macroinvertebrates and fish communities during and following major disturbances (i.e. drydown events) is required to provide an array of foraging patches across the landscape for wading bird populations. Preservation of landscape spatial heterogeneity and configuration, as represented by a cross-section of high to low elevation wetland types, extending across topographical contours from the peripheral edges of such systems inward to central or deep water flowways, is therefore critical to maintaining a wetland system's trophic structure, if a functional landscape mosaic, as demonstrated in this study, is to be maintained. Therefore, a greater need exists for a better definition of a core zone in preserve design approaches for systems in which spatial heterogeneity and the configuration of terrain patterns are important, if such a functional landscape mosaic is to be preserved. A central core-corridor design in such situations cannot substitute for the preservation or restoration of a functional landscape mosaic. A prerequisite for ecosystem integrity is the existence of a functional landscape. For the Everglades, we have defined a functional landscape as one of sufficient areal extent, heterogeneity, configuration, connectivity, and hydrologic periodicity to support native plant and animal assemblages, particularly the top-level carnivores that comprise the higher trophic levels of the system.

3.4. INCREASED TEMPORAL VARIABILITY AS A RESULT OF POST-DRAINAGE LANDSCAPE CHANGES

The above discussion describes the general effects of spatial heterogeneity and temporal variability on the reproductive success of aquatic and terrestrial species. In the Everglades, spatial heterogeneity is provided by a mosaic of wetland types, varying from low elevation, long-hydroperiod central sloughs or flowways and interior wetlands to high elevation, short-hydroperiod marshes along the edges of these interior wetlands. Temporal variability in the Everglades is primarily related to the seasonal and annual variation in rainfall and related changes in marsh water depths.

Post-drainage changes in these characteristics that have resulted in a loss of landscape heterogeneity create resource bottlenecks for key consumer groups. In the pre-drainage landscape, the adverse affects of extreme temporal events, i.e., droughts or floods which produce prolonged high or low water level conditions, on any particular consumer group were attenuated by a greater diversity of wetland types that provided increased options for individuals in which to meet their daily or seasonal needs. For example, high elevation, short-hydroperiod wetlands would have provided extensive shallow water habitats required by the more terrestrial species (wading birds, deer) even during wet season months of high rainfall years. Slower changes in seasonal marsh water depths under pre-drainage conditions (Fennema et al. 1994) also provided more persistent dry season flows required by more aquatic species (fishes, alligators) within the central sloughs and interior wetlands of the system, particularly during lower rainfall years. The loss of short-hydroperiod wetlands (decrease in spatial heterogeneity) was concurrent with other pronounced seasonal changes in marsh water depths resulting from water management practices. To promote drainage of the Everglades, water managers eliminated wetlands by diverting flow into remaining marshes and implementing regulatory schedules to provide flood control, causing pulsing of flows and concentrating releases during a relatively short time period. This produced reduced wet-season runoff retention in storage areas for dry season water supply leading to more pronounced drawdowns during dry season months (Leach et al. 1972, Johnson and Van Lent 1992, Van Lent et al. 1993, Fennema et al. 1994, Light and Dineen 1994).

These pronounced seasonal changes and the increased frequency of major drydowns in central sloughs (increased temporal variability) that characterize post-drainage conditions represents a system in which the effects of temporal events are more dominant upon the system's trophic structure and dynamics than in pre-drainage conditions. Figure 2 shows schematically the effects of reduced spatial heterogeneity. When spatial heterogeneity was high in the Everglades (before extensive drainage), landscape conditions as a whole were generally favorable for both aquatic and terrestrial species over a wide range of rainfall conditions. Because spatial heterogeneity in the post-drainage Everglades is much lower, landscape conditions are favorable for the terrestrial species only under low rainfall conditions and favorable for aquatic species only under high rainfall conditions.

Figure 2. Schematic Diagram Indicating That, When a Wetland Ecosystem Such as the Everglades Has High Landscape Diversity, It Can Still Be Favorable for Both Terrestrial and Aquatic Species. When this heterogeneity is greatly decreased, aquatic species have favorable conditions only at the high rainfall extreme and terrestrial species have favorable conditions only at the low rainfall extreme.

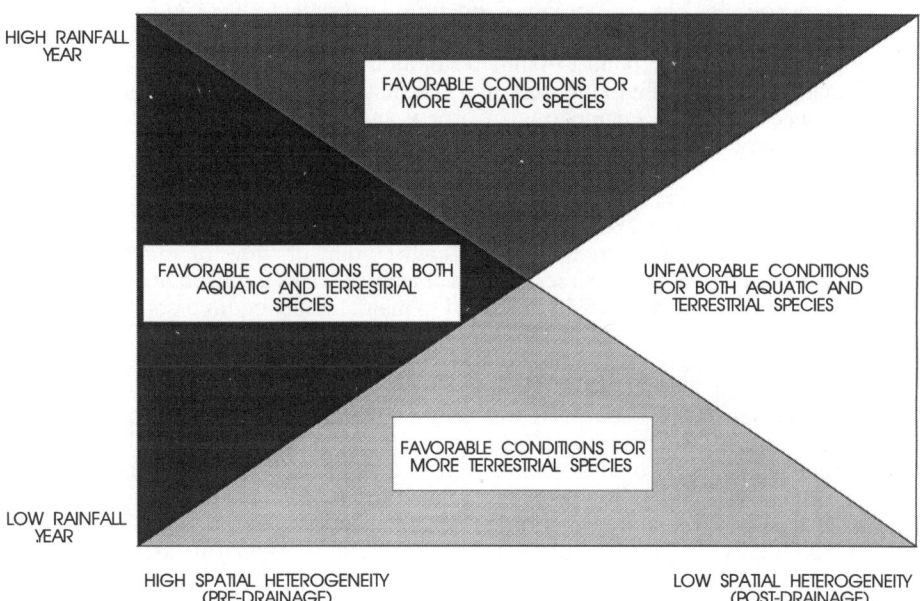

The resource bottlenecks that occur for a key consumer group such as wading birds, as a result of diminished spatial heterogeneity and increased temporal variability in the post-drainage Everglades landscape, demonstrate the need for restoration of a more functional landscape mosaic of adequate areal extent, heterogeneity, configuration, connectivity, and hydrologic periodicity. Priority restoration elements include:

1. restoration of uninterrupted sheet flow across the entire remaining natural areas of the catchment basin to:
   a. restore sheet flow to overdrained short-hydroperiod wetlands on the eastern periphery of the Everglades to increase landscape (wetland) heterogeneity;
   b. reduce impoundment of flows within northern reaches and between these impounded marshes and downstream reaches of the southern Everglades (to provide continuous, small-scale spatial connectivity required for the persistence of aquatic organisms);
2. restoration of persistent, dry season flows from upper to lower reaches of the area to:
   a. reduce temporal variability in seasonal marsh water depths and related occurrences of frequent major drydowns in central sloughs and interior wetlands of the system (to restore hydrologic periodicity and related aquatic productivity);
   b. increase freshwater dry season flows into downstream coastal estuaries (to reduce hypersalinity and hypothesized declines in estuarine productivity) (Walters et al. 1992).

4. Restoring and Preserving Biodiversity in Disturbed Landscapes: Theory and Practice

There are a number of possible alternative ways to restore an environmental system. Principal restoration approaches include:
1. point-to-point: the restoration of an entire landscape to its original condition at a selected point in time;
2. rehabilitation: the point-to-point restoration of selected original attributes of a landscape important to the structural and functional integrity of an ecosystem; and
3. enhancement: the engineering of a new and different system from its original characteristics or attributes (Rowe and Sheard 1981).

The advantages versus disadvantages of these conceptual approaches, in terms of preserving the native animal assemblages within a protected or targeted area, necessarily involve a comparison of each approach from the point of view of practicality, initial investment required to accomplish restoration objectives, risks involved in each restoration approach, and management intensity involved to maintain the required conditions.

Where spatial extent, heterogeneity, configuration, and connectivity are critical for maintaining desired biodiversity, the point-to-point restoration of the landscape, although requiring a high initial investment in some situations, would provide the lowest risk in achieving management objectives, as well as the lowest risks in committing irreversible management errors. Such large scale, intact natural landscapes require less intensive management and support populations least affected by the usual problems associated with habitat fragmentation, including problems related to environmental stochasticity, demographic stochasticity, social dysfunction, or genetic deterioration (Noss 1987).

A rehabilitation approach, or the restoration of selected original attributes important to system structure and function, would be more practical and require less initial investment than a point-to-point approach (in which the entire system is to be restored). The risks involved in this approach, however, would be greater as well as the management intensity required to achieve management objectives, as landscape size, heterogeneity, and the natural configuration and connectivity of terrain patterns are only partially restored or protected.

While enhancement approaches (redesigning/engineering of a new and different system from its original characteristics or attributes) are more practical to implement, and require a relatively lower degree of effort for implementation (compared to point-to-point and/or rehabilitation approaches that seek to restore many of the system's original characteristics), the subsequent risk in achieving management objectives is high. Such an approach usually does not seek to restore a natural landscape mosaic of adequate size, heterogeneity, spatial configuration and connectivity, but seeks instead to create an "optimum" habitat condition for a particular species (USFWS 1980) or to restore functional aspects of a system at the expense of its structural integrity (Walters et al. 1992). This approach would necessarily require a high degree of understanding and control of the natural processes that occur within a landscape and their causal links and mechanisms, over short, as well as long, time scales. Over-coupling of predator/prey populations as well as sudden and unpredictable shifts in habitat conditions may occur in such artificially structured and brittle systems (Holling 1973, DeAngelis and Waterhouse 1987).

Westra (1994) contrasts the natural "autonomous" behavior of a system with that of "willful, active intervention." We have argued above that restoration and management of the Everglades ecosystem must aim at creating an autonomous system that can maintain itself within the normal ranges of natural environmental variability. A necessary condition for such autonomy is sufficient landscape extent, heterogeneity, and, particularly the natural spatial configuration of terrain patterns that influence the patterning of resource availability, as well as connectivity and natural hydrologic periodicity. Attempts to maintain the system through external intervention that do not restore these attributes are unlikely to preserve structural integrity in the form of the diversity of higher order consumers. For example, attempts to address restoration issues for the natural areas of the Everglades by manipulation or rearrangement of terrain patterns that do not functionally mimic those of the natural, historical landscape will not adequately address the resource bottlenecks identified in our simulation study and others (as discussed in the preceding section, the unique characteristics and therefore importance of the eastern, high elevation wetlands and the proposals not to restore sheet flow to these wetlands but to substitute other marsh areas instead). These types of proposals arise from the need to meet present water supply and flood control demands. These ecosystem strategies attempt to integrate conflicting human land-use practices into a natural ecosystem. However, these attempts to "integrate" due so at the expense of restoring a system's landscape integrity.

If preservation of native animal assemblages is a priority within a natural area, then restoration approaches need to be evaluated on a case by case basis, using technical procedures that assess the minimal landscape size, heterogeneity, and spatial configuration requirements of top-level carnivores comprising the trophic communities of targeted areas. Spatially explicit, rule-based models, that include the complex decision-making processes and responses of individuals comprising such populations to changes in these landscape characteristics, provide the capability to identify potential habitat or resource bottlenecks in proposed restoration and/or natural reserve designs.

The persistence of such populations of top-level carnivores is also dependent upon the food webs in which they occur. However, trophic interactions between predator and prey populations are processes that are manifested at variable scales. Primary producers and lower trophic levels are directly driven by biochemical processes while population dynamics and individual behavior tend to dominate the dynamics of upper trophic levels or top-level carnivores. Thus the use of a single modeling method is not appropriate to simulate the hierarchy of interactions that occur among primary producers, lower trophic levels, and top-level carnivores, and across spatial and temporal scales at which these interactions actually occur within an ecosystem.

A modeling approach is required that integrates an appropriate set of spatially explicit, lower and intermediate trophic-level models and individual-based models of higher consumers, coupled across a spatially explicit landscape model (Fleming et al. 1994b). The application of such a landscape-scale, ecosystem-level modeling approach to the design, restoration, and management of such natural areas would assist in avoiding past errors in the identification of core zones for protection. This would be particularly true within large scale, oligotrophic systems such as the Everglades, where prey availability for higher consumer groups is naturally low, and where alterations in landscape heterogeneity have further reduced seasonal prey availability, leading to severe declines in historical numbers of colonial wading birds and other top-level carnivores of the system.

A conceptual core restoration zone must provide a more functional landscape mosaic of adequate extent, heterogeneity, spatial configuration, connectivity, and natural hydrologic periodicity than present post-drainage conditions to address seasonal habitat requirements for wading birds and other top-level carnivores within the Everglades. The core zone proposed in this study (Figure 3) should be viewed as a "minimum dynamic area" for the Everglades that would not preclude any larger scale restoration plans for the entire central and south Florida region. Buffer zones are required to enable full restoration of sheet flow within the "minimum dynamic area" or core zone. Essential restoration steps to achieve this core zone would include:

1. the raising of water levels (storage capacity) within Lake Okeechobee up to their historical levels to meet hydrologic requirements for restoration of sheet flow to

downstream reaches of the Everglades catchment area. Restoration of the lake's historical littoral zone, however, would be required to restore the structural integrity of this lake ecosystem;

2. the elimination of impounded flows within northern reaches of the catchment area and between these impounded marshes and downstream reaches by the removal of internal levees, the causewaying of roads traversing interior marshes, etc;

Figure 3. Depiction of a Conceptual, Core Restoration Zone or "Minimum Dynamic Area" for the Everglades Including an Eastern Buffer Zone.

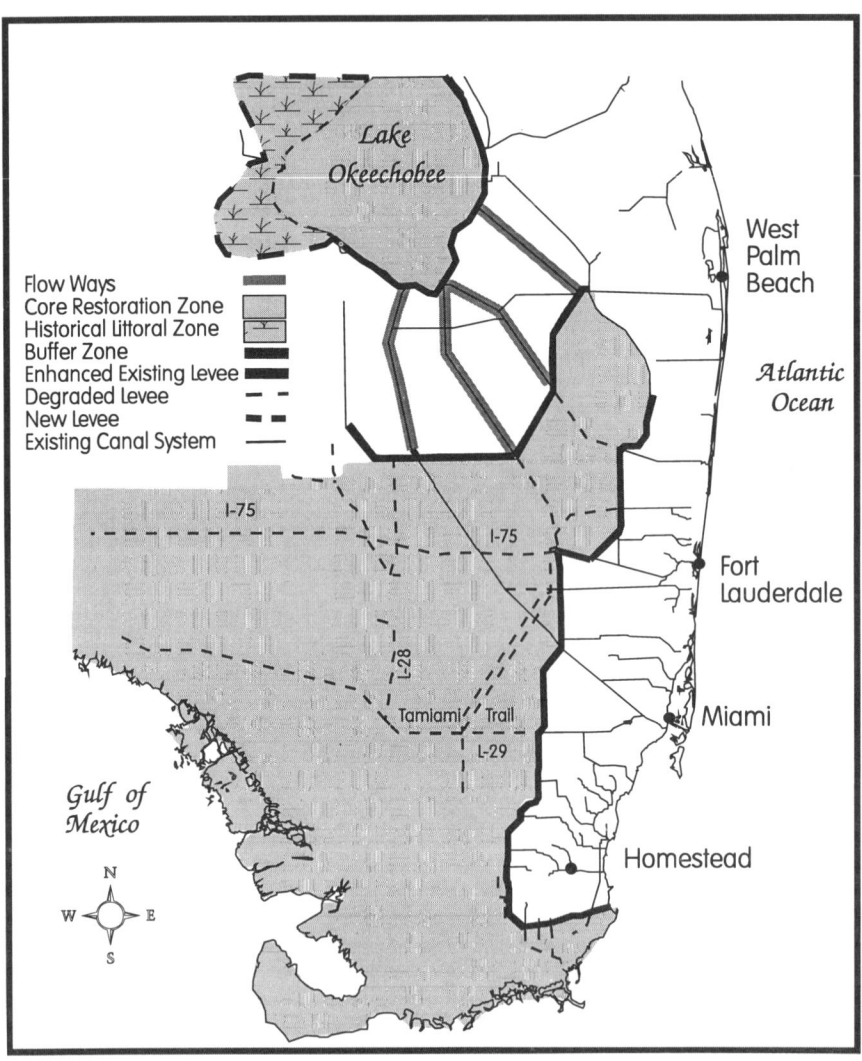

3. restoration of sheet flow to the historical central flowway of the southern Everglades and to short-hydroperiod marshes to the east and west of this central flowway, and
4. restoration of more persistent, dry season flows throughout downstream reaches of the southern Everglades.

Restoration of uninterrupted sheet flow from upstream to downstream reaches as proposed above would restore the small spatial scale connectivity of contiguous marsh areas. This connectivity is required especially during and following drydown events to maintain aquatic productivity through the unimpeded dispersal and recolonization of affected marsh areas by aquatic organisms. However, restoration of such uninterrupted sheet flow by the removal of interior canals and levees will cause overdrainage of current marsh areas in the Everglades during dry season months unless additional storage capacity is attained (South Florida Water Management District 1993). The only reasonable place for obtaining this additional storage is the lake where such storage historically occurred.

5. Conclusions

We have addressed herein the problem of severe declines in wading bird breeding success at traditional colony sites in the southern Everglades, which we argue is the sign of the loss of ecosystem integrity. We argue that a prerequisite for both functional and structural integrity is the existence of a functional landscape mosaic. The existence of such a mosaic ensures that individual species can successfully carry out their life cycles despite natural year-to-year fluctuations in environmental conditions. The system will thus retain its structural integrity of species diversity and "retain the ability to continue its ongoing (autonomous) change and development" (Westra, this volume). We used this concept of a functional landscape mosaic to propose specific restoration measures for the Everglades.

6. Acknowledgments

This paper was accomplished with assistance from the National Park Service, U.S. Department of the Interior. The statements, findings, conclusions, and recommendations in this report are solely those of the authors and do not necessarily reflect the views of the U.S. Department of the Interior, National Park Service. Thanks are extended to Robert Johnson, Robert Fennema, and David Buker for reviewing this paper. Further appreciation is extended to Trupti Bhatt and Fred James for preparing the graphics.

7. References

Cain, M.L. 1985. Random Search by Herbivorous Insects: A Simulation Model. *Ecology* 66: 876-888.
Colinvaux, P.A. 1978. *Why Big Fierce Animals Are Rare: An Ecologist's Perspective.* Princeton University Press, Princeton, New Jersey.
DeAngelis, D.L., and J.C. Waterhouse. 1987. Equilibrium and Nonequilibrium Concepts in Ecological Models. *Ecological Monographs* 57: 1-21.
den Boer, P.J. 1981. Spreading the Risk and Stabilization of Animal Numbers. *Acta Biotheoretica* (Leiden) 18: 165-194.
Diamond, J.M. 1975. The Island Dilemma: Lessons of Modern Biogeographic Studies for the Design of Natural Reserves. *Biological Conservation* 7: 129-146.
Diamond, J.M., and R.M. May. 1976. Island Biogeography and the Design of Natural Reserves. In *Theoretical Ecology: Principles and Applications*, R.M. May, ed. W.B. Saunders, Philadelphia, pp. 163-186.

Fahrig, L., and J. Paloheimo. 1988. Effect of Spatial Arrangement of Habitat Patches on Local Population Size. *Ecology* 69: 468-475.
Fahrig, L., and G. Merriam. 1985. Habitat Patch Connectivity and Population Survival. *Ecology* 66: 1762-1768.
Fennema, R.J., C.J. Neidrauer, R.A. Johnson, T.K. MacVicar, and W.A. Perkins. 1994. A Computer Model to Simulate Natural Everglades Hydrology. In *Everglades: The Ecosystem and Its Restoration*, S. Davis and J. Ogden, eds. St. Lucie Press, Delray Beach, FL, pp. 249-289.
Fleming, D.M., J. Schortemeyer, W. Hoffman, and D.L. DeAngelis. In press a. Colonial Wading Bird Distribution and Abundance in the Pre- and Post-Drainage Landscapes of the Everglades. *Oecologia*.
Fleming, D.M., J. Schortemeyer, W. Hoffman, and D.L. DeAngelis. In press b. Colonial Wading Bird Nesting in the Pre- and Post-drainage Landscapes of the Everglades. *Oecologia*.
Fleming, D.M., W.F. Wolff, and D.L. DeAngelis. 1994a. The Importance of Landscape Heterogeneity to Wood Storks in the Everglades. *Environmental Management* 18: 743-757.
Fleming, D.M., D.L. DeAngelis, L.J. Gross, R.E. Ulanowicz, W.F. Wolff, W.F. Loftus, and M. Huston. 1994b. ATLSS: Across Trophic-Level System Simulation for the Freshwater Wetlands of the Everglades and Big Cypress Swamp. Draft National Biological Survey Technical Report. 76 pp.
Harris, L.D. 1984. *The Fragmented Forest: Island Biogeography Theory and the Preservation of Biotic Diversity*. University of Chicago Press, Chicago.
Hilborn, R. 1975. The Effect of Spatial Heterogeneity on the Persistence of Predator-Prey Interactions. *Theoretical Population Biology* 8: 346-355.
Holling, C.S. 1973. Resilience and Stability of Ecological Systems. *Annual Reviews of Systematics and Ecology* 4: 1-22.
Johnson, R.A. and T.J. Van Lent. 1992. Restoring Flows to the Shark Slough Basin, Everglades National Park. In *Effects of Human-Induced Changes on Hydrologic Systems*. American Water Resources Association, pp. 465-474.
Kareiva, P. 1983. Local Movement in Herbivorous Insects: Applying a Passive Diffusion Model to Mark-Recapture Field Experiments. *Oecologia* 57: 322-327.
Leach, S.D., H. Klein, and E.R. Hampton. 1972. Hydrologic Effects of Water Control and Management of Southeast Florida, Report of Investigation No. 60, Bureau of Geology, Florida Department of Natural Resources, Tallahassee.
Light, S.S. and J.W. Dineen. 1994. Water Control in the Everglades: A Historical Perspective. In *Everglades: The Ecosystem and Its Restoration*, S. Davis and J. Ogden, eds. Saint Lucie Press, Delaray Beach, FL, pp. 47-84.
Loftus, W.F., R.A. Johnson, and G.H. Anderson. 1992. Ecological Impacts of the Reduction of Groundwater Levels in Short-Hydroperiod Marshes of the Everglades. In *Proceedings of the First International Conference of Ground Water Ecology*, J.A. Stanford and J.J. Simons, eds. American Water Resources Association, Bethesda, MD, pp. 199-208
Loftus, W.F. and A.S. Eklund. 1994. Long-Term Dynamics of an Everglades Small-Fish Assemblage. In *Everglades: The Ecosystem and Its Restoration*, S. Davis and J. Ogden eds. Saint Lucie Press, Delaray Beach, FL, pp. 461-483.
Noss, R.F. 1983. A Regional Landscape Approach to Maintain Diversity. *BioScience* 33: 700-706.
Noss, R.F. 1987. Protecting Natural Areas in Fragmented Landscapes. Natural Area Journal 7: 2-13.
Noss, R.F. and L.D. Harris. 1986. Nodes, Networks, and MUMs: Preserving Diversity at All Scales. *Environmental Management* 10: 229-309.

Pearson, S.M., M.G. Turner, R.H. Gardner, and R.V. O'Neill. In press. In *Biodiversity in Managed Landscapes: Theory and Practice*, R.C. Szaro, ed. Oxford University Press.

Pickett, S.T.A., and J.N. Thompson. 1978. Patch Dynamics and the Design of Nature Reserves. *Biological Conservation* 13: 27-37.

Risch, S. 1981. Insect Herbivore Abundance in Tropical Monocultures and Polycultures: An Experimental Test of Two Hypotheses. *Ecology* 62: 1325-1340.

Rowe, J.S., and J.W. Sheard, 1981. Ecological Land Classification: A Survey Approach. *Environmental Management* 5: 451-464.

Soulé, M.E. 1987. *Viable Populations for Conservation*. Cambridge University Press.

South Florida Water Management District. 1993. Draft Working Document in Support of the Lower East Coast Regional Water Supply Plan. South Florida Water Management District, West Palm Beach, FL.

Stanton, M.L. 1982. Searching in a Patchy Environment: Foodplant Selection by *Colias p. eriphyle* Butterflies. *Ecology* 63: 839-853.

U.S.F.W.S. 1980. Habitat Evaluation Procedures (HEP). Manual Release 2-80.

Van Lent, T.A., R.A. Johnson, and R.J. Fennema. 1993. Water Management in Taylor Slough and Effects on Florida Bay. South Florida Research Center, Everglades National Park, Technical Report 93-3.

Walters, C.J., L.H. Gunderson, and C.S. Holling. 1992. Experimental Policies for Water Management in the Everglades. *Ecological Applications* 2(2): 189-202.

Westra, L. 1994. *An Environmental Proposal for Ethics: The Principle of Integrity*. Rowman & Littlefield Publishers, Inc., Lanham, MD.

Wilson, E.O. and E.O. Willis. 1975. Applied Biogeography. In *Ecology and Evolution of Communities*, M.L. Cody and J.M. Diamond, eds. Belknap Press of Harvard University, Cambridge, MA, pp. 522-534..

Wolff, W.F. 1994. An Individual Oriented Model of a Wading Bird Nesting Colony. *Ecological Modelling* 72: 75-114.

Chapter 14
INTEGRITY, SUSTAINABILITY, BIODIVERSITY AND FORESTRY

Peter Miller[1]

1. Introduction: Is Integrity Enough?

According to Leopold (1949), "A thing is right when it tends to preserve the integrity, stability, and beauty of the biotic community. It is wrong when it tends otherwise." Leopold's list of desirable ecosystem attributes has undergone various additions and subtractions. Rodman (1983) added diversity, complexity, harmony, and scarcity, to this multifaceted norm, but removed the notion of beauty—perhaps because he thought that it existed more in the eye of the beholder than in nature itself. Contemporary ecologists raise questions about whether stability is a necessary feature of natural ecosystems, given their dynamical character. For example the fire ecology and disease patterns of most boreal forests in Canada have the result that virtually no forest stands reach 200 years of age without a natural catastrophic destruction. On the other hand, biological diversity and the sustainability of humanly exploited ecosystems are norms that are more recently in focus. Suppose, now, we omit mention of all of the qualities of the biotic community except integrity; will we then have a sufficient principle to guide our conduct affecting ecosystems or nature at large? Can we dispense with the other qualities proposed by Leopold, Rodman, sustainable development theory and others as redundant additions to the norm of integrity?

I suspect that we cannot dispense with these other principles and will attempt to show why. Nonetheless, one has to admire the compelling project of Westra (1994) and her numerous collaborators to show that, from a variety of angles, we need something like a principle of integrity for ecosystems. Her work is directed to theory, policy and application and invites a response in kind, an essay that spans theory and its implementation. In this paper, after exploring several theoretical considerations related to the concept of integrity, I shall turn in application to forest policy and practice in Canada.

I wish first to situate Westra's integrity proposal with respect to a dominant tradition in modern philosophy which I call "Descartes' legacy," but which can as well be called "Locke's legacy" in its political and economic dimensions. This context is important, I believe, because it signifies the cultural forces that we must work with and against as we mend our ways environmentally. Then I shall locate integrity's niche within several types of contemporary environmental ethical theory as a variety of biocentric holism.

Integrity also, as Westra notes, has a prominent place in environmental legislation and policy, which puts it in the same arena as the related norms of sustainable development and biodiversity. Brief examinations of the complementary nature of these policy norms and of

[1]University of Winnipeg, 515 Portage Ave., Winnipeg, Manitoba R3B 2E9; Canada

the concept of integrity itself converge on a common conclusion in answer to our opening question: Is integrity a sufficient fundamental environmental norm for public policy? No it is not, I argue; we need to supplement it with, at least, the norms of biodiversity in living systems and sustainability in human practices. With these theoretical conclusions in hand, I turn next to consider recent developments in Canadian forest policy and practice to illustrate the prescriptive force that these theoretical conclusions can have.

2. Situating Integrity: A Counterweight to Descartes' Legacy

One broad characterization of the modern era of Western philosophy is that it plays out a set of ideas crystallized and set in motion by Rene Descartes and successors like John Locke, which have been highly influential in shaping our understanding of epistemology, metaphysics, axiology, ethics, and socio-political theory. I call this set Descartes' legacy (Miller 1989, 1991a). Some of the elements of that legacy are: (1) a solipsistic immediate self-certainty coupled with a broad skepticism about everything else, including other persons, with which we might be experientially acquainted; (2) a faith in the power of an individual's abstract reason employing general principles, formal concepts, and rigorous quantitative methods as the most reliable means for arriving at truths about the world in the face of skepticism; (3) the homogenization of nature as mere matter; (4) the reduction of humanity to a set of mutually independent conscious minds; (5) the founding of all values in individual egos and their minds, pleasures and wants; (6) the founding of social existence on the coercive restriction of self-interested egos (note that the coercion may be internalized as reason or a social alter ego constraining and prescribing to the appetitive self); (7) the instrumentalization of nature's value while denying it intrinsic value; and (8) the reduction of nature's instrumental value to that of a raw material requiring human cultivation and manufacture to become useful and hence valuable.

There are, of course, many benefits that are associated with Descartes' legacy, such as individual liberty; freedom of expression; the exploration of alternate life-styles; democracy and civil rights; scientific and technical advancement; the empowerment of individuals to exercise their own capacity for critical reflection, analysis, and judgment; and so forth. Moreover, much can be done in environmental ethics without radically altering the traditional assumptions of Descartes' legacy. As we discover the numerous causal linkages in nature affecting human well- and ill-being, new imperatives emerge. "Thou shalt not kill" implies "Thou shalt not pollute food, land, air, and water with toxins."

But whatever the merits and benefits of our modern civilization, which is, in numerous respects, a product of Descartes' legacy, there is growing evidence that this outlook is also, in many ways, theoretically distorting and practically perverse. Pragmatically, it has been a recipe for the destruction of nature, human community, and psychophysical health; biologically, it fails to reflect evolutionary, ecological, and comparative biological reality; axiologically, it is insensitive to the beauty, sublimity and productive creativity of nature, to our ability to value nature for what it is intrinsically, and to our own worth as interdependent parts of a natural whole.

My favorite example of the devaluation of nature in the modern tradition is Locke's (1947) observation:

> Land that is left wholly to Nature, that hath no improvement of Pasturage, Tillage, or Planting, is called, as indeed it is, waste, and we shall find the benefit of it amount to little more than nothing.

Locke further calculated that improved land derives 99 to 99.9 percent of its value from the labour of cultivation rather than from the land resource itself (Hargrove 1983). Whether one's measure is economic rent, absolute utility, productive processes, or appreciation of inherent richness and integrity, Locke's reckoning of the value of land is preposterous and value-blind, leading to disastrous consequences in our mistreatment of the land that sustains us. Even the most noble and sincere moral good-will, accompanied by value-blindness, can be terribly destructive.

The fact that people are willing to pay for timber and mineral rights and undeveloped land indicates that even in an economic sense the land has value prior to the mixing of labour with it. But economic rent, since it is a function of scarcity, is a poor indicator of the absolute utility of the materials nature provides for us. The resource we draw on continuously with least labour is also the most vital to our survival, the free air we breathe. Looking at the land resource itself, we can now recognize that it embodies a multitude of geophysical, chemical, and biological processes and a very long evolutionary history, all marvellous in their own right, to produce the conditions that enable our food to grow. These conditions are vulnerable to mistreatment and ruin.

Consequently, contemporary environmentalism has frequently joined other self-consciously "post-modern" currents of thought in seeking to go beyond the task of assimilating new causal information into the old Cartesian scheme. One or more of the assumptions of the legacy have been brought under scrutiny and criticism and replaced with an alternative.

For example, the strong sense of individual isolation and separateness from others in the Cartesian model has been tempered by observations of the abilities of affect and culture to bond us with others. The powerful "affiliative need" may lead to conformism and a desire for solidarity with others (often with negative consequences, unfortunately, as "groupthink" suspends individual critical judgment, promotes us/them polarization, and joins persons in destructive causes (Janis 1982). More positively, the altruistic sentiments of compassion, sympathy, benevolence, and love draw our concerns beyond ourselves to others. The message of Wilson and others, and of many hunter/gatherer cultures is that such altruistic and social sentiments can take the form of a "biohumanism" or "biophilia" that links us to a wider community of nature (Wilson 1984, Kellert and Wilson 1993).

Likewise the abstract autonomy of the human mind is tempered by psychoanalytic, phenomenological and anthropological reflections that our conscious and rational experience has physical roots as the experience of a "lived body." Our reason also has collective roots as a participant in a shared language and culture. We are thus both inheritors and transformers of a cultural dialogue in which the natural world of which we are a part already has various kinds of significance interwoven with our personal and cultural identities. Sagoff (1974, 1989) and Norton (1984, 1987) in particular stress the prospects of a "weak anthropocentrism" that underscores the cultural importance of preserving the natural contexts of human life. More recently, McKibben (1990) in *The End of Nature* has commented on the profound sense of loss and regret he finds at the thought that wild nature no longer exists, since, due to humanly induced global changes, there no longer exists a corner of nature untouched and unaltered by our hand. We have made the entire planet into a human artifice.

Finally there are those critiques of the Cartesian legacy, often joined with or supplemental to the above, that reject the simplification and homogenization of nature and the setting of humanity over against nature. Instead they seek to understand human life and value as a part of larger natural world that is itself already rich in value. For example human sexual coupling is an instance of sexual reproductive strategies extending far back in evolutionary history and shared broadly by a tremendous diversity of life forms. Does value arise only as sexual coupling becomes a human phenomenon or does it exist already in the mating of non-human animals? And what of sexual reproduction in plants? An extended value naturalism affirms and develops the thesis that as human existence is naturalized and biologized in our understanding, the values that we find there are best conceived as specifications of a broader realm of values (Rolston 1988). How, then, may we conceive such non-anthropocentric values? To do so, we must break with the Cartesian anthropocentric and mentalistic confinement of values.

This brings us to a number of non-Cartesian (i.e., "pre-modern" and "post-modern") attempts to get beyond atomistic psychological value categories. Natural teleology or conativism (the purposiveness of living things and their organic parts) is one of these; self-actualization (the realization of mature and more complete states of being) is another (Aristotle, Taylor 1986). Both of these pertain primarily to individual organisms. Values that apply also at the level of biological communities include ecosystem richness (Miller 1982), health (Miller 1981, Costanza et al. 1992) and integrity (Westra 1994). What is characteristic of all of these generic conceptualizations of natural value is that they enable value identifications and value distinctions in ways compatible with widespread usage that do not limit the value phenomena described to states of consciousness. An acorn pursues its telos in putting down roots and pushing up stem and leaves, and the air, sun, soil, and water are instrumentally good for it, enabling it eventually to actualize its mature state and flourish as a mighty oak tree. But no one supposes that it does any of this consciously. A flourishing natural ecosystem has integrity and richness in that it has unity, complexity, diversity, and a creative potential even when no being is conscious of the system.

Thus we can understand the integrity of ecosystems, along with organic teleology, self-actualization, richness, flourishing and health, as belonging to a cluster of value concepts that run counter to the Cartesian devaluation of nature and psychologically rooted axiology. Westra's elaboration and defence of the principle of integrity is a variety of biocentric holism, one of several significant types of position in environmental ethics today. To these we now turn.

3. Situating Integrity: Biocentric Holism in Ethics

Having located the principle of integrity amidst other alternatives to the subjectivist axiologies inspired by Descartes, I shall now try to map its place within current ethical theory. The most inclusive ethical principle that I can think of is, as St. Thomas noted, to do good and avoid evil (Aquinas). Ethics, as I understand it, has the job of reflecting on human practice— on the many activities we engage in individually and collectively. It does so in order to identify some of the most egregious harms we are prone to commit and the most basic goods we should strive to protect or realize, so that we may better fulfill the broad mandate to do good and avoid evil. Of course no one can avoid inflicting some harms in life and our capacities for doing good are often quite limited, so many of the interesting questions of ethics are concerned with the selection and priorization of what is good or bad, i.e., with decision, choice and trade-offs. Nonetheless, undergirding decision and choice are the goods and evils, the positive and negative values, that give meaning to choice and action.

Environmental ethics, in turn, specializes the broad mandate of ethics by trying to identify salient features of our natural environment that bear on doing good and avoiding evil. Such features may be identified and evaluated within a variety of frameworks, of which I shall single out four: an informed humanism, an enriched humanism, an organismic biocentrism, and an holistic biocentrism. I call an ethical view humanistic if the goods and evils that it prescribes for us to do or avoid happen only to human beings. In contrast, a biocentric view counts goods and evils occurring to other forms of life than the human as ethically important too. Organismic biocentrism focuses on the welfare of individual organisms; whereas holistic biocentrism attends to values rooted in ecosystems and the biosphere. It is noteworthy that the recent *Canada Forest Accord* takes a biocentric stance when it declares that "our goal is to maintain and enhance the long-term health of our forest ecosystems, for the benefit of all living things both nationally and globally..." (CCFM 1992).

The further contrast between informed and enriched humanism (in relation to our natural environment) I draw as follows. Informed humanism makes no new departures in identifying what is intrinsically good and evil in human affairs. Generally speaking, life, health, the development of our faculties, fulfilling human relationships, happiness and the like are good; death, sickness, incapacitation, social isolation and destructive relationships, suffering and despair are bad. Doing good means we should try to promote these human goods and to avoid, mitigate or alleviate the evils. Within the framework of informed humanism, as we become more knowledgeable about the causal processes in nature, we learn, for example, of new pathways by which human activities cause harm to other humans. Lead in our gasoline incapacitates by causing neurological damage; carcinogens kill. Thus on traditional humanistic ethical grounds (as we noted earlier), taking into account new scientific information, the activities which generate these environmentally mediated harms are wrong, other things being equal.

Enriched humanism goes beyond informed humanism in recognizing new ranges of value, of goods that might be enjoyed or destroyed by humans. For example, the *Canada Forest Accord* includes the recognition that "our forest heritage is part of our past, our present and our future identity as a nation," and that "the spiritual qualities and the inherent beauty of our forests are essential to our physical and our mental well-being" (CCFM 1992). These summary claims can be elaborated in a multitude of ways as the religious, cultural, emotional, aesthetic, recreational, educational, and scientific dimensions of our interactions with the natural world are explored. We need to take account of a vast range of "psycho-spiritual utilities" that nature provides us in addition to her "material-economic utilities" (Callicott 1985). Let us call a humanism that is both informed and enriched enlightened humanism.

We should distinguish as well a class of goods that are equally recognizable within several frameworks. Primary examples of these are the environmental services and the productive and generative capabilities of ecosystems. Ecosystems sustain the biophysical conditions of life by moderating climate, cycling water, maintaining a balance of atmospheric gases, providing food and habitat to all species, decomposing wastes and recycling nutrients, and enabling the evolutionary development of species. A holistic biocentrism might take these functions to be intrinsically valuable qualities of ecosystems. Organismic biocentrism and informed humanism would value them instrumentally for their contributions to the existence and well-being of organisms generally or humans in particular.

Broad environmental services like these are an important instance of value convergence in environmental ethics. When values identified from the standpoints of different ethical frameworks converge, there is a possibility of developing a common public policy despite disagreements about the ethical foundations for the policy. Such ethical redundancy is not superfluous in practical contexts of consensus-building but an important asset for defining a broadly acceptable "common good." Ethical casuistry in developing public policy, as, for example, in the areas of agriculture, forestry, fisheries and aquatic resources, should be prepared to appeal to a plurality of value frameworks.

The moral promiscuity of appealing to multiple value frameworks is one of the most significant differences of practical or applied ethics from theoretical ethics. It recognizes that decisions will be made and actions taken before or apart from the achievement of a complete consensus in principle or in conscience among the affected parties, if that were ever possible. Thus both for strategic reasons and out of respect for persons as moral agents operating from personal convictions, one must be flexible in identifying the appropriate justificatory patterns that will enable different parties to concur on the rightness of a course of action. This, of course, is not to say that there will not be points where the consensus becomes unhinged as different moral convictions and priorities are seen to have different practical implications. For that *practical* reason, we must also attend to ethical theory and ethical education in order to try to critique, improve and legitimate ethical frameworks and enhance the quality of ethical reflection, research and practice. Laura Westra's integrity project nicely expresses this practical imperative to become theoretically clearer about the increasingly invoked norm of integrity.

Up to this point I have situated Westra's principle of integrity amidst a cluster of norms characterizing natural values that are supposed to be independent of human states of mind. These values thus represent a part of a more general critique and rejection of "modernity" or what I have called "Descartes' Legacy." In particular, amongst contemporary versions of environmental ethics, the principle of integrity is a form of biocentric holism. But enlightened humanistic justifications of the primacy of integrity can also be invoked, as Westra does, since integral ecosystems serve as our life-support systems, and thus as a precondition of our lives and whatever is valuable in and for them (Westra 1994, Ch. 2).

An assumption of the foregoing, and of Westra's integrity project, is that the concept of integrity is both normatively and empirically meaningful and that one can develop criteria by which different states of affairs can be distinguished as having more or less integrity and thus be evaluated as better or worse, other things being equal. The most significant challenge to the integrity project, I believe, is to refine and operationalize the definition of the concept of integrity. Later, I wish to explore some of the difficulties that need to be addressed. But first let us consider integrity in relation to other environmental norms in public policy.

4. Situating Integrity in Public Policy

According to Westra, it was the entrenchment of the principle of integrity in international legislation and public policy, namely the Great Lakes Water Quality Agreement (GLWQA) of 1978, which provided the starting point for the integrity project (Westra 1994, pp. xv, 24). As Westra and Regier point out, the U.S./Canada GLWQA contains as a regulatory goal that "The Purpose of the Parties is to restore and maintain the chemical, physical and biological integrity of the waters of the Great Lakes Basin Ecosystem..." (Westra and Regier 1992, p. 4).

Even more widely accepted than the norm of ecological integrity is the philosophy of "sustainable development" advocated by the Brundtland Commission report *Our Common Future* (WCED 1987) and widely promoted by the United Nations, especially at UNCED, the United Nations Conference on Environment and Development held at Rio de Janeiro in 1992. Canada in particular, in rhetoric, policy and consultative structures, such as a network of federal and provincial "round tables" across the country, has made sustainable development a cornerstone of environmental and economic policies.

Many biologically informed individuals and environmentalists are quite skeptical of the sustainable development philosophy and consider the term an oxymoron that fails to recognize the biological limits to growth. There is good reason for skepticism and vigilance, I think, but in Canada and other nations which have espoused the philosophy, the concept's political potency and the consensual agreement of diverse parties to explore the idea together recommend that we give it serious attention as a social rallying point that helps us to define a common social good. My promiscuously pragmatic casuistical tendencies as a practical ethicist lead me in that direction, and I shall indicate shortly where it has taken us in Manitoba in the creation of a body for researching and potentially managing a Model Forest.

As indicated above, the concepts of "integrity" and "sustainable development" both lie in the arena of public policy. They thus are subject to analysis and debate that have political qualities in addition to whatever philosophical and scientific character they might have. Westra and Regier are concerned that the concept of "integrity" shall not be an operationally empty symbol but have real force in guiding the restoration of the Great Lakes ecosystem. And many of us who have been involved in sustainable development debates are concerned that the philosophy not be bent to reduce all environmental values to economic ones (as is the predisposition of some government and business proponents), that sustainability shall take priority over development, and that development shall be sharply distinguished from growth in material throughput for the economy and even from economic growth. Combining the two concepts of "integrity" and "sustainable development," we might urge the following social imperative:

> Whatever human activities take place under the rubric of "development," they ought not to compromise nature's capacities to sustain these and other activities required to meet human needs nor ought they to compromise the integrity of the biosphere or its constituent ecosystems.

With qualifications, I am willing to defend such a social imperative that gives priority to sustainability and the preservation of ecological integrity. It certainly is stronger than the standard of sustainability alone because it makes explicit a biocentric norm of ecosystem integrity, which is important in its own right and as a condition for the long-range sustainability of human activities. But also, I would argue, the joint imperative is stronger than the standard of integrity alone, since it directs attention to a principle for the self-management of human affairs, which is a necessary condition for the preservation of the integrity of ecosystems. Without some consensus on environmentally sensitive human ideals to shape our lives, our communities, our economies, and a global world order, we shall be unable to achieve the "mutual coercion, mutually agreed upon" needed to prevent the human destruction of the integrity of ecosystems (Hardin 1968). Each ideal, integrity and sustainability, requires the other for its realization. Together they are stronger, if properly articulated, than either apart.

However, for both humanistic and biocentric reasons, I do not find even the combined norm of sustainability and integrity to be sufficient or adequate as an informed, enriched, and enlightened environmental imperative. Humanistically, it is troubling because it does not offer an account of human needs and one fears that many of the enriched "psycho-spiritual utilities" derived from nature will be given short shrift. Biocentrically it is troubling because the notion of the "integrity" of ecosystems or the biosphere is sufficiently abstract, formal and systemic that it might be compatible with numerous and significant species losses. The result could be a reduction in the mix of species while maintaining sufficient "integrity" of natural ecosystems to provide the essential environmental services that ensure our survival. If Grizzly Bears vanish from the Rockies or Caribou from Canada's north or Great Blue Whales from the oceans, it is not obvious that the integrity of these ecosystems or of the biosphere is compromised (unless, of course, one explicitly incorporates biodiversity as an aspect of ecosystem integrity). The essential environmental processes that maintain these ecosystems would, in all likelihood, adjust to the change in species mix and continue to sustain what remains of life on earth, including human life.

Shall we then abandon the norm of ecosystem integrity? I think not, for that would undermine the significant accomplishment of establishing an ecosystemic standard in legislation, just as the abandonment of the concept of sustainable development would undermine the significant international consensus achieved by the Brundtland Commission. Rather we should frankly acknowledge the plasticity of these concepts and the political struggles to which they are prone and proceed to define them and build upon them in ways that can strengthen public commitments to the human, natural and ecosystemic values we seek to protect. In particular, as the previous paragraph suggests, we need to incorporate into the mix of broad environmental norms a standard of biodiversity as well (Norton 1987). This is particularly important now that it too has entered the international policy arena through the

Biodiversity Convention signed at Rio (UNEP 1992; BSAT 1994). In keeping with this recommendation, we may revise the above social imperative as follows:

> Whatever human activities take place under the rubric of "development," they ought not to compromise nature's capacities to sustain these and other activities required to meet human needs, compromise the integrity of the biosphere or its constituent ecosystems, or reduce global and regional biodiversity.

I have claimed that a generic concept of integrity applied to ecosystems is insufficient to protect basic environmental values and that it needs supplementation, at a minimum, by environmentally sensitive humanistic norms and by a standard of biodiversity protection. The next section elaborates that claim.

5. The Concept of Integrity

In the abstract, the concept "integrity" connotes a unity of parts that constitute a whole. The integers, for example, represent an enumeration of multiple whole units, whether they be chairs or people. The verbal form, "to integrate," connotes making, strengthening, or adding to a whole by bringing appropriate parts into appropriate union with one another. There is also a value side to the concept, since integrity is viewed as a desirable characteristic in human personalities, in engineered (and natural) structures, in aesthetic creations, and, as has been proposed, in ecosystems, the biosphere, and nature as a whole. What lacks integrity may be dishonest, corrupt, incoherent, ugly, degraded, or ready to fail or disintegrate. Thus "integrity," I think, can be explicated in terms of its several aspects or dimensions of unity, completeness, and value. It is presumably the value dimension of the concept as intrinsic to natural wholes that makes it of interest to environmental ethics. How successful is the concept of integrity in identifying what ought to be of concern to us in environmental decision-making? To answer that we must look at both its descriptive adequacy or applicability to the realm of nature as well as its axiological and normative significance for value analysis and ethics.

Although the path is somewhat murky, the route I propose to follow is to use the unity in integrity as a criterion to identify potential domains of application and then to examine how the other two dimensions, completeness and value, might obtain in the situation. A preliminary hypothesis is that the concept of integrity designates in part the "supervenient" value of wholes. That is, a whole is valuable, i.e., has an abstract and concrete "integrity-value"), in being something that is (relatively) unitary and complete (abstract integrity-value) and in being the particular whole that it is (concrete integrity-value).

A further preliminary hypothesis is that the integrity-value of wholes varies with the degree of their unity and completeness, i.e., other things being equal, wholes become more valuable as they increase in unity and completeness. Since wholeness is resolvable into the two dimensions of unity and completeness, the integrity of a whole may fail through lack or loss of either unity or completeness. A partially assembled automobile may be bolted firmly together with parts in their proper places but lack doors, seats and dashboard instrumentation. The parts that are there are unified in an appropriate way but the whole is incomplete. On the other hand, a disassembled automobile with parts strewn all over the garage floor is complete in the sense of having all its parts, but lacks the appropriate unity. Both of these are inferior in integrity-value, other things being equal, to a properly assembled complete automobile.

The concept of integrity connotes unity, completeness, and value. Normally we use it as a contrastive term identifying a valuable quality that some things and persons have and others lack. But if we speak of the integrity of nature, its domain is regarded as universal, applying to nature at large. This linking of unity and value with the whole of existence echoes a theme of classical and medieval philosophy, which identified "being," "unity," and "good" as three coextensive but distinct transcendental categories that apply in one way or another to everything whatsoever that exists. I think there is a truth to these traditional claims, but the trick is to see the many different ways in which the categories apply. To see this in specific terms, let us look at some of the different forms of unity that can be identified, note (in parentheses) the contrasting disunity, and then ask of each how the unitary entity might be more or less complete, i.e., approaching wholeness.

1. Extensive forms of unity, such as spatio-temporal continuity (vs. discontinuity) and causal connection (vs. causal disconnection). These are basic forms of unity that encompass the whole of nature to make it a universe. If we interpret degrees of completeness in terms of the extent encompassed of the space-time continuum with its causal network, then larger and more enduring items are more complete than smaller and more transitory items. By this criterion, everything within nature and even the "whole" of nature is always incomplete, because temporal and causal process is never complete. By this logic too, the biosphere, amongst living systems, is more complete than its component ecosystems and organisms, simply because it is more extensive than they. By the same logic, however, the larger intergalactic universe is more complete yet, and the living portion may fade into relative insignificance according to the criterion of extensiveness.

2. Qualitative unity, i.e., the sharing of some common characteristic or feature such as being red, being a cube, being a horse, or being alive (vs. lack or loss or impurity of the quality). One might speak of completeness in terms of the set of all items in the universe that possess the quality in question. Subsets of this most inclusive set would be incomplete in varying degrees. If it is true that the surface layers of the Earth provide the only locus of living organisms and systems, then once again the biosphere stands out as the largest causally connected system of life that contains all the instances of life. Again, the aggregate whole is greater (thus, per hypothesis, more valuable) than any constituent parts.

3. Qualitative continuity, i.e., a transformation along a continuous qualitative dimension such as the colors of a rainbow (vs. a discontinuous succession). For the realm of living things, qualitative continuity is particularly significant along the temporal axis as:

4. Developmental continuity, i.e., continuous transformations of an organism or system over time (vs thwarted development or death). A particular species of organism has a typical maximum life-cycle of transformational stages that it passes through, even if most members of the species have their life-cycles interrupted or thwarted before completion. Particular ecosystems likewise trace a developmental course of seral stages, which may be reset to early stages from natural or humanly induced catastrophes prior to completion of the more mature stages. However the system of life as a whole appears to have a more linear history of evolutionary diversification and speciation. Human cultural history too may be less cyclical and more linear (although not necessarily progressive) in character. Other things being equal, continuing further along the course of life is more complete than early terminations.

5. Contained unity, i.e., enclosure and containment by some type of boundary that separates what is inside from what is outside (vs. a breach or dissolution or penetration of the boundary). Boundaries may be just interfaces between two media, like water and air, or integuments like skin or cell membranes, and they may have various degrees and kinds of thickness, texture, and definiteness. Many human artifacts, such as fences, walls, ship hulls, and the many varieties of receptacles or vessels, are also designed to contain. Such containment is structurally typical of

organisms but not of ecosystems, which may bleed into one another along physical gradients.
6. Harmony, i.e., mutual complementation of parts to produce a holistic gestalt (vs. disharmony, dissonance, incoherence). Such mutual complementation of parts might be grounds for attributing, with Aristotle and Leopold, an objective beauty to organisms and ecosystems. This is not to say that there are not many disharmonies and underlying chaos in ecosystems as well.
7. Structural integrity, i.e., maintaining the connectedness and appropriate functioning of the parts so as to be capable of withstanding various forms of stress (vs. weakening, breaking up or losing a structural part). Structural integrity appears to be a quite significant evaluative dimension of integrity, indicating a kind of strength permitting entities to preserve themselves under stress. It is an important value for engineered products, such as bridges, buildings, and aircraft. A "strong constitution" maintains or reconstitutes health in the face of disease and injury. Moral integrity makes one less susceptible to wrong actions in the face of temptation. And the resilience of ecosystems enables recovery from destructive or stressful events.
8. Functional/dynamic coordination, i.e., an interactive system with feedback mechanisms which preserve certain features or control a directional transformation - dynamic equilibrium (vs. functional failure and uncoordinated processes).
   a. mechanical coordination, i.e., part of a larger whole which reliably produces certain outputs when it receives certain inputs, as when pressure on the brake pedal slows the car and increased concentrations of carbon dioxide in the lungs stimulate a breathing reflex (vs. mechanical breakdown).
   b. communicative coordination, i.e., the successful transmission and interpretation of signals (vs. communicative failure).
   c. organic coordination, i.e., possessing properties of organisms such as self-development, self-governance, self-repair, reproduction, etc. (vs. disease, disorganization, organic failure or death).
   d. ecosystemic coordination, i.e., characteristic communities of organisms and interactive processes including autopoietic capabilities for self-repair, development and evolution (vs. degradation).
9. Human integrity
   a. Personal integration, i.e., consistency between words, beliefs, feelings, and deeds and coherent motivation (vs. inconsistency, incoherence of personal traits and self-defeating behavior).
   b. Integrity of character, i.e., personal integration in morally significant ways, such as honesty (correspondence between beliefs and words) and fidelity in feeling and action towards moral principles, the deliverances of conscience, and moral commitments and relationships.
   c. Biohumanistic integrity harmonizing one's life, sensibilities, and commitments to one's bioregion and the biosphere.

The above list is intended merely to suggest a range of possible applications of the concept of integrity in ways that are both empirically and normatively meaningful. Note, however, the great variety of features of the world to which the concept can be applied. The abstract notion of integrity is quite flexible and polymorphous, but not, thereby, meaningless. It is, however, in need of further specification as to the type of integrity under consideration and the relevant features that are subject to integration.

Although integrity does appear to be a value that belongs to entities in virtue of their unity and completeness as some kind of whole (akin to the convergence of the transcendental categories of being, unity, and goodness in classical and mediaeval philosophy), integrity by itself is insufficient to establish a norm to guide our behavior or measure its impacts on the environment. At first blush it would seem to provide clear direction. An injunction to respect the integrity of nature would tell us to respect what has integrity-value, i.e., to protect, preserve and restore rather than destroy natural wholes. But there are problems with this norm as it stands, which lead me to conclude that integrity by itself is not enough. There are several reasons for this conclusion.

1. Some forms of integrity are non-contingent, e.g. the universal space/time continuum with various sorts of causal interconnection. But morality is concerned with contingent goods and evils, which our actions can produce or affect for better or for worse. If integrity in the form of the universal space/time continuum or causal network is inevitable, we do not need to worry about preserving or destroying it and thus it is not morally significant. What, then, is morally significant?
2. The fact that there are many forms which integrity can take means that something beyond generic integrity is needed to differentiate and identify the forms which should be the focus of concern. In ethical contexts, these will be valuable contingent instances or contingent degrees of integrity.
3. Because the contingent integrities of nature are multitudinous, not singular, there are often conflicts between them. In order to utilize other living things for food, we must, for example, destroy their organismic integrity in order to sustain our own. Or, at the level of ecosystems, one type, with its peculiar mix of species and other biophysical properties, can displace another from natural or anthropogenic causes. For example, either beaver or humans can dam a stream and flood low-lying forest land. Once stabilized, the pond ecosystem may be equivalent in integrity to the previous stream ecosystem, although the pond will favor algal species and carp, while the stream favors trout. Thus one cannot simply be on the side of integrity but must be able to discriminate and comparatively evaluate different forms and instances of integrity.

In sum, although integrity does have prima facie normative force in identifying a value in natural wholes (and artificial wholes as well) which is destroyed when they are destroyed, it does not capture all of the values of those entities. Nor does it provide a broad enough basis for a comparative evaluation when choices must be made. In short, integrity is not enough. Thus the challenge for the Westra integrity project is to pursue a combination of strategies that do the following:

1. Clarify and distinguish one or more normative concepts of integrity applicable to ecosystems.
2. Operationalize the above conceptions of integrity so as to enable comparative judgments of degrees of integrity in various contexts. Otherwise "integrity" cannot serve as a guide to policy and practice.
3. Identify other evaluative and normative concepts that pertain to ecosystems with some indication of how they too may be applied.
4. Indicate the relationships between the different norms identified in 1-3 above and how they can be integrated with one another or priorized in application. In particular indicate what extra content or function, if any, is provided by the concept of integrity over and above the other normative concepts identified.
5. Provide justifications for the above steps that can legitimate them for the various stakeholders.

In response to these challenges, one can envisage two principal definitional strategies, the monistic and the pluralistic, as well as several hybrids. A monistic strategy attempts to find some singular essence to the notion of integrity, as applied to ecosystems, from which all the core values of ecosystems can be deduced. That is to say the monistic position finds all primary ecosystemic values to be already implicit in the value of integrity—a unitary whole which, when you unpack it, reveals its many facets. A pluralistic strategy, on the other hand, like Leopold's and Rodman's formulations cited at the beginning of this chapter, may simply list an array of ecosystem values, including, perhaps, integrity, without any claim that all are derivable from one.

A third strategy is a stipulative (and political) hybrid of monism and pluralism. According to this political umbrella strategy, the concept of integrity embodied in legislation and policy is an abbreviated code word which means, roughly, the state of well-being or excellence of ecosystems, which can only be specified by a plurality of other concepts. In other words, whichever intrinsic properties of ecosystems we believe to be worth protecting we shall argue to belong on the list covered by the concept of integrity enshrined in legislation and policy.

Finally, stemming from the need to operationalize the political umbrella approach, a mensurational monistic strategy attempts to develop a formula for measuring and aggregating various ecosystem values to produce a single score for the integrity or health of an ecosystem. The most widely used (for aquatic ecosystems) is Karr's (1986, 1992) Index of Biotic Integrity. Costanza (1992) has produced a similar proposal for ecosystem health.

A key issue for mensurational monism is the rationale for the aggregating formula that is supposed to provide an index for ecological integrity. That rationale will depend upon one's theoretical and empirical understanding of valuable key features of ecosystems, which themselves may be measured by a variety of distinct indices. If these features co-vary (as a monistic view might argue), then any one variable might be an index of the rest and of the integrity of the whole. But if the valuable features vary independently or even inversely from one another (as a more pluralistic conception of ecosystem integrity might claim), then decisions must be made about how the different variables are to be weighted when aggregated and what is the rationale for a particular weighting.

Noss adopts the above umbrella approach in his recent monograph on protected areas prepared for the World Wildlife Fund (Noss 1995). Unlike Westra, Noss interprets integrity to mean a conceptual, rather than a biological integration. "Integrity...is an integrating and holistic concept; it pulls together many related ideas" (p. 20). He regards ecosystem health, biodiversity, a dynamical form of stability or homeorhesis, sustainability, naturalness, wildness, and beauty as values encompassed by integrity. The advantage he claims for this explicit listing of component aspects of integrity is that we have a better idea of how to operationalize them and can identify criteria and indicators for their presence. Unlike Karr and Costanza, he does not produce a mensurational summing of diverse indicators.

Noss's approach to ecological integrity in the context of protected reserves is analogous to the Canadian Criteria and Indicators Initiative for Sustainable Forest Management; both are irreducibly pluralistic in their criteria. The Canadian initiative is linked to an international effort to develop suitable criteria and indicators to monitor the condition and management of the world's forests. Besides economic and social criteria and indicators, they have formulated a set of environmental criteria and indicators under the general headings of (1)biodiversity, (2)forest condition and ecosystem productivity, (3)soil and water conservation, and (4)global ecological cycles (CFS 1994). It remains to be seen whether or not they define their criteria and indicators well, but they have, I believe, adopted the right approach at the present juncture, as I shall argue below.

Westra's approach is more ambivalent than Noss's or the Canadian forest initiative. She too produces an "umbrella" definition of integrity having multiple parts, but she is also strongly attracted to a paradigm of integrity in organisms and presses harder in the monistic direction to integrate the subordinate concepts and derive some aspects of integrity from other aspects. For example, a key part of her definition of ecosystemic integrity is that "the system's optimum capacity for the greatest possible on-going developmental options within its time/location remains undiminished" (Westra 1994). From this key feature of integrity she derives the principle of biodiversity as a foundation of genetic potential and relational information needed to preserve the system's developmental capacities.

I will not pass judgment as to the ultimate satisfactoriness of these different definitional strategies in the long run, since I believe they are all still under development, and I expect other contributors to this volume to further that developmental task. I will only insist that the key concept of natural biodiversity must occupy a central place amongst biocentric norms either as a derivation from integrity, an explicitly listed component of integrity, or as an independent auxiliary concept. Like integrity, biodiversity itself comes in different forms, including genetic diversity within and between species, diversity of ecosystem types, and structural and age diversity within an ecosystem. The preservation of the forms of ecosystem diversity is critical to the preservation of genetic diversity of the multitude of organisms fitted to different niches. Thus the preservation of biodiversity at the ecosystemic level is a very important norm relating, in particular, to forest practices and policies.

Although it may be premature to evaluate the potential for success in the long run of the different approaches to defining integrity, I believe that a relatively pluralistic approach to ecosystem values is the most practical at the present time and a stepping stone to further development. A pluralistic approach might take advantage of the political weight of the concept of integrity through an umbrella strategy for defining integrity, such as Noss pursues, or simply offer separate definitions of a plurality of environmental indicators without reference to integrity, as found in the Canadian Criteria and Indicators Initiative (CFS 1994). In either case, the focus is on the conceptualization and measurement of a set of distinct ecosystemic variables. There are several reasons for favoring this pluralistic approach:

1. As previously argued, the concept of integrity is too abstract and polymorphous to capture all important ecosystem values unless they are expressly and stipulatively identified.
2. A separate identification of important ecosystem values permits (a) a distinct theoretical rationale for each, explaining its importance and justifying its inclusion, and (b) the formulation of specific criteria and measurable indicators for empirically identifying the presence and degree of each value.
3. The accomplishment of 2 (a) and (b) permits further exploration of the connections and relations of the theoretical and empirical variables identified, to gain a better understanding of the workings of ecosystems.
4. If one is seeking to develop an integrative rationale, based on ecosystems theory, for a more monistic concept of ecosystem integrity, then the rationales for the distinguishable properties of ecosystems developed in 2 (a) will provide a test of adequacy for various integrative proposals.
5. If one is pursuing the strategy of mensurational monism, then the theory and measurements from 2 (a) and (b) and 3 provide both measurements to be integrated into a single index and the rationale for constructing the index.
6. Because a unitary index of ecological integrity is derivative from its elements and their rationales, it is important that decision-makers not lose sight of the severally important elements when they are combined into a single index. A single index has the potential to ride rough-shod over some values, as experience with GNP as a measure of human welfare demonstrates.

6. Forestry in Canada: Principle, Policy and Practice

In the preceding sections, I gave a qualified endorsement of the principles of integrity and sustainability as norms for human activities affecting the biosphere, but also urged that they be explicitly supplemented by the multi-layered norm of biodiversity. In this section, I consider these norms in relation to forest policy, with illustrations from experience in Canada.

Canadian forestry, like forestry in many parts of the world, including the U.S., is profoundly controversial and conflict-ridden and has been for much of its recent history. A large proportion of the Canadian economy is based upon natural resource extraction and processing with forest products at the lead in export earnings. Thus there are powerful economic interests (in which, to some degree, most Canadians have a stake) that drive the maintenance and expansion of industrial forestry.

Against the economic benefits from extractive industry is arrayed a tremendous range of environmental costs. Our forest industry has been the greatest polluter of air and water in Canada, ahead of mining, although both sectors are now showing improvements (Sinclair 1988). A second count against the pulp and paper section of the forest industry is that its products are the largest contributors to the industrial world's garbage glut. The responses to that glut in turn contributed to an economic crisis within the Canadian pulp and paper industry, from which it is only now beginning to recover. As waste reduction and recycling programs kicked in, many with legislative support in the form of legally prescribed proportions of recycled content in paper, the demand for Canadian virgin fibre products declined. This decline, in combination with the recent recession at a time of significant overcapacity in the industry, caused a number of pulp and paper mills to close. Thirdly and finally, added to the problems of pollution and garbage is the obvious short-term, and often long-term, devastation to the landscape created by forest operations, an assault on the integrity of forest ecosystems.

Although industrial pollution certainly raises questions about attacks on the "integrity" of aquatic ecosystems, and the garbage glut raises questions about the "sustainability" of wasteful practices in urban centers, I want to restrict attention for now to policies pertaining to forest ecosystems themselves and human utilization of them.

There is, I think, an emerging consensus in Canada of where, in principle, we should be going. How far we can go in practice is another matter. Agreement in principle does not prevent profound disagreements in judgment on how to priorize and realize these principles in policy and in practice, particularly since some of the principles tug in contrary directions. For example, the previously cited *Canada Forest Accord*, which summarizes a more substantial document *Sustainable Forests: A Canadian Commitment*, contains a variety of elements that reflect all of the ethical orientations identified earlier. How the different anthropocentric and biocentric ethical considerations and the economic objectives in this eclectic document are to be reconciled and integrated is not altogether clear. The document calls for the protection of biodiverse ecosystems in harvested areas plus increased preservation of unharvested old growth forests, in line with previously cited enriched humanistic and biocentric objectives. But the same document also has an economic agenda of improving "the quantity, quality and continuity of supply of forest resources," and intensive silvicultural practices are a part of that program (CCFM 1992).

Intensive forest management, as presently practiced, subsumes forestry under the agricultural model, as indicated by the title of a Canadian Pulp and Paper Association booklet *Farming Canada's Forests*. The conversion process from forest to tree farm is described by Maser (1988) in his book *The Redesigned Forest*:

> As we liquidate the ancient forests, we are redesigning the forests of the future. In fact we are redesigning the entire world, and we are simultaneously throwing away nature's blueprint. Nature designed a forest as an experiment in unpredictability; we are trying to design a regulated forest. Nature designed a forest over a landscape; we are trying to design a forest on each hectare. Nature designed a forest with

diversity; we are trying to design a forest with simplistic uniformity. Nature designed a forest of interrelated processes; we are trying to design a forest based on isolated products. Nature designed a forest in which all elements are neutral; we are trying to design a forest in which we perceive some elements to be good and others bad. Nature designed a forest to be a flexible, timeless, continuum of species; we are trying to design a forest to be a rigid time-constrained monoculture. Nature designed a forest of long-term trends; we are trying to design a forest of short-term absolutes. Nature designed a forest to be self-sustaining and self-repairing; we are designing a forest to require increasing external subsidies—fertilizers, herbicides and pesticides.

Note that in Canada, at least in the West, we are, for the most part, still planting and tending our first "crop" of trees. The ones we cut are still in original natural forests, but in ever more remote locations. This has two consequences: (1)the replacement process described by Maser is spreading ever farther afield into the interior of B.C. and the more northerly boreal forests of the so-called "prairie" provinces of Alberta, Saskatchewan and Manitoba with a consequent steady loss of virgin forest wilderness, and (2)a process of "highgrading" at the landscape level is occurring, whereby the stands that are most valuable because of their accessibility, the quality of their fibre, and the volume of wood per hectare are harvested first leaving the commercially less valuable natural stands for the future (Hammond 1991). We are still engaged in a mining operation, but with agricultural aspirations. At least in the coastal areas of B.C., there is no hope or intention of ever again replacing the volumes and quality of wood of the original ancient forests that have been harvested, much less their climax ecosystems. I know of no forest management plan that intends to institute a 1000-year rotation.

Once we have cut as much of the ancient forests of B.C. as we are willing to cut, a massive "falldown" in quality and quantity of harvestable timber resource will occur as a consequence of our biologically unsustainable cutting rates and the resulting soil degradation, loss of biodiversity, macro and micro climate change, poor regeneration, and the like. We shall be left with much impoverished and diminished younger plantations that cannot yield the same high grade products as in the past. For example, according to an analysis by Forintek, second growth Douglas Fir has problems with structural strength, warping and workability. In order to be able to mill high grade lumber from it, the trees require pruning during growth and an extension of rotation age from 60 to 90 years (which would require a reduction in Annual Allowable Cut). Because such cultivation requirements appear to be uneconomic, the industry's future is projected to lie increasingly in the production of chips that can be pressed into boards and structural materials and pulped for paper (Kellog 1989).

At an Ottawa conference on resources, I tried to highlight this tension between the joint commitments to ecological protection and increased extraction with the following question put to commentators from government and industry:

> How do you propose that we increase the quantity and quality of timber and fibre from our forests in the face of the predicted "falldown" of the same resulting from the exhaustion of the old growth "mine" across Canada and the new environmental agenda of increased protection of wild lands and biodiversity? (Miller 1992)

I received only hand-waving in reply. Since then, the British Columbia government has begun a land re-allocation process based on regional consultations that has assigned increased lands for preservation with a consequent reduction in Annual Allowable Cuts. British Columbia mills, in turn, have gone out of province into Alberta and Saskatchewan to augment their wood supply, often by stripping private lands. The Alberta government is reportedly planning measures to curtail the outflow and excessive harvest levels (Nikiforuk 1994).

Because of tensions like this in the eclectic documents that are currently being cobbled together, sustainable development, which attempts to bridge economic and environmental concerns, will probably remain a very political concept in forestry and other sectors. Within this political context, the project of spelling out the practical implications of environmental norms like integrity, biodiversity, and sustainability is extremely important. These norms must also be related to more standard humanistic norms and to the processes required to implement them.

I also believe that, although important policy generalizations can be made, the values and issues vary somewhat, for both human and natural reasons, depending upon where one is on the landscape. Sagoff (1992) has written that the introduction of concerns for ecological "health" or "integrity":

> ...does not entail...that the integrity of ecological systems requires they remain pristine or protected absolutely from human use. Such a criterion—historical authenticity—if used as the measure of ecological health, would doom that concept to irrelevance.

He cites the Lake District in England as a healthy, but humanly altered landscape.

Surely Sagoff is right that we need measures of health and integrity that can apply to humanly altered and humanly exploited landscapes, and that to some degree most landscapes on earth bear marks of human impacts, even when they are not directly visible. But that does not mean that the historical authenticity of natural ecosystems is itself irrelevant to determinations of integrity or an unimportant value in itself. Indeed it becomes increasingly important as one moves west and north on the North American continent, particularly in Canada, because there remnant original ecosystems with lesser industrial and agricultural impacts are still to be found. And in more easterly and disturbed portions of the continent, restoration ecology must draw on relatively undisturbed pockets for the pools of natural biodiversity with which to recolonize larger parts of a more natural landscape.

In addition to the degree and nature of human colonization and disturbance of the landscape, there are more natural ecosystem properties that also vary with the landscape and that should be reflected in policies and practices. One of the most significant is whether or not the landscape is characterized by ecosystems in a climax state or not. For example, the coastal temperate rain forest in British Columbia is a climax ecosystem. Individual trees die, fall and decay to provide a structural base and nutritional source as so-called "nurse logs" for other vegetation, including a new generation of trees. However these climax stands (unlike most of the boreal forests in Canada) are generally not subject to catastrophic destruction through fire or insects to set the whole ecosystem back to an early successional stage. Where periodic wholesale natural destruction occurs, controlled clearcuts might mimic aspects of the disturbance with a greater chance of preserving ecosystem functions, structure and biodiversity over the landscape. Thus one could hypothesize (based on this consideration alone) that the impact on natural integrity of large scale industrial logging could be much more significant for coastal rainforests than for boreal forests. We need a good deal more research, though, to test such hypotheses and to evaluate various logging and silvicultural prescriptions.

In the remainder of this paper, I wish to introduce several other statements of policy which I believe might contribute to the task of reforming forest practices. One of the more inspired programs in Canada's recent Green Plan is the Canadian Forest Service's support of a string of 10 Model Forests across Canada (and a growing number of partners in other lands, including Russia, Mexico, and Malaysia), designed to bring together usually suspicious and often antagonistic parties for the joint research and management of a chunk of Canada's forests. As a representative of a coalition of environmental groups called T.R.E.E. (Time to Respect Earth's Ecosystems), I had the privilege of helping to create Manitoba Model Forest, one of Canada's winning proposals. Although the proposal itself is extensive and detailed, it includes the following statement of a Vision and Principles to which the partners have subscribed (MAMFPP 1992):

> Building upon the special natural and human features of the region and linkages beyond, we intend to create an operationally-viable, ecologically-sustainable model of forest management and use in which a wide spectrum of interests and values are represented. We shall achieve this vision by integrating these values into a harmonious partnership working towards wiser forest management practices based on an improved understanding of Canada's and Manitoba's Boreal forest ecosystems, their benefits and values, and human impacts on them. This vision is based on the following principles and values:
> 1. Respect and care for the community of life of which we are a part and exploration and conservation of the ecological processes and biodiversity of our shared forest and wildlife heritage.
> 2. Respect for the diverse streams of human experience, learning and culture that shall guide a multi-valued appreciation and wise and equitable use of the forest.
> 3. A candid and open problem-solving approach that faces squarely potential negative impacts and conflicts and attempts to resolve them in accordance with the principles of respect for the community of life and the human community.
> 4. Reliance upon human ingenuity working creatively with nature's inventions and productivity to solve problems, pursue new opportunities, and improve environmental and economic benefits and efficiency.

Although we speak of integrating values and interests on the management side, our principles do not use the expression "ecological integrity." However in the first, and most fundamental, principle of respect and care for the community of life, we follow the two World Conservation Strategies (IUCN 1980, 1991) and speak of valuing and conserving ecological processes and biodiversity, which I would take to be a semantic alternative to "integrity" in its umbrella usage. What is, and has to be, basic in such policy statements is to record and reinforce the importance of forests as sustainers of life, including our own, and their importance as objects of love, respect and understanding, as well as utility, to which we are related on many different levels. For better and for worse, our relations to our forests color our relations to one another. These two relations, to our forests and to one another, and the responsibilities they entail define the ethical context for forest policies and practices.

The implications of these commitments are further spelled out by the objectives of the T.R.E.E. coalition which include (1)Sound Forest Policy, (2)Adequate Wilderness Preservation, (3)Comprehensive Environmental Assessments, (4)Forest Research, (5)much broader Forest Inventories, and (6)Public Education. That is, T.R.E.E. advocates the protection and preservation of the health and multiple values of natural forest ecosystems, which shall be achieved through the determination of a sound, sustainable forest policy for Manitoba to provide a proper framework for land and resource allocations, protection, regulation, and management. Two major prongs of this policy should be a *wild* forests policy that protects from exploitation sufficiently large and interconnected wilderness reserves representing the full range of Manitoba's ecosystems as well as special natural features, and a *Harvested Forests* policy that protects as much of the natural features and biodiversity of the exploited forests as possible. A sound forest policy, once formulated, shall provide a basis for comprehensive environmental assessments of all facets of the forest industry, which shall guide forest resource allocations, management and monitoring programs. Environmental assessments as well as forest management, monitoring, and land-use planning must be based on an adequate knowledge base and multifaceted valuing of our forests, and these in turn require forest-related research, improved databases and inventories, and public education (Miller1991b).

Note that the policies for both Wild Forests and Harvested Forests reject the intensively managed agricultural model primarily on the grounds that it is an enemy of biodiversity. Non-commercial as well as commercial species must flourish and significant portions of our forests must be allowed to grow old in order to guarantee the full range of ecological niches and habitat.

Finally I would like to cite a remarkable document, *A Wildlife Policy for Canada*, which was signed in September, 1990 by Canada's ministers responsible for wildlife, many of whom, as natural resource ministers, later signed the Forest Accord. The policy's goal is "to maintain and enhance the health and diversity of Canada's wildlife, for its own sake and for the benefit of present and future generations of Canadians. This requires: (1)maintaining and restoring ecological processes, (2)maintaining and restoring biodiversity, and (3)ensuring that all uses of wildlife are sustainable" (WMCC 1990).

These are not entirely new themes in policy at a generic level, but this policy does contain some new points of departure. Most fundamental is the broadening of the definition of wildlife to include:
>all wild organisms and their habitats—including wild plants, invertebrates, and microorganisms, as well as fishes, amphibians, reptiles, and the birds and mammals traditionally regarded as wildlife.

Thus our forests are not only habitats for wildlife but themselves wildlife and subject to this policy in their own right.

Note also an expanded value base for the policy as stated in the Guiding Principles. I shall single out two: "(1)wildlife has intrinsic, social, cultural, and economic values; and (2)the maintenance of viable natural populations of wildlife always takes precedence over their use by people." The latter principle is a clear statement of priority that should be applied to any threatened population of wildlife and their habitat such as the Woodland Caribou in Manitoba's Nopiming Park. Also relevant to Woodland Caribou and other species that might be threatened are the directives on wildlife populations, which note that:
>Within-species genetic diversity is essential for each species to adapt and survive and is the raw material of plant and animal breeding. It is a matter of priority to prevent species extinctions and to conserve as much within-species diversity as possible.

>Wildlife should be managed to maintain a diversity of species broadly distributed throughout their range at levels of abundance sufficient to enable them to adapt to environmental changes and to meet management objectives. Local and regional population changes may also need to be addressed (WMCC 1990).

Finally I call your attention to the fact that this policy document supports and elaborates the objectives of what is popularly known as the "Endangered Spaces" campaign for preservation of representative natural ecosystems.

> Governments will complete and maintain comprehensive systems of protected areas, through legislation and/or policy, that include representative ecological types and give priority to the protection of endangered or limited habitats. To allow species to change their local and regional distributions in response to climate change and other factors, the protected area systems must be designed to: (1)protect the diversity of Canada's physical environments, (2)contain a range of environments within each protected area, and (3)link protected areas by corridors of suitable habitat (WMCC 1990).

Remember that the wildlife here discussed is natural indigenous flora and fauna including macro- and micro-organisms of every kind.

What emerges as an ideal is a reverse image of the policies of the past, which countenanced a few islands of protection in a sea of development. Instead, were these policies followed, there would be pockets of development, themselves as ecologically sound as possible, surrounded and buffered by a connected network of wild lands, a swiss cheese landscape. That is the picture of sustainable forest practices which would flow from the principles of integrity, sustainability and biodiversity as interpreted by *A Wildlife Policy for Canada*. It resembles very much the image of protected areas and corridors advocated by The Wildlands Project for ecological recovery in the United States, for which Reed Noss is the scientific director (Noss 1993). The difference is that in Canada, such a vision is official policy, signed by representatives of all of Canada's governments, although not enacted in legislation. However, many of us in the environmental movement suspect that the ministers did not know what they were signing and will never enshrine it in legislation. At least they do not act as though they will in their allocations of forests for harvest (with the possible exception of British Columbia after major protests and massive arrests in that province).

The natural swiss cheese vision of the Canadian national wildlife policy, it is fair to say, is countered by another interpretation of the same principles of integrity, sustainability and biodiversity by the forest industry (Booth 1993). Industry is prepared to manage forests for diverse values on an integrated basis by treading more lightly in their forest operations and thus obviating the need for extensive protected areas to maintain integrity and biodiversity. Whether this can be done or not is a matter for research, but that research itself will require more natural and humanly undisturbed areas as a baseline for comparison. Thus on either scenario, there is a case to be made for protected areas and for softened impacts from forest operations.

One final point to note is that both the industrial and the *Wildlife Policy for Canada* scenarios share at least a verbal commitment to the principles of integrity (or ecosystem health), sustainability and biodiversity and both recognize the dynamical character of ecosystems that are subject to catastrophic perturbations and seek to make provision for such disturbances, with their displacement of ecosystems and wildlife, at the landscape level. Integrity, then, as an umbrella or summation of ecosystemic values, must become operational at the level of bioregional landscapes and not just in isolated local ecosystems. Many other factors, both natural and socio-political—stemming from the whole gamut of humanistic and biocentric values, will determine what compromises are made and what forms of implementation will occur in any particular region.

7. References

Aquinas, St. Thomas. *Summa Theologiae* Ia2ae, 94, 2.
Aristotle. *Physics*. Book II.

Booth, D., D.W.K. Boulter, D.J. Neave, A.A. Rotherham, and D.A. Welsh. 1993. Natural Forest Landscape Management: A Strategy for Canada. *The Forestry Chronicle* 69: 141-145.

(BSAT) Biodiversity Science Assessment Team. 1994. *Biodiversity in Canada: A Science Assessment for Environment Canada*. Environment Canada. Ottawa.

Callicott, J.B. 1985. Intrinsic Value, Quantum Theory, and Environmental Ethics. *Environmental Ethics* 7(3): 262.

(CCFM) Canadian Council of Forest Ministers. 1992. *Canada Forest Accord* and *Sustainable Forests: A Canadian Commitment*. Hull, Québec.

(CI) Canadian Criteria and Indicators Initiative for Sustainable Forest Management. 1994. *CI Newsletter* I:3 (August/September). Canadian Forest Service, Hull, Québec.

Costanza, R., B.G. Norton, and B.D. Haskell, eds. 1992. *Ecosystem Health: New Goals for Environmental Management*. Island Press, Washington, D.C.

Costanza, R. 1992. Toward an Operational Definition of Ecosystem Health. In *Ecosystem Health: New Goals for Environmental Management*, R. Costanza, B.G. Norton, and B.D. Haskell, eds. Island Press, Washington, D.C. pp. 239-256.

Hammond, H. 1991. *Seeing the Forest Among the Trees: The Case for Wholistic Forest Use*. Polestar Press, Vancouver, BC. pp. 87-90.

Hardin, G. 1968. The Tragedy of the Commons. *Science* 162: 1243-48.

Hargrove, E. 1983. Anglo-American Land Use Attitudes. In *Ethics and the Environment*, D. Scherer and T. Attig, eds. Prentice-Hall, Englewood Cliffs, N.J. pp. 96-113.

IUCN/UNEP/WWF. 1980. *World Conservation Strategy: Living Resource Conservation for Sustainable Development*. International Union for Conservation of Nature and Natural Resources, United Nations Environment Programme and World Wildlife Fund, Gland, Switzerland.

IUCN/UNEP/WWF. 1991. *Caring for the Earth: A Strategy for Sustainable Living*. International Union for Conservation of Nature and Natural Resources, United Nations Environment Programme and World Wildlife Fund, Gland, Switzerland.

Janis, I.L. 1982. *Groupthink: Psychological Studies of Policy Decisions and Fiascoes*. 2nd edition. Houghton Mifflin, Boston, MA.

Karr, J.R., K.D. Fausch, P.L Angermeier, P.R. Yant, and I.J. Schlosser. 1986. Assessment of Biological Integrity in Running Water: A Method and Its Rationale. Publication 5. Illinois Natural History Survey, Champaign, Illinois.

Karr, J.R. 1992. Ecological Integrity: Protecting Earth's Life Support Systems. In *Ecosystem Health: New Goals for Environmental Management*, R. Costanza, B.G. Norton, and B.D. Haskell, eds. Island Press, Washington, D.C. pp. 223-238.

Kellert, S.R. and E.O. Wilson. 1993. *The Biophilia Hypothesis*. Island Press, Washington, D.C.

Kellog, R.M., ed. 1989. *Second Growth Douglas-fir: Its Management and Conversion for Value*. Special Pub. No. SP-32. The Douglas-Fir Task Force, Forintek Canada Corp., Vancouver, B.C.

Leopold, A. 1949. *A Sand County Almanac*. Oxford University Press, New York, N.Y.

Locke, J. 1947. Second Treatise. In *Two Treatises of Government*, ed. T.I. Cook. Hafner Press, New York and London. sec. 42-43. Cited in Hargrove 1983. p. 109.

(MAMFPP) Manitou Abi Model Forest Project Partnership. 1992. *Manitou Abi Model Forest Proposal*. Pine Falls, Manitoba.

Maser, C. 1988. *The Redesigned Forest*. E. Miles, San Pedro, California.

McKibben, W. 1990. *The End of Nature*. Anchor Books, New York.

Miller, P. 1981. Is Health an Anthropocentric Value? *Nature and System* 3: 193-207.

Miller, P. 1982. Value as Richness: Toward a Value Theory for an Expanded Naturalism in Environmental Ethics. *Environmental Ethics* 4:101-114.

Miller, P. 1989. Descartes' Legacy and Deep Ecology. *Dialogue* XXVII: 183-202.
Miller, P. 1991a. Integrity is Not Enough: Remarks on the Theme 'The Integrity of Creation.' Presented at the Central Division of the American Philosophical Association. April 26.
Miller, P. 1991b. A Brief from T.R.E.E. (Time to Respect Earth's Ecosystems) on the Environmental Assessment of the Abitibi-Price 8 Year Forest Management Plan.
Miller, P. 1992. Towards a Forest Policy for Canada: New Directions from Environmental Ethics. Presented at (IREE) Institute for Research on Environment and Economy workshop on An Environmental Ethics Perspective on Canadian Policy for Sustainable Development. University of Ottawa, Ottawa, Ontario. October 2.
Nikiforuk, A. 1994. The Great Alberta Timber Rush: A Hot Lumber Market Cuts a Swath through Alberta's Privately Owned Forests. *Environment Views.* Winter 1994. Green Thread Publishing, Edmonton, Alberta. pp. 11-14.
Norton, B.G. 1984. Environmental Ethics and Weak Anthropocentrism. *Environmental Ethics* 6: 131-48.
Norton, B.G. 1987. *Why Preserve Natural Variety?* Princeton University Press, Princeton, N.J.
Noss, R.F. 1992. The Wildlands Project: Land Conservation Strategy. *Wild Earth* (Special Issue): 10-25.
Noss, R.F. 1995. *Maintaining Ecological Integrity in Representative Reserve Networks.* World Wildlife Fund, Toronto.
Rodman, J. 1983. Four Forms of Ecological Consciousness Reconsidered: Ecological Sensibility. *Ethics and the Environment*, D. Scherer and T. Attig, eds. Prentice-Hall, Englewood Cliffs, N.J. pp. 88-92.
Rolston, H. 1988. *Environmental Ethics: Values in and Duties to Nature.* Temple University Press, Philadelphia.
Sagoff, M. 1974. On Preserving the Natural Environment. *The Yale Law Journal* 84: 245-267.
Sagoff, M. 1989. *The Economy of the Earth.* Cambridge University Press, Cambridge, England.
Sagoff, M. 1992. Has Nature a Good of Its Own? In *Ecosystem Health: New Goals for Environmental Management,* R. Costanza, B.G. Norton, and B.D. Haskell, eds. Island Press, Washington, D.C. pp. 57-71.
Sinclair, W.F. 1988. *Controlling Pollution from Canadian Pulp and Paper Manufacturers: A Federal Perspective.* Environment Canada, Ottawa.
Taylor, P. 1986. *Respect for Nature.* Princeton University Press, Princeton, N.J.
Westra, L. 1994. *An Environmental Proposal for Ethics: The Principle of Integrity.* Rowman and Littlefield, Lanham, MD.

Chapter 15
THE GLOBAL POPULATION, FOOD, AND THE ENVIRONMENT

David Pimentel[1]

1. Introduction

The world's human population is currently more than 5.6 billion, projected to reach nearly 8.4 billion by the year 2025 and may reach a disastrous 15 billion by 2100 (PCC 1989). Presently a quarter million humans are added each day. Many leading scientists and public organizations are concerned about the rapid growth in population numbers and the deterioration of natural resources and the environment caused by human numbers and activities (CEQ 1980, Keyfitz 1984, Hardin 1986, Demeny 1986, Ehrlich and Ehrlich 1990, Holdren 1992). As populations and their consumerism increase basic resources are depleted, this leads to environmental degradation while freedom of individual choice and quality of life decline (Durning 1989, Durham 1992). At present, from 1.2 billion (Durning 1989) to 2 billion people (Abernethy, Vanderbilt University, PC, 1992) worldwide are living in poverty, malnourished, diseased, and experiencing short life-spans. In the United States 36 million now are living in poverty (USBC 1994).

The natural resources required to sustain human life include ample supplies of fertile land, forests, water, energy, and diversity of natural biota. The interdependencies of these resources and their current and projected future status are analyzed in this paper. An optimum population for the United States and the world is proposed based on a high standard of living while maintaining the sustainability of renewable resources and the environment. The goal is to determine the population size that will insure the possibility of individual prosperity for everyone while maintaining a quality environment. This information will assist the public and governments to make thoughtful decisions that lead to reducing population numbers and consumption levels while effectively managing natural resources and the environment to sustain future generations.

2. Ecosystem Integrity

An ecosystem is a network of energy and mineral flows in which the major integrated components are populations of plants, animals, and microorganisms (Westra 1994a). Each organism performs different specialized functions in the system. All self-sufficient, integrated ecosystems consist of producers (plants), consumers (animals and microbes), and reducers or decomposers (microbes and animals). Macro- and microscopic plants collect

---

[1]Department of Entomology and Section of Ecology and Systematics, Cornell University, Ithaca, NY 14853-0901, U.S.A.

solar energy and convert it into chemical energy via photosynthesis. Plants use this energy for growth, maintenance, and reproduction. In turn, these plants serve as the primary energy source for all the other living organisms in the ecosystem. Animals and microbes consume plants, animals eat other animals, reducers feed on both plants and animals and recycle, thus conserving the vital chemicals required by life to be used once again by plants. Thus, consumers, reducers, and decomposers all depend, directly or indirectly, on plants as their food source.

The exact number of species and biodiversity needed for a particular integrated and self-sufficient ecosystem depends upon many physical and chemical factors, as well as temperature and moisture conditions, and the diverse types of species that make up the ecosystem. At present, our knowledge is insufficient to predict accurately how many and what kinds of species are necessary for the biodiversity of the ecosystem. In the United States, approximately 500,000 species of plants and animals are vital to the environment and the integrity of the ecosystem. No one knows how many of these species can be eliminated before the quality of the ecosystem will be diminished.

Elton (1927) pointed out that the "whole structure and activities of the community are dependent upon questions of food supply". Plants are nurtured by the sun and by the vital chemicals of life that they obtain from the atmosphere, soil, and water. The remainder of the species in the ecosystem depend on living and dead plants.

Several chemical elements, including carbon, hydrogen, oxygen, nitrogen, phosphorus, potassium, and calcium, are essential to the integrity of ecosystems. Various biogeochemical cycles have evolved to insure that plants, animals, and microbes have suitable amounts of these vital chemical elements. Biogeochemical cycles both conserve the vital elements and keep them in circulation in the ecosystem. Indeed, the mortality of living organisms keeps the vital elements in circulation, enabling the system to evolve and adapt to new and changing ecosystems. These biogeochemicals themselves are a product of evolution of the living ecosystem. If the living system had not evolved this integrated way of keeping vital chemicals in circulation and conserving them for use in the ecosystem, it would have become extinct long ago.

To obtain food, humans manipulate natural ecosystems and alter its biodiversity. In altering the natural system to produce vegetation and/or animal types (livestock) different from that typical of the natural systems, a certain amount of human and fossil energy input is necessary. In principle, the greater the change required in the natural ecosystem to produce crops and livestock, the greater the energy and labor that must be expended. This same principle applies in reverse to minimizing the energy inputs into all agricultural systems. That is, the more closely the agricultural system resembles the original natural ecosystem, the fewer the inputs of energy and other factors that will be required in an agricultural ecosystem. Equally important, the closer the agricultural system is to the natural ecosystem, the more sustainable it is. This is because the ecosystem has great ecological integrity.

Because the world population numbers nearly 6 billion and we add about a quarter million people to the population daily, more food must be produced. In the simplest terms, the production of more food requires more land, water, energy, and biological resources. As the world population grows and stresses the ecosystem, its ecological integrity is reduced. No one knows how much ecosystem integrity can be reduced and the system continue to function. Therefore, there is a clear need for humans to reduce their numbers to about 2 billion on earth.

3. Populations and Consumption of Resources

Human behavior demonstrates a strong will to survive, to reproduce, and to achieve some level of prosperity and quality of life. However, individuals as well as societies differ in their view of what they consider a satisfactory life. Contrasting some aspects of life in the United States, China, and world reveals disparities in lifestyles which most often are functions of the natural resources available per person. Furthermore, most of these basic resources (land, water, energy, and biota) are finite and not unlimited in their supplies and as human populations continue to grow, prosperity and quality of life can be expected to decline (Fornos 1987, UNFPA 1991).

The present population of the United States is 256 million and is growing at a rate of 1.1 percent per year (USCB 1992). If the numbers of immigrants are increased as proposed by the President and Congress, then the rate of U.S. population growth will increase at a greater rate. In contrast, China already has a population of 1.2 billion, and despite the governmental policy of permitting only one child per couple, it is growing at a rate of 1.4 percent (PRB 1991). The world population is now 5.5 billion and growing at a rate of 1.7 percent. Based on these data, the world population is expected to double in 41 years and the U.S. population to double in 63 years.

Each American consumes about 23 times more goods and services than the average third-world citizen. Also each person in the United States consumes about 53 times more goods and services than a Chinese citizen (PRB 1991). Achieving the U.S. standard of living is impossible for the rest of the world, based both on projections of future resource availability and population growth. The excessive consumption levels characteristic of Americans depend on the importation of natural resources from other countries (USBC 1991) and are reflected in the highest debt of any nation in the world.

Since the 1850s, Americans have relied increasingly on energy sources other than human power for their food and forest products. The relatively cheap and abundant supplies of fossil fuel have been substituted for human and draft animal energy. Commercial fertilizers and pesticides as well as machinery have enabled U.S. farmers diminish the level of human energy they must expend to farm the land. Chinese farmers use as much fertilizer and pesticides per hectare as American farmers. But they also depend on about 1,200 hrs/ha per year of human labor for grain production, compared with only 10 hrs/ha per year in the United States (Wen and Pimentel 1984).

Industry, transportation, home heating, and food production account for most of the fossil energy consumed in the United States (DOE 1991a). In China most fossil energy is used by industry and a lesser amount for food production (Kinzelbach 1983, Smil 1984). Per capita use of fossil energy in the United States is 10,000 liters of oil equivalents per year or almost 14 times the level in China (Table 1). U.S. per capita energy consumption is nearly 7 times the world average level. The relative affluence presently enjoyed by Americans has been made possible by our abundant supplies of fertile cropland, water, and fossil energy per capita. As our population continues to grow (Figure 1), we will inevitably experience resource shortages similar to those now being experienced by China and other nations (Tables 1 and 2).

4. Status of World Environmental Resources

What standard of living will be experienced by each person in the United States and the world in the future? We have already suggested that this depends on population numbers and the quality and quantity of land, water, and energy as well as of biological resources and the technologies employed to manage these resources. The U.S. population currently has 256 million consumers of these vital resources, many of which are being depleted, with no hope of renewal after the next hundred years. Reports indicate that the average standard of living in the United States began to decline during the last decade (Fuchs and Reklis 1992) and is projected to continue to decline if the U.S. population doubles its numbers during the next

Table 1. Resources Used Per Capita Per Year in the United States, China, and the World to Supply Basic Needs.

Resources	USA	China	World
Land			
Cropland (ha)	0.52[a]	0.13[b]	0.28[g]
Pasture (ha)	1.3[a]	0.35[b]	0.58[g]
Forest (ha)	1.3[a]	0.15[b]	0.76[g]
Total (ha)	3.12	0.63	1.62
Water (liters × 10^6)	1.9[a]	0.43[c]	0.66[h]
Fossil Fuel			
Oil equivalents (liters)	10,000[e]	700[f]	1,500[i]
Forest Products (kg)	1,400[e]	40[d]	70[e]

[a]USDA (1990); [b]Shi Yulin (1991); [c]Sun Julin (1990), Water Use in China from Wen Dazhong, Inst. of Appl. Ecology, Shenyang, China, PC, 1992; [d]SSBPRC (1991); [e]USBC (1991); [f]SSBPRC (1990); [g]Buringh (1989); [h]WRI (1991); [i]UNEP (1985).

Figure 1. Rapid Growth in the U.S. Population From 1800 to Date. At the Current Growth Rate, the U.S. Population is Projected to Double in 63 Years (USCB 1992).

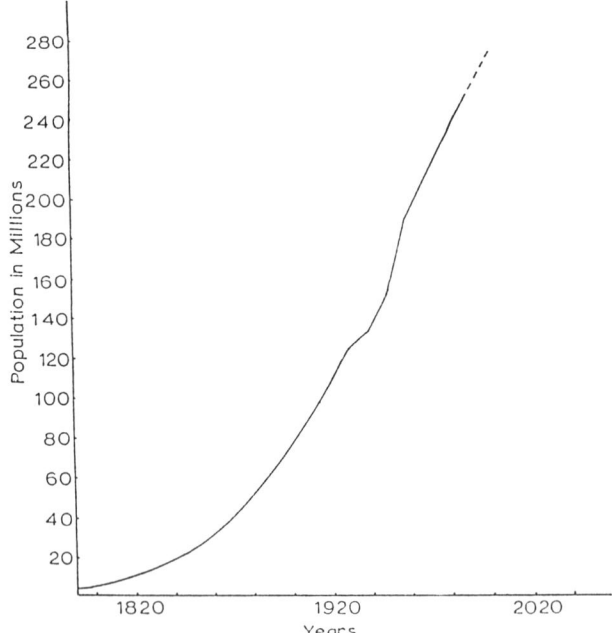

Table 2. Foods and Feed Grains Supplied Per Capita (kg) Per Year in the United States, China, and the World.

Food/Feed	USA[1]	China[2]	World[3]
Food grain	77	265	201
Vegetables	129	180	130
Fruit	46	15	53
Meat and fish	88	32	47
Dairy products	258	4	77
Eggs	14	6	6
Fats and oils	29	6	13
Sugar and sweeteners	70	7	25
Total food	711	515	552
Feed grains	663	70	166
Grand Total	1374	585	718
Kcal/person/day	3600	2500	2667

[1] Putnam and Allshouse (1991).
[2] All food item data, except vegetables, are from AMPRC (1989); the vegetable data are from D. Wen, Institute of Ecology, Shenyang, China, PC, 1991. Feed grains are from Ding Junsheng (1988).
[3] FAO (1991), except for feed grain data which is from FAO (1989).

63 years (USCB 1992). The world population, as mentioned, is projected to double in just 41 years (PRB 1991) and already shortages of fertile land, water, and fossil energy exist in many regions (WRI 1991, Worldwatch Institute 1992).

4.1. ECOLOGICAL INTEGRITY

Food supplies for humans and all other terrestrial organisms depend on a fertile land, freshwater, energy, and biodiversity. If any of these factors is missing, the food and ecological systems will be non-functional. Thus, the vital role played by the principle of *ecological integrity* that is so well explained by Westra (1994b) should be taken into account when assessing problems of food supplies. She points out that a shortage of any one of these factors becomes the prime limiting factor for the ecosystem.

The estimates for an optimum population of 200 million for the U.S. and 2 billion for the world are based on which of the four factors is limiting for various regions of the earth. The item that is most difficult to assess as a limiting factor is biodiversity (Westra 1994c). In North America there are an estimated 500,000 plant and animal species and we depend upon most of these organisms for our agriculture, forestry, and quality of the environment (Pimentel et al. 1992a). No one knows which species or how many species can be exterminated before the functioning of the ecosystem is threatened and/or its function significantly reduced. Thus, a reason for limiting the numbers of humans and protecting our vital environmental resources and all the species in the system.

4.2. LAND RESOURCES

More than 98 percent of world food comes from the terrestrial environment and the remaining small percentage comes from ocean, lake, and other aquatic ecosystems (Pimentel and Hall 1989). Worldwide, food and fiber crops are grown on 12 percent of the earth's total land area (Buringh 1989). Another 24 percent of the land is used as pasture to graze livestock that provide meat and milk products. Forests cover an additional 31 percent (Buringh 1989). The small percentage of forestland and grassland set aside as protected national parks to conserve biological diversity, amounts to only 3.2 percent of the total terrestrial ecosystem (Reid and Miller 1989). Most (34 percent) of the remaining portion of land area is unsuitable for crops, pasture, and forests because it is too cold, too dry, steep, stony, or wet, or the soil is too shallow to support plant growth (Buringh 1989).

To provide a diverse nutritious diet of plant and animal products about 0.5 ha of cropland per capita is needed (Lal 1989). The U.S. is at this level now, but the world average is only 0.28 ha of cropland available per capita, or nearly one-half this optimum value (Table 1). This shortage of productive cropland is in part the cause of food shortages and poverty that many humans are experiencing today.

Currently, a total of 1,374 kg of agricultural products are produced annually to feed each American while the Chinese's supply averages only 585 kg/capita/yr. Note that the world average value is 718 kg/capita/ yr. Based on available data each person in China eats essentially a vegetarian diet. Further they have reached the carrying capacity of their agricultural system, even with huge inputs of fossil energy now used on Chinese farms (Wen and Pimentel 1990).

Escalating land degradation threatens most crop and pasture land throughout the world (Lal and Pierce 1991). The major types of degradation include water and wind erosion, salinization, and water-logging of soils (Mabbutt 1989). Indeed, more than 10 million hectares of productive arable land are severely degraded and abandoned each year (Pimentel et al. 1992a). Moreover, each year an additional 5 million hectares of new land must be put into production to feed the 92 million humans added yearly to the world population. Most of this total of 15 million ha needed for replacement and expansion is coming from the world's forests. The urgent need for agricultural land accounts for 80 percent of the deforestation now occurring worldwide (Myers 1990).

Soil erosion, the single most serious cause of soil loss and land degradation, is more intense than ever before in history (Pimentel and Hall 1989, WRI 1991, Pimentel 1993). In Africa during the past 30 years, the rate of soil loss has increased 20 times (Tolba 1989). Wind erosion is so serious in China that Chinese soil can be detected in the Hawaiian atmosphere when planting starts in China (Parrington et al. 1983). Similarly in 1992, soil eroded from Africa was detected in Florida and Brazil (Simons 1992). Soil erosion on cropland ranges from about 16 t/ha/yr in the USA to 40 t/ha/yr in China (USDA 1991, Wen Dazhong 1993, McLaughlin 1993). Soil erosion worldwide is about 30 t/ha/yr (Pimentel 1993). This magnitude of erosion is of particular concern because of the slow pace of soil formation; it takes approximately 500 years for 2.5 cm of topsoil to form under agricultural conditions (OTA 1982, Elwell 1985, Troeh et al. 1980). Thus, topsoil is being lost 20 to 40 times faster than it is being replaced. Erosion adversely affects crop productivity by reducing water availability, water-holding capacity, soil nutrients, soil organic matter, and soil depth. Estimates are that agricultural land degradation can be expected to depress world food production between 15 percent and 30 percent during the next 25-year period (Buringh 1989). The arable land currently used for crop production includes some marginal land which is highly susceptible to degradation. When such changes occur crop production is depressed and the requirement for fossil energy inputs in form of fertilizers, pesticides, and irrigation increased in an effort to offset some degradation (OTA 1982, Follett and Stewart 1985, Pimentel 1993).

4.3. WATER RESOURCES

The present and future availability of adequate supplies of fresh water is frequently taken for granted. Natural collectors of water such as rivers and lakes vary in distribution throughout the world and frequently are shared by several countries. All water supplies, especially in arid regions are diminished by evaporation. Reservoir water experiences an average yearly loss of about 24 percent (Meyers 1962).

All vegetation requires and transpires massive amounts of water during the growing season. For example, a corn crop that produces about 7,000 kg/ha of grain will take up and transpire about 4.2 million liters/ha of water during the growing season (Leyton 1983). To supply this much water to the crop, not only must 10 million liters (1,000 mm) of rain fall per hectare, but a significant portion must fall during the growing season.

The greatest threat to maintaining fresh water supplies is overdraft of surface and groundwater resources to supply the needs of the rapidly growing human population and of the agriculture which provides its food. Agricultural production "consumes" more fresh water than any other human activity (Falkenmark 1989). Worldwide about 87 percent of the fresh water is "consumed" (non-recoverable) by agriculture (Postel 1989), while in the United States this figure is about 85 percent (NAS 1989). An individual requires nearly 3 liters/day of fresh water for drinking, but needs a minimum of 90 liters/day for cooking, washing, and other domestic needs (Brewster 1987). Each American uses about 400 liters/day for domestic needs (USBC 1991).

As the world's population grows, so do its water needs. To provide the ever increasing amount of water required to meet human needs is resulting in increased demand for surface water and groundwater resources. For example, by the time the Colorado River enters Mexico it has literally disappeared because of the excessive removal of its water by the states of California, Arizona, and Colorado (Sheridan 1983). Veltrop (1991) calculates that if the world's population increases about 20 percent, the demand for water will double.

Surface water and groundwater each supply half of the freshwater supply in the world (Wolman 1986, Falkenmark 1989). Groundwater resources are renewed at various rates, but usually at the extremely slow rate of about 1 percent per year (CEQ 1980). Because of this slow recharge rate, groundwater resources must be carefully managed to prevent overdraft. Yet humans are not effectively conserving groundwater resources and overdraft is a serious problem worldwide. For example, in Tamil Nadu, India groundwater levels declined 25 to 30 m during the 1970s because of pumping for irrigation (Postel 1984, UNFPA 1991). Beijing, China records a decline in its groundwater table of about 1 m/yr; and in Tianjin, China it drops 4.4 m/yr (Postel 1984). In the United States overdraft averages 25 percent higher than replacement (USWRC 1979). But in some locations, like the U.S. Ogallala aquifer, annual overdraft is 130 percent to 160 percent above replacement (Beaumont 1985). If this continues, this vast aquifer is expected to become non-productive in about 40 years (Soule and Piper 1992). Loss of available water limits the option of irrigation in arid regions. The irrigation area worldwide is now declining per capita because of salinization, waterlogging and population growth (Postel 1989).

Another major threat to maintaining ample fresh water resources is pollution caused by people and industries. Considerable water pollution is documented in the United States (USBC 1991), but it is more serious in developing countries. For example, in Latin American countries, untreated urban sewage is often dumped into rivers and lakes (WRI 1991), resulting in fecal-coliform bacterial counts higher than 100,000 per ml of water (0.1/ml is the maximum acceptable level for U.S. drinking water). Pesticides, fertilizers, and sediments pollute water resources as they accompany eroded soil; industries dump toxic chemicals untreated into rivers and lakes (WRI 1991). Pollution by sewage, as well as various chemical wastes, makes water unsuitable for human drinking and for application to crops.

4.4. BIOLOGICAL RESOURCES

In addition to land, water resources, and crop and livestock species, humans depend on the millions of other species that exist in agroecosystems and nature (Pimentel et al. 1992a). Humans have no technologies that can substitute for the service provided by wild biota. In the United States there are approximately 500,000 species of plants, animals, and microbes that provide many essential functions for humans including: pollination of crop and wild plants; recycling manure and other organic wastes; degrading chemical pollutants; and purifying water and soil (Pimentel et al. 1992a). These diverse species also serve as a vital reservoir of genetic material for future development of agriculture and forestry. Yet the world is losing about 150 species per day because of human activities of deforestation, pollution, applying pesticides, urbanization, etc. (Reid and Miller 1989).

Ecologists have reported that if sufficient natural biological diversity is to be maintained to ensure a quality environment, then about one-third of the terrestrial ecosystem should be preserved as natural vegetation (Odum 1971). This biomass is essential to provide food, shelter, and protection for these valuable species and ensures the preservation of adequate biodiversity (Pimentel et al. 1992a).

Clearly humans need these organisms as well as their livestock and crop species. For example, honey bees and wild bees play an essential role in pollinating about $30 billion worth of U.S. crops annually in addition to pollinating natural plant species. It has been calculated that honey bees and wild bees in New York State on a bright, sunny day in July pollinate 10^{12} blossoms (Pimentel 1994). Humans have no technology to substitute for this natural service supplied by wild biota.

4.5. ENERGY RESOURCES

Some form of energy is expended to provide humans with all their needs. About 369 quads from all energy sources per year are used worldwide, the amount directly related to the rapid growth in the world population and the environmental degradation imposed by human activity (Pimentel and Pimentel 1979) (Table 3). Although worldwide about 50 percent of all solar energy captured by photosynthesis is used by humans, it is inadequate to meet their needs of food and forest products (Pimentel 1989, Pimentel and Pimentel 1991). To make up the addition, about 319 quads (10^{15} BTU or 337×10^{18} Joules) of fossil energy are utilized worldwide each year (UNEP 1985, IEA 1991), of which 79 quads are consumed in the United States (DOE 1991a). These 79 quads represent nearly 3- times the 28 quads of solar energy harvested as crop and forest products, and about 40 percent more energy than is captured by U.S. vegetation. Fossil energy has also been used to fuel a wide array of human activities including industrial production, fuel for automobiles and trucks, highway construction, heating and cooling of buildings, and packaging of all goods.

Fossil energy is used to feed an increasing number of humans as well as improve the quality of life in many basic ways, such as protecting humans from numerous diseases. For example, delivering clean water has helped to eliminate a wide array of disease organisms that are transmitted in polluted water.

Developed nations annually consume about 80 percent of the fossil energy worldwide while the developing nations, which have about 75 percent of the world population, consume only 20 percent (UNEP 1985, DOE 1991a). The United States consumes about 25 percent of the world's fossil energy annually.

Several developing nations that have high rates of population growth are increasing the use of fossil fuels in their agricultural production. For example, since 1955 there has been a 100-fold increase in the use of fossil energy in Chinese agriculture (Wen and Pimentel

Table 3. Fossil and Solar Energy Use in the U.S. and World.

	USA[1]		World[2,3]	
	Quads	Percent	Quads	Percent
Total energy	85.1	100	368.9	100
Fossil energy	78.5	92.3	319.2	86.5
Solar energy	6.6	7.7	49.7	13.5
Hydropower	3.0	3.5	21.2	5.7
Biomass	3.6	4.2	28.5	7.8

[1]DOE 1991a, [2]DOE 1991b, [3]UNEP 1985.

1984). Similarly, fossil energy use in different U.S. economic sectors has increased 20- to 1,000-fold in the past 3 to 4 decades, attesting to our heavy reliance on this finite energy resource (Pimentel and Hall 1989).

Projections of the availability of fossil energy resources are discouraging. A recent report published by the U.S. Department of Energy (DOE 1991a) based on current oil-drilling data indicates that the estimated amount of national oil reserves has plummeted. This means that instead of the 35-year supply of U.S. oil reserves that was projected about 4 years ago, the current known and discoverable potential oil reserves are now limited to a 10- to 13-year supply at present rates of pumping (DOE 1990, Lawson 1991). Since the United States is now importing more than half its oil, a serious problem already exists (Gibbons and Blair 1991).

The world supply of oil is greater than that of the United States and is projected to last about 35 years at current pumping rates (Matare 1989). Both in the United States and the world, the natural gas supply is adequate for about 35 years and coal for about 100 years (Matare 1989). Other estimates range as high as 150 years for total fossil energy, primarily coal (BP 1991). However, these estimates are based on current consumption rates and current population numbers. If all people in the world enjoyed a similar standard of living and energy consumption as the U.S. average, and the world population continued to grow at a rate of 1.7 percent, the world's fossil fuel reserves would last a mere 20 years.

At present about 34 percent of total U.S. energy consumption is electricity and nuclear energy provides 18 percent of these electric needs (USBC 1991). Nuclear energy production of electricity has some advantages over fossil fuels because it requires less land than coal-fired plants, causes fewer human injuries and deaths, and its use does not contribute to acid rain and global warming (Holdren 1991, Meeks and Drummond 1991). However, there are several limitations to the expansion of the use of nuclear fission and fusion energy in the future.

First, uranium resources are limited worldwide and are expected to last about 100 years, without nuclear breeder reactors (Hafele 1991). Second, the risks of disposing radioactive wastes and lack of public acceptance for storage of wastes may influence the widespread use of both fission and fusion energy (Hafele 1991). Fusion technology will require a great many years of research for development before it will be ready for use (Matare 1989).

Both nuclear fission and fusion technologies produce enormous amounts of waste heat, which is a serious environmental pollutant (Bartlett 1989). For example, it has been estimated that if the number of nuclear power plants in the U.S. were increased from the current 108 to 1,500, the temperature of aquatic ecosystems in the United States would increase about 10°C (Kendall 1992). This degree of heat pollution would cause a major loss of biological diversity in aquatic systems and would also alter existing climate patterns which influence agricultural and forestry production.

5. Improved Use of Resources

The prime resources—land, water, energy, and biological resources—function interdependently and each can be manipulated to a degree to make up for a partial shortage in one or more of the others. For example, to bring desert land in agricultural production, it can be irrigated. This can occur only if groundwater or river water is available, if sufficient fossil energy is available to pump and move the water, and if the soil is suitable for irrigation and fertile to support crop growth. Because the availability of these essential resources is fast diminishing, the options for substitution are also diminishing. This emphasizes the need to examine alternative strategies.

Large quantities of fossil based fertilizers are major sources of nutrient enhancement of agricultural soils throughout the world. Yet in the United States about $18 billion per year of fertilizer nutrients are lost as they are eroded along with soils (Troeh et al. 1980). Further, U.S. livestock manures, which have an amount of nitrogen equal to that in commercial nitrogen fertilizer applied to agriculture each year, are underutilized, and wasted. Significant quantities of fossil energy could be saved if effective soil conservation methods were implemented, and if manures were used more extensively as a substitute for commercial fertilizer (Pimentel et al. 1989a, 1989b).

Pesticides are also fossil based in their production and are wasted (Pimentel 1990). Since 1945 the use of synthetic pesticides in the United States has grown 33-fold, yet crop losses to pests continue to increase (Pimentel et al. 1991). For example, despite a 1,000-fold increase in the use of insecticides on corn, corn losses to insects have risen nearly 4-fold (Pimentel et al. 1991). Pesticide use has increased because agricultural technologies have been changed. For some major crops like corn, crop rotations have been abandoned. Now about 40 percent of U.S. corn land is used to grow corn continuously as a monoculture. This has caused an increase in the number of corn pests and in pesticides required to protect the crop (Pimentel et al. 1991). Adopting sustainable and environmentally sound agricultural technologies, including a return to crop rotations, would stem soil erosion, conserve fertile land, reduce water requirements for irrigation, decrease pesticide and fertilizer use, and thus save fossil fuel, soil, and water resources (Pimentel et al. 1989a, 1989b).

The use of more land to produce food reduces the total energy inputs necessary for crop production and would lead to greater solar energy dependence and sustainability in agriculture. This, of course, assumes the availability of sufficient land, halving crop yields per hectare, but maintaining the same total amount of food produced.

6. Conclusion

Does human society want 10 to 15 billion humans living in poverty and malnourishment or 1 to 2 billion living with abundant resources and a quality environment? Citizens of the United States and the world must support their leaders in making these critical decisions for the future. This fundamental commitment to move toward a sustainable-sized population and an energy-secure future must include the active political participation of all people.

Given the present level of fertility and immigration, the U.S. population will double in 63 years to more than half a billion or roughly half the size of present day China. Comparisons to the problems now being experienced in China emphasize why the United States will be unable to maintain its present level of prosperity and relatively high standard of living, unless population growth is controlled.

For Americans to continue to enjoy a high standard of living and for society to be self-sustaining in renewable energy and food and forestry products, given U.S. land, water and biological resources, the optimum U.S. population is about 200 million—significantly less than the current level of 256 million. However, with one billion people as now live in China, the U.S. population could be sustained but in relative poverty. Sometime soon the United States needs to determine its population policy and vision for the future.

At present the pressure imposed by the large and expanding world population is more serious than that being experienced in the United States. The world population is 5.7 billion with about 1.0 billion humans now malnourished and from 1.2 to 2 billion living in poverty. Fertile cropland, fresh water, and fossil energy resources are now in serious short supply in many regions of the world. Their scarcity accounts for inadequate food and forest production, a deteriorating environment, and a diminished standard of living for most people. At current use levels most oil, natural gas, and coal reserves will be used up within the next century, with actual rates of consumption driven by population growth and rising consumer expectations. In addition, soil degradation is intensifying, water shortages and pollution increasing, forests are being removed, and more biological species are being destroyed than ever before.

Thus far, the Americans as well as world citizens appear unwilling to deal with the growing imbalances of human population and the energy and environmental resources that support all life. Humans have a disappointing record of effectively managing and protecting their essential resources and the environment from over-exploitation in the face of rapidly growing population. World leaders seem not to understand or acknowledge the interdependencies existing among individual standard of living, population density, availability of life-supporting resources, and the quality of the environment. Local, national, and global problems exist because governments have not tried to develop cohesive and cooperative policies that recognize how supplies of the natural resources are affected by human numbers and consumption levels.

Decision making tends to be based on crises; decisions are not made until catastrophe strikes. Thus, decisions are ad hoc, designed to protect and/or promote a particular resource or aspect of human well-being instead of examining the problem in a holistic manner. Based on past experience, we expect that leaders will continue to postpone decisions concerning human carrying capacity of the world (Fornos 1987), maintenance of a standard of living, conservation of resources, and the preservation of the environment until the situation becomes intolerable, or worse still, irreversible.

Starting to deal with the imbalance of the population-resource equation before it reaches a crisis level is the only way to avert a real tragedy for our children's children. With equitable population control that respects basic individual rights, sound resource management policies, support of science and technology to enhance energy supplies and the environment, and with all people working together, an optimum population can be achieved. With such cooperative efforts we would fulfill fundamental obligations to generations that follow—to ensure that individuals will be free from poverty and starvation in an environment that will sustain human life with dignity.

7. References

AMPRC. 1989. *Agricultural Statistical Data of China in 1988*. Agricultural Ministry of PRC, Agricultural Press, Beijing, China.

Bartlett, A.A. 1989. Fusion and the Future. *Physics and Society* 18(3): 11.

Beaumont, P. 1985. Irrigated Agriculture and Groundwater Mining on the High Plains of Texas, USA. *Environmental Conservation* 12: 11.

BP. 1991. *British Petroleum Statistical Review of World Energy.* British Petroleum Corporate Communications Services, London, UK.
Brewster, J.A. 1987. *World Resources 1987. A Report by the International Institute for Environment and Development and The World Resources Institute.* Basic Books, Inc., New York.
Buringh, P. 1989. Availability of Agricultural Land for Crop and Livestock Production. In *Food and Natural Resources,* D. Pimentel and C. W. Hall, eds. Academic Press, San Diego, pp. 69-83.
CEQ. 1980. *The Global 2000 Report to the President of the U.S. Entering the 21st Century.* Pergamon Press, New York.
Demeny, P.G. 1986. *Population and the Invisible Hand.* Center for Policy Studies Paper No. 123. Population Council, New York.
Ding, Junsheng. 1988. Some Inspirations From the Comparison of Food Consumptions Between China and Other Countries. *The People's Daily* July 25: 5.
DOE. 1990. *Annual Energy Outlook.* U.S. Department of Energy, Washington, DC.
DOE. 1991a. *Annual Energy Outlook with Projections to 2010.* U.S. Department of Energy, Energy Information Administration, Washington, DC.
DOE. 1991b. *1989 International Energy Annual.* U.S. Department of Energy, Washington, DC.
Durham, D.F. 1992. Cultural Carrying Capacity: I = PACT. *Focus* 2(3): 5-8.
Durning, A.B. 1989. *Poverty and the Environment: Reversing the Downward Spiral.* Worldwatch Institute, Washington, D.C.
Ehrlich, P.R., and A.H. Ehrlich. 1990. *The Population Explosion.* Simon and Schuster, New York.
Elton, C.S. 1927. *Animal Ecology.* London, Sidgwick and Jackson, LTD.
Elwell, H.A. 1985. An Assessment of Soil Erosion in Zimbabwe. *Zimbabwe Science News* 19: 27-31.
Falkenmark, M. 1989. Water Scarcity and Food Production. In *Food and Natural Resources,* D. Pimentel and C. W. Hall, eds. Academic Press, San Diego, CA, pp. 164-191.
FAO. 1989. Aspects of the World Feed Livestock Economy: Structural Changes, Prospects and Issues. Economic and Social Development Paper #80. Food and Agriculture Organization of the United Nations, Rome, Italy.
FAO. 1991. *Food Balance Sheets.* Food and Agriculture Organization of the United Nations, Rome, Italy.
Follett, R.F. and B.A. Stewart. 1985. *Soil Erosion and Crop Productivity.* American Society of Agronomy, Crop Science Society of America, Madison, WI.
Fornos, W. 1987.*Gaining People, Losing Ground: A Blueprint for Stabilizing World Population.* The Population Institute, Washington, DC.
Fuchs, V.R. and D.M. Reklis. 1992. America's Children: Economic Perspectives and Policy Options. *Science* 255: 41-46.
Gibbons, J.H. and P.D. Blair. 1991. U.S. Energy Transition: On Getting From Here to There. *Physics Today* 44: 22-30.
Hafele, W. 1991. Energy From Nuclear Power. *Scientific American* September: 137-144.
Hardin, G. 1986. Cultural Carrying Capacity: A Biological Approach to Human Problems. *BioScience* 36: 599-606.
Holdren, J.P. 1991. Energy in Transition. In *Energy for Planet Earth,* J. Piel, ed. W.H. Freeman Co., New York, pp. 119-130.
Holdren, C. 1992. Population Alarm. *Science* 255: 1358.
IEA. 1991. *Energy Statistics of OECD Countries.* International Energy Agency, Paris.
Kendall, H. 1992. Personal conversation. Department of Physics, Massachusetts Institute of Technology.

Keyfitz, N. 1984. Impact of Trends in Resources, Environment and Development on Demographic Prospects. In *Population, Resources, Environment and Development.* New York: United Nations, pp. 97-124.
Kinzelbach, W.K. 1983. China: Energy and Environment. *Environmental Management* 7: 303-310.
Lal, R. 1989. Land Degradation and Its Impact on Food and Other Resources. In *Food and Natural Resources,* D. Pimentel, ed. San Diego, CA: Academic Press, pp. 85-140.
Lal, R. and F.J. Pierce. 1991. *Soil Management for Sustainability.* Soil and Water Conservation Society in Cooperation with World Association of Soil and Water Conservation and Soil Science Society of America, Ankeny, Iowa.
Lawson, R.L. 1991. The U.S. Should Increase Its Use of Coal. In *Energy Alternatives,* C. P. Cozic and M. Polesetsky, eds. Greenhaven Press, San Diego, pp. 41-45.
Leyton, L. 1983. Crop Water Use: Principles and Some Considerations for Agroforestry. In *Plant Research and Agroforestry,* P. A. Huxley, ed. International Council for Research in Agroforestry, Nairobi, Kenya, pp. 379-400.
Mabbutt, J.A. 1989. Impacts of Carbon Dioxide Warming on Climate and Man in the Semi-Arid Tropics. *Climatic Change* 15: 191-221.
Matare, H.F. 1989. *Energy: Fact and Future.* CRC Press, Boca Raton, FL.
McLauglin, L. 1993. Soil Erosion and Conservation in Northwestern China. In *World Soil Erosion and Conservation,* D. Pimentel, ed. Cambridge University Press, Cambridge, UK. pp. 87-107.
Meeks, F. and J. Drummond. 1991. Nuclear Power Can Be Environmentally Safe. In *Energy Alternatives,* C. P. Cozic and M. Polesetsky, eds. Greenhaven Press, San Diego, CA. pp. 46-52.
Meyers, J.S. 1962. *Evaporation from 17 Western States.* Geological Survey Professional Paper 272-D, Washington, DC.
Myers, N. 1990. Mass Extinctions: What Can the Past Tell Us About the Present and Future? *Global and Planetary Change* 2: 82.
NAS. 1989. *Alternative Agriculture.* National Academy of Sciences, Washington, DC.
Odum, E.P. 1971. *Fundamentals of Ecology.* W.B. Saunders Co., Philadelphia, PA.
OTA. 1982. *Impacts of Technology on U.S. Cropland and Rangeland Productivity.* U.S. Congress Office of Technology, Washington, DC.
Parrington, J.R., W.H. Zoller and N.K. Aras. 1983. Asian Dust: Seasonal Transport to the Hawaiian Islands. *Science* 246: 195-197.
PCC. 1989. *Population.* Population Crisis Committee, Washington, DC.
Pimentel, D. 1989. Ecological Systems, Natural Resources, and Food Supplies. In *Food and Natural Resources,* D. Pimentel and C. W. Hall, eds. Academic Press, San Diego, pp. 1-29.
Pimentel, D. 1990. Environmental and Social Implications of Waste in U.S. Agriculture and Food Sectors. *Journal of Agriculture Ethics* 3: 5-20.
Pimentel, D., ed. 1993. *World Soil Erosion and Conservation.* Cambridge University Press, Cambridge, UK.
Pimentel, D. 1994. Arthropods and Their Biology. In *The Encyclopedia of the Environment.* Houghton Mifflin Co., Boston, in press.
Pimentel, D. and C.W. Hall, eds. 1989. *Food and Natural Resources.* Academic Press, San Diego.
Pimentel, D. and M. Pimentel. 1979. *Food, Energy and Society.* Edward Arnolds, London, UK.
Pimentel, D. and M. Pimentel. 1991. Land, Energy and Water: The Constraints Governing Optimum U.S. Population Size. *Focus* 1: 9-14.

Pimentel, D., Armstrong, L.E., Flass, C.A., Hopf, FW., Landy, R.B. and M.H. Pimentel. 1989a. Interdependence of Food and Natural Resources. In *Food and Natural Resources,* D. Pimentel and C. W. Hall, eds. Academic Press, San Diego, pp. 31-48.

Pimentel, D., Culliney, T.W., Butler, I.W., Reinemann, D.J. and K.B. Beckman. 1989b. Ecological Resource Management For a Productive, Sustainable Agriculture. In *Food and Natural Resources,* D. Pimentel and C. W. Hall, eds. Academic Press, San Diego, pp.301-323.

Pimentel, D., McLaughlin, L., Zepp, A., Lakitan, B., Kraus, T., Kleinman, P., Vancini, F., Roach, W.J., Graap, E., Keeton, W.S. and G. Selig. 1991. Environmental and Economic Impacts of Reducing U.S. Agricultural Pesticide Use. In *Handbook of Pest Management in Agriculture,* D. Pimentel, ed. CRC Press, Boca Raton, pp. 679-718.

Pimentel, D., Stachow, U., Takacs, D.A., Brubaker, H.W., Dumas, A.R., Meaney, J.J., O'Neil, J., Onsi, D.E. and D.B. Corzilius. 1992a. Conserving Biological Diversity in Agricultural/Forestry Systems. *BioScience* 42: 354-362.

Postel, S. 1984. *Water: Rethinking Management in an Age of Scarcity.* Worldwatch paper no. 62. Worldwatch Institute, Washington, DC.

Postel, S. 1989. *Water for Agriculture: Facing the Limits.* Worldwatch Institute, Washington, DC.

PRB. 1991. *World Population Data Sheet.* Population Reference Bureau, Washington, DC.

Putnam, J.J. and J.E. Allshouse. 1991. *Food Consumption, Prices, and Expenditures, 1968-89.* ERS, Statistical Bulletin No. 825. US Dept. of Agriculture, Washington, DC.

Reid, W.V. and K.R. Miller. 1989. *Keeping Options Alive: The Scientific Basis for Conserving Biodiversity.* World Resources Institute, Washington, DC.

Sheridan, D. 1983. The Colorado—An Engineering Wonder Without Enough Water. *Smithsonian* February: 45-54.

Shi, Y. 1991. The Land Resource of China. *Chinese Statistical Press* January 11: 2.

Simons, M. 1992. Winds Toss Africa's Soil, Feeding Lands Far Away. *New York Times* October 29: A1, A16.

Smil, V. 1984. *The Bad Earth, Environmental Degradation in China.* M.E. Sharpe, Inc., Armonk, NY.

Soule, J.D. and D. Piper. 1992. *Farming in Nature's Image: An Ecological Approach to Agriculture.* Island Press, Washington, DC.

SSBPRC. 1990. *The Yearbook of Energy Statistics of China in 1989.* State Statistical Bureau of PRC, Chinese Statistical Press, Beijing, China.

SSBPRC. 1991. The Statistical Announcement of the Economical and Social Development in China in 1986—1990. *Guangming Daily* March 14: 3.

Sun J. 1990. *Water Use in China* (in Chinese).

Tolba, M.K. 1989. Our Biological Heritage Under Siege. *BioScience* 39: 725-728.

Troeh, F.R., Hobbs, J.A. and R.L. Donahue. 1980. *Soil and Water Conservation for Productivity and Environmental Protection.* Printice-Hall, Englewood Cliffs, N.J.

UNEP. 1985. *Energy Supply Demand in Rural Areas in Developing Countries.* Report of the Executive Director. United Nations Environment Programme, Nairobi.

UNFPA. 1991. *Population and the Environment: The Challenges Ahead.* United Nations Population Fund, New York.

USBC. 1992. *Current Population Reports,* January. U.S. Bureau of the Census, Washington, DC.

USBC. 1994. *Statistical Abstract of the United States 1991.* U.S. Government Printing Office, Washington, DC.

(USDA) U.S. Department of Agriculture. 1990. *Agricultural Statistics.* Government Printing Office, Washington, DC.

(USDA) U.S. Department of Agriculture. 1991. *Agricultural Resources: Cropland, Water, and Conservation Situation and Outlook Report.* Economic Research Service, AR-23. U.S. Department of Agriculture, Washington, DC.

(USWRC) U.S. Water Resources Council. 1979. *The Nation's Water Resources 1975 - 2000. Second National Water Assessment.* USWRC, Washington, DC.

Veltrop, J.A. 1991. There Is No Substitute for Water. *Water International* 16: 57.

Wen, D. 1993. Soil Erosion and Conservation in China. In: *World Soil Erosion and Conservation,* D. Pimentel, ed. Cambridge University Press, Cambridge, pp. 63-85.

Wen, D. and D. Pimentel. 1984. Energy Inputs in Agricultural Systems of China. *Agriculture, Ecosystems and Environment* 11: 29-35.

Wen, D. and D. Pimentel. 1990. Ecological Resource Management for a Productive, Sustainable Agriculture in Northeast China. In *Agricultural Reform and Development in China,* T. C. Tso, ed. IDEALS, Beltsville, MD, pp. 297-313.

Westra, L. 1994a. *An Environmental Proposal for Ethics: The Principle of Integrity.* Lanham, MD: Rowman and Littlefield.

Westra, L. 1994b. *An Environmental Proposal for Ethics.* Rowman & Littlefield Publishers, Inc., Lanham, MD.

Westra, L. 1994c. Biodiversity and Food Production: The Perspective of Ecosystem Integrity. Paper presented at the AAAS meeting, San Francisco, CA, Feb 21, 1994.

Wolman, M.G. 1986. Consensus, Conclusions, and Major Issues in Water Resources. In *Managing Water Resources,* J. Cairns and R. Patrick, eds. Praeger, New York, pp. 117-126.

Worldwatch Institute. 1992. *State of the World.* Worldwatch Institute, Washington, DC.

WRI. 1991. *World Resources 1990-91.* New York: Oxford University Press.

Chapter 16
SUSTAINABLE DEVELOPMENT AND ECONOMIC GROWTH

Joel E. Reichart[1]
Patricia H. Werhane[1]

1. Introduction

Since its introduction into mainstream environmental thought via the Brundtland Commission report (Brundtland 1987), the meaning of the term 'sustainable development' has seemingly acquired several connotations based on its broad definition. One is that, to many environmentalists, sustainable development is desirable only if it means economic development without growth. This position is exemplified by economist Daly (1993) who writes:

> In its physical dimensions the economy is an open subsystem of the earth ecosystem, which is finite, nongrowing, and materially closed. As the economic subsystem grows it incorporates an ever greater proportion of the total ecosystem into itself and must reach a limit at 100 percent, if not before. Therefore its growth is not sustainable....The term 'sustainable development' therefore makes sense for the economy, but only if it is understood as development without growth-i.e., qualitative improvement of a physical economic base that is maintained in a steady state by a throughput of matter-energy that is within the regenerative and assimilative capacities of the ecosystem.

Sustainable development is connected with the notion of environmental sustainability or environmental health. The goals of environmental sustainability are to create a balanced ecosystem whereby: (1) pollution and waste are permitted if they can be absorbed without further degrading the environment; (2) natural resources are used only to the extent that renewable resources are regenerated at the same rate, and nonrenewable resources are equally replaceable with renewable substitutes. The goal is that, given the present state of affairs, we sustain resource *status quo* while arresting further degradation of the environment.

The thesis that sustainable development should be equated with zero or even negative economic growth and the goal of environmental health form the starting point for a more radical concept of ecosystem integrity recently proposed by Westra (1994). Ecosystem integrity goes beyond pleas for sustainable development, containing economic growth and restoring environmental health. Its aims are to restore the biosystem to its former diversity and integrity, to protect all species from further biological harm, and to provide sustenance

---

[1]Darden Graduate School of Business Administration, University of Virginia, P.O. Box 6550, Charlottesville, VA 22906-6550, U.S.A.

for present and future populations world wide. To achieve these ends, it is argued, we need to rethink our ecosystem priorities. In brief, according to principle of ecosystem integrity, this entails restoring and preserving large wilderness areas and protective buffer zones, reintroducing native species of plants and animals, eliminating transgenic and exotic species, forgoing the use of pesticides, PCBs, and other biotechnological substances, and reordering, even dismantling of the economic life-style priorities of those of us who live in developed countries.

These conclusions, that sustainable development should be equated with zero or even negative economic growth and that the restoration of biointegrity requires a parsimonious life style and reduction of economic largess enjoyed by citizens in developed countries, provide the ammunition for much of the criticism of the environmental movement. Speaking about what she calls 'green ideology,' Postrel (1990) rails against environmentalists for their unrealistic conceptions of a desirable world existing in economic stasis. Postrel writes:

> Take electricity. Environmentalists, of course, rule out nuclear power, regardless of the evidence of its safety. But then they say coal-powered plants can cause acid rain and pollution so they're out, too. Oil-fired plants release greenhouse gases (and are expensive, too). Hydroelectric plants are no good because they disrupt the flow of rivers.
>
> Solar photovoltaic cells have always been the great hope of the future. But making them requires lots of nasty chemicals, so we can expect solar cells to be banned around the time they become profitable. Pretty soon, you've eliminated every conceivable source of electricity. Then your only option is to dismantle your industry and live with less...
>
> ...Many ordinary human beings would like a cleaner world. They are prepared to make sacrifices-*tradeoffs* is a better word-to get one. But ordinary human beings will not adopt the Buddha's life without desire, much as E. F. Schumacher might have ordained it.

This brief snapshot of a critique of environmentalism presents an opinion that is shared by a large proportion of individuals in the business and political world and has contributed to a cynical view of the environmental movement. This view is reinforced by a number of negative implications of ecosystem integrity that preclude the acceptance of this idea on at least two important deeply held moral grounds. First, Westra's (1994) position appears to equate human survival and well-being with that of all species, and indeed, she argues, restoring ecosystem integrity is a goal that is primary, overriding even questions of human rights and justice. The ecosystem as a whole, not human beings, are of infinite worth and the primary focus of concern. Second, in order to achieve ecosystem integrity, a common and overriding good according to this line of argument, one would have to sharply limit and control not only human preferences, but what most of us on this planet have come to value most, human freedom to choose and control our own lives. Given these alternatives, many of us will opt for continuing our present haphazard environmentally questionable customs. While this view is frankly anthropocentric, it challenges defenders of ecosystem integrity to present strong counterarguments to what we have come to regard as having intrinsic value, namely, human rights, fairness, and most particularly, freedom. Further, if developed countries and their citizens are asked to dismantle their life styles, reduce economic growth, exchange already developed areas for wilderness, and live under more restrictive regimes that enforce these outcomes, most of us will choose otherwise. Indeed, the inherent danger of radical environmental arguments is that they can create cynicism that leads some business people and politicians to separate environmental from economic and political priorities, if not discount environmental concerns altogether.

Conversely, many environmentalists also decouple the two priorities. As Sandbrook (1991) points out:
> Members of the "green" lobby, or at least some who purport to represent it, seem to sympathize with this [decoupled] view, emphasizing more the plight of nature than that of humankind, as if people were not a part of the natural world....Dealing with environmental issues in isolation [from economics and politics] is bound to fail in all but the most unusual circumstances; thus, those concerned with global environmental change must be concerned with development too.

These debates do not address what Daly (1993) and others take to be the central crisis: the fact remains that at some point world quantitative economic growth will eventually have to either level-off or recede since infinite growth appears at least intuitively to be impossible. Daly notes that, "The earth will not tolerate the doubling of even one grain of wheat 64 times, yet in the past two centuries we have developed a culture dependent on exponential growth for its economic stability." Therefore at some future point the carrying capacity of the earth, that is the earth's ability to both provide natural capital and absorb our waste, will be maximized or surpassed. Evidenced by the extent of natural resource depletion and ecosystem damage, some would even argue that, given our past patterns of population growth, consumption, and technological choices, the carrying capacity of the earth has already reached the critical threshold. As Brown (1994) notes:
> The earth's environmental assets are now insufficient to sustain both our present patterns of economic activity and the life-support systems we depend on....Since mid-century, three trends have contributed most directly to the excessive pressures now being placed on the earth's natural systems-the doubling of world population, the quintupling of global economic output, and the widening gap in the distribution of income.

Nowhere is the threat of impending environmental catastrophe more readily apparent then in lesser developed countries (LDC's). The exponential population growth currently being experienced in many LDC's is one aspect of a continuing cycle that encompasses poverty and environmental destruction. Growing populations of poor are forced to choose between urban migration and subsistence agriculture for mere survival. The former usually creates massive unemployment in urban centers as well as uncontrolled environmental degradation. The migration of the population to urban environments is a necessity due to limited or non-existent opportunities in rural agriculture, but this migration is the principal cause of unemployment in urban centers, particularly among the young. To meet the demands of the burgeoning population as urban LDC societies expand and grow, they will experience increasing pressures on their industrial infrastructure, compounded by export needs to meet debt service requirements. Corresponding increases in energy and raw material use, toxic and non-toxic waste emissions and other secondary effects of industrialization can therefore be expected.

At the same time subsistence agriculture results in an inefficient overuse of natural resources. The Brundtland Commission (Brundtland 1987) reports that:
> The woman who cooks in an earthen pot over an open fire uses perhaps eight times more energy than an affluent neighbor with a gas

stove and aluminum pans...[S]hortage of money...[forces them] to use 'free' fuels and inefficient equipment because they do not have the cash or savings to purchase energy-efficient fuels and end-use devices. Consequently, collectively they pay much more for a unit of delivered energy-services.

But it would be unconscionable to stop subsistence farmers from increasing their "wealth" of children and trying to provide for their families. They have little alternative but to continue along their current path of migration or environmental degradation. We are left with population explosion or poverty as likely starting points for dealing with LDC issues. Yet these two cannot be decoupled nor separated from environmental concerns. According to Sagoff (1991):

Poverty is one of today's greatest environmental and ecological problems. This is because people who do not share in the wealth technology creates must live off nature; in their need to exploit the natural commons, they may destroy it. Analogously, in an urban context, poor people have had to send their children to work in sweat shops -to survive....Accordingly, I question the adequacy of the argument environmentalists often make that we must protect nature to provide for the welfare of human beings. I think it is also true that we must provide for the welfare of human beings if we are to protect the natural environment.

We appear to be in a hopeless dilemma. Without economic growth and technological development the condition of LDCs will only decline, environmental degradation in these regions will increase, and environmental health of the world will be further threatened. Yet both economic growth and technological development appear to threaten the aims of sustainable development, environmental sustainability, and ecosystem integrity. And while we are urged to think of the world as one community, the transfer of capital, wealth and resources to LDCs with zero-sum economic growth, even if morally defensible, will not be accepted by those with capital, power, and the ability to execute such transfers, the citizens, businesses, and governments in developed countries.

2. Another Perspective

In what follows we want to suggest that these issues need to be reformulated. The first horn of the dilemma, LDC poverty, ignores environmental issues, the second, the demand for zero-growth ecosystem integrity, refuses to consider economic and political ones. But if Sagoff (1991) is correct that these are inseparably interrelated, we need to rethink the concept of ecosystem integrity in light of some obvious facts and some pressing moral and practical imperatives.

It is highly unlikely that we can restore the entire earth to its original or pristine condition nor that we can even restore parts of it to that condition. This is because humankind as an intrinsic component of nature has been influencing and shaping the earth to meet our needs ever since the first person scratched on cave walls, cut down trees, picked berries, built fires, grew plants, and tamed, grazed, and bred animals. We have humanized the earth to such an extent that it is impossible to restore or resurrect its original condition. Indeed, what we call "nature," "earth," "biodiversity," "ecosystem," or even a "natural or original condition" are humanized concepts. It is impossible to make sense of nature apart from our conceptions of it. So when one demands "integrity," for instance, that demand carries with it the baggage of a particular human perspective. The way in which we conceive and think about the ecosystem is so socialized through language and culture, that the meanings of these terms, by their very uses in language, cannot escape the humanized socio-political economic framework in which they are defined.

The focus then, should be on how we can preserve and restore what we call "nature" (but not to the original condition) based on a number of sound moral reasons, reasons that make sense to economists, politicians, and business people as well as to philosophers and ecologists. Valuing nature or the beauty of nature for its own sake is a good reason to preserve and restore wilderness areas. Indeed, companies like McDonald's who have invested in a rain forest in Puerto Rico or Conoco who created a comprehensive plan of environmental and indigenous people protection plan to develop oil in Ecuador (Salter 1993) illustrate how business can and does value the environment for its own sake. Moreover, there is evidence that preserving biodiversity is critical not only for species survival, but also because of the possibility of finding new uses for natural resources, possibilities that may not be realized except by future generations.

Along with this it is obvious for the sake of self-preservation and the survival of future generations that we need to continue to seek ways to clean up or eliminate past and present pollution and waste emissions. This sort of goal of environmental health, however, requires technology, advanced technology that either supersedes our present knowledge or is available but largely unknown, too expensive or too complex to be utilized to a great extent. Returning to an earlier, less resource consuming lifestyle cannot be a return that precludes the continued exploration of these new technologies. That would be to our peril, because undoubtedly we have created some mess or other, not yet recognized, that will require new technological means to clean it up. So the survival of future generations requires not just wilderness preservation but advancing technology. Similarly it is critical that we learn to recycle or replace resources, none of which is of infinite quantity. But part of that learning is through advanced technology, as is clearly illustrated in the development of solar energy on a practical scale.

Finally, it is unconscionable to stop economic growth if that will preclude LDC development. One needs to recouple the notions of sustainable development with technological development and economic growth, and question the conclusion that economic growth always and exclusively entails depleting natural resources or harming the environment. One should recall from Economics 101 or from reading Adam Smith, that one can take a limited population, limited resources, and create new economic value by reorganizing that population and its resources through industrialization, advanced technology, etc. And with advanced environmental technologies one can achieve positive economic growth without environmental harm. Indeed, as companies such as the Body Shop have demonstrated, one can create economic value, reduce poverty, and improve the environment. How is this possible?

3. Technology—The Answer?

The ideal of sustainable development, when considered in light of the need for continued economic development, appears to require a definition of sustainability that goes beyond the admirable goal of zero net depletion in natural resources. Ecological damage from economic development is largely the result of two phenomena: patterns of resource consumption and the kinds of technology employed to facilitate that consumption. Reducing personal consumption in the developed world through lifestyle modification is one alternative that certainly has found many adherents in the environmental community. But, barring some form of unprecedented and politically acceptable mass ecological awakening of the vast majority of individuals residing in developed countries, this is, at best, unlikely to happen any time soon. A renewed emphasis on technology would therefore be the logical focal point for achieving sustainable development within the constraints of integrity. Like consumption reduction, the call for technological advancement is not new. Brown (1990) writes:

> Advances in technology...offer at least a partial way out of our predicament. The challenge of finding ways to meet the legitimate needs of our growing population without further destroying the natural resource base certainly ranks among the greatest missions humanity has ever faced. In most cases, 'appropriate' technologies will no longer be engineering schemes, techniques, or methods that enable us to claim more of nature's resources, but instead systems that allow us to benefit more from the resources we already have.

What is now needed is a view of technology that is dynamic versus one that is static. According to Postrel (1990), the static view, espoused by many environmentalists, envisions technology as some sort of evil to be avoided at all costs. Technology is the great destroyer that will some day lead to our ultimate doom. Because of the inherent dangers, we need to heavily regulate industry and provide other regulatory roadblocks to discourage development and limit growth.

Conversely, an absolute devotion to technology as the world's savior (environmentally or otherwise), is another version of the static view. This 'blind faith' view refuses to acknowledge the many negative aspects involved in using unproven or environmentally destructive technologies and can lead to complacency and/or apathy toward one's contribution to environmental problems. It is simply too easy for those devoted to technology to assume that, since all of the world's environmental ills will be somehow, some day solved, one need not care about the current destruction of the environment. This aspect of the static position is problematically similar to the anti-technology view in that it is based on an unrealistic position that leads one to believe a completely unregulated business environment is to be preferred. This is just as unwelcome since it does not discriminate between environmentally destructive versus salutary technology.

4. Ecoefficiency

A dynamic view argues that sustainable development will not become a reality without advanced technology and the encouragement of new environmentally sound products and processes. The story one tells is that of ecoefficiency realized in new products, new building and community designs, and new services, all of which can be produced both in developed countries and in the emerging markets of LDCs.

Nowhere is this more important than at the point where the concept of ecosystem integrity and development meet. The requirements of ecosystem integrity, particularly those of wilderness and buffer zone land allocation, would place additional burden on the world's human population to make more efficient use of the earth's remaining natural resources. Therefore, the concept of 'ecoefficiency,' which advocates new approaches to designing buildings, processes, products and services that emulate the ideal of sustainability, needs to be more thoroughly explored and information about newly discovered methods must be made available to business and governments.

Business, itself, is beginning to recognize the importance of maintaining the environment to their long-run survival. The Business Council for Sustainable Development (Dechant and Altman 1994) argues that sustainability is:

> [a redefinition] of the rules of the economic game in order to move from a situation of wasteful consumption and pollution to one of conservation, and from one of privilege and protectionism to one of

fair and equitable changes open to all....For business the implication is that environmental common sense will stimulate growth and competitiveness while the lack of environmental concern will act as a brake on a firm's progress.

Some of the forces impacting the business community include: Staying ahead of regulatory issues, competitive pressures, and, perhaps most importantly, stakeholder activism from employees, customers and communities. Heretofore, however, business, or at least United States and Canadian corporations, have focused primarily on environmental health. Many companies have become proactive in reducing pollution emissions and dumping of dangerous waste materials, increasing recyclable products, actively participating in recycle programs, educating their employees and customers and working to reduce uses of non-renewable natural resources. Many multinational companies such as Levi-Strauss, DuPont, Motorola and GE require the same environmental standards in their multinational operations as they do in the United States.

What is less common is the development of new, "ecoefficient" technologies, and in particular, ecoefficient technologies that would be fiscally appropriate and adaptable in LDCs. According to the Brundtland Commission (Brundtland 1987), appropriate technologies, "...are (those that are) more efficient in terms of resource use,...generate less pollution and waste,...are based on the use of renewable rather than non-renewable resources, and...minimize irreversible adverse impacts on human health and the environment." With this in mind, we now present a case example that we believe exemplifies the ideal of ecoefficiency in light of the concept of integrity. This case provides a small, but extremely vital, illustration of how technology and innovative ideas can benefit the entire world within a framework of sustainability and economic development.

5. McDonough Architects

If we understand that design leads to the manifestation of human intention, and if what we make with our hands is to be sacred and honor the earth that gives us life, then the things we make must not only rise from the ground but return to it, soil to soil, water to water, so everything that is received from the earth can be freely given back without causing harm to any living system. This is ecology. This is good design (McDonough 1993a).

The preceding quotation embodies the design philosophy of William McDonough, founder of William McDonough Architects. According to McDonough (1993a), the design world's reliance on two Industrial Revolution technologies, namely, cheap energy and the large sheet of glass, has led architects and designers to subvert nature rather than symbiotically work within nature's constraints. Because of the advent of these two technologies, "...architects no longer rely upon the sun for heat or illumination....Our culture has adopted a design stratagem that essentially says that if brute force or massive amounts of energy don't work, you're not using enough of it." The modern result is enclosed buildings that are heated via the sun and lack of ventilation and subsequently need to be cooled with refrigeration and energy. And in not a few cases the many toxic chemicals that make up a modern structure's building materials, furnishings, and fixtures lead to unhealthy indoor pollution levels.

But why start with design and architecture? McDonough recounts an analogy used by Peter Senge, a systems theorist at MIT, who asks visiting CEO's at his leadership laboratory, "Who is the leader on a ship crossing the ocean?" The obvious answers are given: the captain, navigator, etc. Senge's response is, "No, the leader of that ship is the designer of that ship, because everyone on that ship is affected by its design." In light of the designer's role, McDonough's concept of sustainable natural design is founded upon three defining characteristics. The first is the 'waste equals food' concept. The raw materials needed already exist in the form of stone, clay, wood, water and air; all of which are given to us by nature and cycle naturally without waste. All natural materials will eventually become food for different living systems as opposed to industrially manufactured materials and waste which can remain intact in nature for hundreds or even thousands of years. The question McDonough would like architects and designers to ask themselves is, "Who's food is in the buildings we make" (Wagner 1993).

The second characteristic is based on the earth's perpetual source of energy, the sun. "Current solar income" is seen by McDonough as nature's "interest" whereas petroleum-based energy is considered the world's "capital reserves." Maximizing use of current solar income in the design of buildings is therefore to be stressed and petroleum reserves should only be used sparingly. Solar income is available to us indirectly in the form of trees and other biomass and directly from the sun. These renewable materials are incorporated into structure design as much as possible and are replenished as they are used. This includes replenishing a building's "embodied energy" which is the energy required to produce the building materials and transport them to the building site. Accordingly, McDonough not only designs his buildings to maximize the sun's natural light and warmth but many times he requires the developer to plant a new forest to offset the total energy used. The Warsaw Trade Center in Poland, for example, will seed 6,400 acres of new forest at a cost of $150,000 which the developer justified because it is, "equivalent to a small fraction of our advertising budget" (Gutfeld 1989).

The final characteristic is the Earth's biodiversity which, "sustains [the earth's] complex and efficient system of metabolism and creation." To foment biodiversity, McDonough (1993a) uses non-traditional, sustainably produced natural materials where possible. He helped create a computer data base called the Forest Resource Information System (FORIS) which will provide useful information regarding thousands of tree species. Information such as availability, potential uses, biological and machining characteristics, harvesting methodology, endangered status, etc., is provided to designers and architects.

The following are three briefly described examples of McDonough's projects each of which incorporates the three concepts of waste equals food, current solar income, and protecting biodiversity, and uses a blend of both old and new technology.

5.1. WAL-MART "ENVIRONMENTAL" STORE, LAWRENCE, KANSAS

- The original design plans were converted from steel construction to wood thus significantly reducing the construction and embodied energy used in the manufacture.
- Six highly illuminative experimental skylights will be installed that are specially designed to spread light evenly throughout the store. The design will require fewer skylights and lower Wal-Mart's electrical usage by 54 percent.
- The heating and cooling system will use hydroflourocarbons which, though not environmentally benign, are significantly less destructive to the ozone layer than the traditionally used chlorofluorocarbon-based systems.
- To minimize future impact on the local community, the entire structure is convertible to two-story apartment housing in the event of Wal-Mart closing operations or moving to another location.

5.2. DAY CARE CENTER, FRANKFURT, GERMANY

- The award winning design uses skylights with track-mounted shutters which can be opened and closed by the children at different times depending on the sun's position and the children's activities, thus providing a connection to the environment and attuning the children to the natural rhythms of the day. Indeed the entire building must be opened and closed by the children each day.
- To compensate for the embodied energy used in construction, the children donated 5,000 trees to Leipzig, Frankfurt's sister city, during the center's opening ceremony.
- The building is designed to be converted into three houses, six flats, or twelve apartments if needed in the future.

5.3. EXPO 2000, THE WORLD'S FAIR, HANNOVER GERMANY

In light of the tentative theme of the next world's fair, "Humanity, Nature, and Technology," the office of William McDonough developed "The Hannover Principles" which it is hoped, "...will help form the foundations of a new design philosophy underlying the future of proposed systems and construction for the City, its region, its global neighbors and partners in the world exposition." The Hannover Principles will act as guidelines for the Fair's many design competitions to "demonstrate that sustainable thinking can be put into practice in the real world" (see Appendix). The following is a very brief sketch of some of EXPO 2000's guidelines.

- All materials should be chosen in light of their waste to food natural sustainability and embodied energy. Materials used shall: not be animal tested, minimize hazardous chemicals, be constructed for reusable recycling, be life cycle analyzed.
- Air quality issues will play a vital role at the fair sites. Both external and internal air quality and related issues must be evaluated and pollution eliminated.
- Solar energy should be evaluated and on-site energy production should be considered. Minimum energy consumption, including embodied energy, is encouraged and structures, "should, wherever possible, be net exporters of energy."
- Decentralization of the Fair throughout the world is encouraged through the use of mass communication high-technologies, such as "tele-presence," to make the Fair available to those in other countries who do not possess the means or ability to journey to Hannover. In this way, "It may then be possible for the wisdom of thousands of encouraged 'sustainable' solutions and examples to be shared and enjoyed among the world's people in a 'sustainable world's fair.'"

6. Conclusion

The examples outlined serve as models of sustainable development and economic growth within the constraints of ecosystem integrity by abstaining from the industrial world's current linear development path and recognizing and working within the substantive cyclical characteristics of nature. Ecoefficient technologies, innovatively applied, can help restore some of the earth's ecosystem, help to protect biodiversity, enhance ecosystem health and provide the impetus and justification for continued but sustainable economic development within both the developed and developing worlds.

We concur with Hawken and McDonough (1993) who wrote, "We believe business is on the verge of a transformation, a change brought on by social and biological forces that can no longer be ignored or put aside, a change so thorough and sweeping that in the decades to come, business will be unrecognizable when compared with the commercial institutions of today." But time is running short, and business is traditionally slow to change. Therefore this and similar stories of socially responsible and innovative entrepreneurs and businesses willing to take risks to advance a development philosophy based on market characteristics and ecologically sound principles need to find their way into the business and political language of the world.

But what about the plight of LDCs? The continuing need for these countries to exploit and deplete their natural capital to meet debt service requirements and the subsistence needs of exponentially growing populations must be addressed and innovative solutions explored. Otherwise LDCs will continue to have no choice but to continue along their current path of environmental degradation, overpopulation and poverty. But if development is going to be beneficial to the developing world it must not only follow McDonough's Hannover Principles, but products and services must be designed for, and produced in, those countries. The best way to address these issues is to design new technologies, and manufacture products and buildings that are both environmentally sound and appropriate for these developing markets. We have not yet seen much in the way of such initiatives, but with environmentally sound designs as models, multinational corporations, always seeking new markets, and developing country governments, ever seeking new economic opportunities, can profitably exploit emerging economic opportunities without further jeopardizing the environment or economic development. Using McDonough's principles and techniques as starting places, the possibilities for LDC viable economic development within the constraints of sustainability and ecosystem integrity are unlimited.

A non-economic growth answer to curbing environmental degradation solves one problem while untenably leaving two equally critical concerns, LDC poverty and overpopulation, begging. Clearly, if we wish to better the existence of the world's population as well as preserve the earth's ecosystem, a viable means to achieve this goal is through increasing the emphasis on prudent technological solutions and disseminating real world stories of ecoefficiency and sustainable development.

7. References

Brown L.R., et al. 1994. *State of the World 1994*. W. W. Norton & Company, London, pp. 4-5.
Brundtland, G.H. 1987. The World Commission on Environment and Development. *Our Common Future*, Brundtland Commission, Oxford University Press, Oxford, pp. 8-9.
Daly, H. 1993. Sustainable Growth: An Impossibility Theorem. *Valuing the Earth: Economics, Ecology, Ethics*, H.E. Daly and K.N. Townsend, eds. The MIT Press, Cambridge, pp. 267-8.
Dechant, K. and B. Altman, 1994. *Environmental Leadership: From Compliance to Competitive Advantage*. Academy of Management Executive, VIII, pp. 7-27.
Gutfeld, R. 1989. Will Poland Plant a Forest to Satisfy a U.S. Architect? *Wall Street Journal* 23 October: 1+.
Hawken P. and W. McDonough 1993. Seven Steps to Doing Good Business. *Inc.* November: 92.
McDonough, W. 1993a. A Centennial Sermon: Design, Ecology, and Ethics. The Cathedral of Saint John the Divine, New York: 2.
McDonough, W. 1993b. A Boat for Thoreau. *Business Ethics* May/June: 26.21.
Postrel, V. 1990. The Environmental Movement: A Skeptical View. *Vital Speeches of the Day* 8: 731-732.
Sagoff, M. 1991. Zuckerman's Dilemma: A Plea for Environmental Ethics. *Hastings Center Report*, September-October: 34.
Salter, M. 1993. Block 16: Conoco's "Green" Oil Strategy (A-D), Harvard Business School Case #9-394-001.

Sandbrook, R. 1991. Development for the People and the Environment. *Journal of International Affairs* 44: 403.
Wagner, M. 1993. Creative Catalyst. *Interiors* March: 55.
Westra, L. 1994. *An Environmental Proposal for Ethics: The Principle of Integrity.* Rowan and Littlefield, Lanham, MD.

8. Appendix

8.1. THE HANNOVER PRINCIPLES

1. Insist on rights of humanity and nature to co-exist in a healthy, supportive, diverse and sustainable condition.
2. Recognize interdependence. The elements of human design interact with and depend upon the natural world, with broad and diverse implications at every scale. Expand design considerations to recognize even distant effects.
3. Respect relationships between spirit and matter. Consider all aspects of human settlement including community, dwelling, industry and trade in terms of existing and evolving connections between spiritual and material consciousness.
4. Accept responsibility for the consequences of design decisions upon human well-being, the viability of natural systems, and their right to co-exist.
5. Create safe objects of long-term value. Do not burden future generations with requirements for maintenance or vigilant administration of potential danger due to the careless creation of products, processes, or standards.
6. Eliminate the concept of waste. Evaluate and optimize the full life-cycle of products and processes, to approach the state of natural systems, in which there is no waste.
7. Rely on natural energy flows. Human designs should, like the living world, derive their creative forces from perpetual solar income. Incorporate this energy efficiently and safely for responsible use.
8. Understand the limitations of design. No human creation lasts forever and design does not solve all problems. Those who create and plan should practice humility in the face of nature. Treat nature as a model and mentor, not an inconvenience to be evaded or controlled.
9. Seek constant improvement by the sharing of knowledge. Encourage direct and open communication between colleagues, patrons, manufacturers and users to link long term sustainable considerations with ethical responsibility, and re-establish the integral relationship between processes and human activity.

The Hannover Principles should be seen as a living document committed to the transformation and growth in the understanding of our interdependence with nature, so that they may adapt as our knowledge of the world evolves.

Chapter 17
ETHICAL OBLIGATIONS OF MULTINATIONAL CORPORATIONS TO THE GLOBAL ENVIRONMENT: THE MCDONALD'S CORPORATION AND CONSERVATION

James D. Nations[1]
Ray Cesca[2]
J. Angus Martin[3]
Thomas E. Lacher, Jr.[4]

1. Introduction

The Greek word *oikos*, or house, is the root of the English words ecology and economy. It is thus ironic that economic development for many years has been viewed as being incompatible with ecological well-being. The world economy has become increasingly more interconnected during the past several decades with the formation of regional and international trading blocks and economic communities. At the same time, we have become more cognizant of the global nature of large-scale environmental change. The connections between global economic activity and global environmental conditions are well defined. In a sense we have come full circle back to the Greek root of a common space, our house.

Global economic activity is less controlled by individual countries, as in the past. Multinational corporations (MNCs) have become major players in international economic activity and the decisions that they make have significant economic, social, and environmental ramifications. We explore the emerging role of MNCs in the global economy, discuss their impacts on and obligations to the environment, and present a case study of the kinds of environmentally sound activities and policies that a MNC can foster.

2. Multinational Corporations and Their Role in the Economic Development of Less Developed Countries

2.1.   THE IMPORTANCE OF MNCS IN THE WORLD ECONOMY

MNCs date back to the colonial era but their significance and importance in the world economy took on new dimensions beginning in the post-war expansion of western economies, and grew tremendously following the transformation in the structure of international finance and production since the 1960s (Dixon et al. 1986).

MNCs are very large corporations with billions of dollars in sales far outreaching the Gross National Product of the majority of developing countries they do business with and as a result exhibit tremendous political power over Less Developed Country (LDC)

---

[1]Conservation International, 1015 18th St., NW, Suite 1000, Washington, DC 20036, U.S.A.; [2]McDonald's Corporation, Kroc Drive, Oak Brook, IL 60521, U.S.A.; [3]Department of Agricultural and Applied Economics, Clemson University, Clemson, SC 29634, U.S.A.; [4]Archbold Tropical Research Center, Clemson University, Clemson, SC 29634-1019, U.S.A.

governments. The case of Chile and the International Telephone and Telegraph Company illustrates the extensive political and economic power of MNCs when the latter was implicated in the removal of Chile's President, Salvador Allende, from power. MNCs, with large market shares in many international industries and finance, dominate the global economy at a time when national borders appear to be disappearing and governments are losing control of their respective economies. Supporters of the 'free-market' economic system see the eminent demise of the nation state as a consequence of recent economic and technological developments. As such, the nation state is an anachronism and the MNCs become the critical transmitter of capital, ideas, and growth (Kefalas 1992).

MNCs are centralized in developed capitalist countries, predominantly the United States, Western Europe and Japan, with subsidiaries in many developing countries. They have evolved from the extraction of raw materials and manufacturing to portfolio investment and foreign direct investments for which they account for over US $10 billion annually.

2.2. THE PROS AND CONS OF THE MNCS ROLE IN ECONOMIC DEVELOPMENT

Over the years, especially in the last three decades, a number of schools of thought regarding MNCs and their role in economic development have proliferated. The debate revolves among two groups, the Dependencia School (the Nationalists) and the 'free-market, laissez faire' supporters. The proponents of MNCs and foreign direct investments contend that the billions of dollars of annual investment in developing countries play a primary role in filling 'gaps' between domestically available supplies of savings, foreign exchange, government revenue and technology. The Dependencia School counters that MNCs and their foreign direct investments reinforce dualistic economies, exacerbate income inequalities, contribute inappropriate capital-intensive technologies of production and influence anti-development policies (Todaro 1982). On an ideological level, the opponents of MNC involvement in LDC development contend that MNCs, because of their power and wealth, have undermined the social and economic stability and independence of sovereign nations in the Third World and their growth must be controlled.

The involvement of MNCs in LDC economies, predominantly as extractors of raw materials and in the manufacturing sector, have resulted in an unequal situation that has contributed to distrust and suspicion. Unethical behavior by some MNCs in developing countries leaves much to be desired as incidents such as Union Carbide chemical spill in Bhopal, India and the Nestlé baby food controversy in Africa illustrate. According to Sen, "In the Third World there is often a deep rooted skepticism of the reliability and moral quality of business behavior. This can be directed both at local businessmen and commercial people from abroad" (Sen 1993).

Yet, MNCs have had positive impacts on the economies of developing countries as is illustrated by the phenomenal growth experienced by the Asian Tigers (South Korea, Taiwan, Hong Kong and Singapore) during the 1970-80s. These countries, with largely multinationally controlled and financed manufacturing sectors, have experienced rapid rates of growth and exports (Dixon, et al. 1986) rivaling developed countries' economies gaining them the name Newly Industrialized Countries (NICs).

2.3. THE FUTURE ROLE OF MNCS IN THIRD WORLD DEVELOPMENT

With extensive capital accumulation MNCs are seen by many western governments and agencies of development as the source of foreign direct investments which developing countries are desperately in need of. It is a model which has had its successes for countries in Asia. Yet, as Todaro points out, MNCs present a unique opportunity and a host of critical problems for those many less developed nations in which they conduct business (Todaro 1982). The decrease in government aid to LDCs from developed countries has made the issue more pressing since MNCs seem more capable of filling the role as investor rather than many developed country governments which are experiencing depressed economies.

It has been traditionally accepted that the MNCs' role in LDC development can be achieved through technology transfer, investment in infrastructure, financial restructuring and export promotion. Yet, this has not been the case for many countries which have created favorable conditions for MNC investment. The arguments against the MNCs past role in the economic and social development of LDCs can easily be substantiated with empirical data. If foreign direct investments by MNCs are to become an important stimulus to LDC development the two have to coordinate their interests. As long as MNCs see their role strictly as profit maximization there will be little gained by LDCs. MNCs have to assume a new role, already recognized by some of its members, which can only be achieved via the long-run view as opposed to a short-run economic view.

The economic instability confronting LDCs and the emergence of MNCs as a primary economic force to bring about badly needed social and economic change have led to the examination of the role of MNCs in the new economic order. Many MNCs have already begun to put aside the sole economic objective of profit maximization to become socially responsible investors in LDC development. MNCs have recognized their important roles in social, economic and political development as was the case with divestiture from South Africa.

3. Multinational Corporations and the Environment: Past and Future Role

3.1. THE ENVIRONMENTAL RECORD OF MULTINATIONAL CORPORATIONS

As the investment of multinational corporations in developing countries has increased, so has the environmental impact of these corporations. The pursuit of economic activities based upon the use of land or natural resources has environmental impacts. The decision to grow bananas or to mine iron ore has, as a given, the fact that forest will be removed, soils will be altered, and the natural environment will be eliminated from the designated area. Without any alteration of the natural environment, these activities cannot be conducted. Objections to the pursuit of these kinds of economic activities are objections to development in general and not specifically to the activities of MNCs. Of course, in many developing countries, only MNCs have access to the necessary capital to carry out large-scale development projects and, as such, are frequently the target of protest over these activities (Redclift 1984). The large capital investment required in many high pollution industries such as the production of chemicals or large-scale mining also means that MNCs are heavily involved in these activities in the developing world.

In addition, MNCs often interact with large national companies in LDCs which are owned by the wealthy. Multinationally controlled development activities therefore also are often cast in the mode of political economy and class conflict (Redclift 1984). There is a belief in many circles that the behavior of multinational corporations is unethical and these companies are often not held to the same high standards of environmental regulations as they are in their countries of origin. Also, increasing levels of environmental degradation have been linked to the expansion of rural poverty (Brandt 1980) thereby increasing the distance between developed and developing country standards of living.

Several cases of apparent neglect or mismanagement by MNCs in high polluting industries have made headlines in recent years. The most serious of these cases was the leak of methyl isocyanate gas from a Union Carbide pesticide plant in Bhopal, India. Over 2,000 (perhaps as many as 5,000) people were killed and 200,000 injured (Gladwin 1987a, Donaldson 1989). The reported continued use of the pesticide DBCP in banana plantations until 1979 by the company Dole Food, Inc. and its Costa Rican subsidiary Standard Fruit Co., after it was banned in the United States by the Environmental Protection Agency in 1977, allegedly caused at least 1,500 plantation workers to become sterile (Thrupp 1990, MacKerron and Cogan 1993).

Even if these cases are isolated or exceptional, they contribute to the belief held by some that corporations invest in developing countries to avoid environmental regulations. Leonard (1984), in a study of the mineral and chemical industries, addresses this issue in detail and concludes that "neither the cost nor the logistics of complying with environmental regulations are emerging as decisive across-the-board factors inducing US manufacturing industries" to relocate to foreign developing countries. Indeed, Leonard asserts that a major effect of environmental regulations has been to speed the decline of obsolescent industries and to provide strong incentives for technological innovation that keeps U.S. industry competitive in the global market.

3.2. THE ROLE OF MNCS IN INTERNATIONAL ENVIRONMENTAL PROTECTION

In order to avoid criticism based upon the perception of dual standards in the level of attention paid to environmental issues, MNCs must address the moral and ethical obligations of doing business in developing countries (Pearson 1987). Donaldson (1989) develops an ethical algorithm for multinational enterprises; he proposes a list of ten fundamental international rights to guide corporate behavior (Table 1). The United Nations Centre on Transnational Corporations (1985) specifically addresses the role of MNCs in environmental protection. The document recognizes that environmental problems are often the result of economic activities and thus MNC investments can conflict with environmental protection. The United Nations has developed a draft code of conduct for transnational corporations with a provision that states that corporations should "carry out their activities in accordance with national laws, regulations, administrative practices and policies relating to the preservation of the environment of the countries in which they operate and with due regard to relevant international standards." The document points out that environmental issues have gained increasing attention in the research and development programs of MNCs and in fact pollution control and prevention is becoming an emerging MNC industry. This supports the contention of Leonard (1984) that environmental regulations can stimulate technological innovation and create jobs and profits.

Table 1. The ten fundamental international rights as proposed by Donaldson (1989).

1. Freedom of Physical Movement
2. Ownership of Property
3. Freedom from Torture
4. Fair Trial
5. Nondiscriminatory Treatment
6. Physical Security
7. Freedom of Speech and Association
8. Minimal Education
9. Political Participation
10. Subsistence

There are other things that multinational corporations can do when addressing the conflict between development and environment. Gladwin (1987b) discusses how MNCs can work to address several issues of environmental concern, including sustainable development, environmental policy, environmental management, and international environmental protection. An important theme is how developing countries can tap into the substantial amount of talent and expertise that large international corporations have in areas of research and development that could assist in the creation of economic activities that are both profitable and more environmentally sound. A major goal of international development agencies is technology transfer; this technology generally rests in the hands of corporations, not governments. The transfer of advanced technologies in the field of environmental protection and management from developed to developing countries via multinational corporations could be a major factor in achieving the transition to a more sustainable development.

3.3. ENVIRONMENTAL ISSUES IN THE TROPICS

Many of the most publicized environmental disasters in developing tropical countries involve accidents or misuse of chemicals, such as the Bhopal incident or the health problems caused by the application of DBCP in Costa Rica; or chronic, excessive industrial pollution such as reported for Cubatão, Brazil, often described as the world's most polluted city (Pimenta 1987, Donaldson 1989). However, these problems are not special to the tropics. Increased public awareness concerning the decline of global biodiversity has made the threat of tropical deforestation one of the most prominent concerns of the contemporary global environmental movement. Thus, the simple act of the conversion of habitat to something other than its natural state, even in the absence of any industrial pollution, is considered unacceptable development to many.

Myers (1980, 1984) publicized the high rates of deforestation of tropical rain forests in two widely cited books. Though other authors have disputed the rates of deforestation and attendant losses of biodiversity as cited by Myers (Lugo, 1988) no one would deny that excessive losses of tropical forests will have substantial impacts not only on global biodiversity, but perhaps on global climate as well (Houghton 1990, Fearnside et al. 1993).

All factors that influence the loss of tropical forests have come under increased scrutiny in recent years, from forestry (Gillis 1987, MacKerron and Cogan 1993) to oil production, mining, agribusiness and ecotourism (MacKerron and Cogan 1993). It is important to note that there are positive stories to report in addition to problems (see Gradwohl and Greenberg 1988, Reid et al. 1988). Nevertheless, all MNCs that do business in countries that contain tropical rain forests will come under intensive scrutiny to evaluate their impact on the deforestation or alteration of natural habitats. The burden is no longer only on those industries traditionally considered to be heavy polluters, but on all businesses that utilize the landscape for commercial activities.

4. A Case Study of Multinational Corporations and Environmental Ethics: The McDonald's Example

4.1. MCDONALD'S CORPORATION AND ENVIRONMENTAL ISSUES

At the height of the environmental activism of the 1970s and 1980s, the idea of a peaceful partnership between corporate America and environmentalists seemed out of the question, especially between an international fast food giant and a green organization intent on taking big business to court. Nevertheless, such a partnership became a reality in 1990 when McDonald's agreed to work with the Environmental Defense Fund (EDF) to help the large multinational fast food chain deal with some of its most challenging environmental issues, from recycling and packaging, to waste reduction and composting.

With 15,000 restaurants in 80 countries, McDonald's high profile can be an inviting target for critics, and that was certainly the case in the 1980s when some activist groups attacked McDonald's for its use of polystyrene packaging. After studying the issue, and battling it in the public arena, McDonald's decided that it was time to pursue a new, bold approach by working with the "other side". Seemingly a simple idea now, it was an aggressive, breakthrough initiative for both organizations that helped to produce tangible and innovative gains on the environmental front, much more than confrontation could have achieved. The EDF was invited into the restaurants, the corporate offices, and behind-the-scenes to see firsthand what the McDonald's hamburger world was about. The goal was to work together to bring a fresh perspective to environmental issues. After an open and candid review, McDonald's and the EDF announced a sweeping package of environmental initiatives, beginning with the news that McDonald's would no longer serve its hamburgers in polystyrene "clamshell" boxes. This announcement not only made news for its green policy decision, but also for the innovative partnership.

4.2. TROPICAL DEFORESTATION AND THE "HAMBURGER CONNECTION"

The McDonald's-EDF collaboration was a model the company would follow again when several radical environmental groups attacked McDonald's and other corporations with misinformation about fast food restaurants and their alleged role in rain forest deforestation.

McDonald's had never knowingly purchased beef raised on recently deforested rain forest land in Central and South America, however, at least one other fast food chain, Burger King, had. That was all the information activists needed to take their case to politicians and the media. Soon, rumors were gaining credibility that every fast food restaurant was buying beef raised on rain forest land that had been converted into cattle pasture. To help transmit this message, a few environmental pranksters took to the streets dressed as trees being pursued by chainsaw-wielding loggers, hoping TV cameras would put their act on the evening news. "Stop importing Central America beef, and deforestation will disappear," the protester's pamphlets read.

In reality, the process of rain forest destruction was much more complex, but the protesters had fallen prey to what social scientists call the "Goliath Syndrome." Having perceived an environmental problem in rain forest destruction, they pointed to some of the most visible symbols of corporate America as the obvious culprits: fast food companies, including Burger King, Wendy's and McDonald's. While protesters had indeed identified a bona fide ecological challenge, in their misguided zeal they had jumped to an overly simple solution. Although only one company had been identified as actually importing beef from Central American rain forest land, other fast food companies, innocent of the charge, were indirectly implicated by the campaign. The industry had to deal with the challenge.

Clearing and burning rain forest to produce beef cattle has been one of the leading causes of forest destruction in Mexico, Central America, and Panama during the past four decades, but the situation is not as simple as some environmental organizations have assumed. In reality, beef cattle production serves as the motive force behind a three-stage pattern of rain forest eradication.

The process begins when timber or oil companies open roads through the rain forest to extract commercial resources. Later, landless peasant families use these roads to filter into the forest in search of land. They clear the remaining vegetation to plant subsistence crops such as maize, manioc, and rice, and low-level cash crops such as coffee, chilies, and bananas.

After one to three years of this production, however, insect plagues, weeds, and soil exhaustion lead the colonists to clear additional forest land for crop production. Rather than allow their previous crop land to regenerate into forest, they seed the area with introduced pasture grasses and begin to produce beef cattle or, in an increasingly common pattern, they sell their cleared forest land to cattle producers who follow in their wake, buying up small plots to convert them into large ranches.

In many regions of Mexico and Central America, government agencies, multilateral development banks, and international development organizations have actively promoted the transformation of rain forest into cattle pasture by providing incentives such as generous loans, new roads, beef packing plants, and pest eradication programs. In most cases, these incentives are designed to increase export earnings by expanding the amount of beef sold to overseas markets, especially western Europe and the United States. In response to this financial and technical support, exports of deboned, frozen beef were the most dynamic sector in Central American trade during the 1960s and 1970s, with a 400 percent increase between 1961 and 1974 alone. Today, although many Latin American countries are phasing out subsidies to the beef cattle sector, Central America as a whole is exporting more beef than ever before (Jukofsky and Wille 1994).

In the importing countries, Mexican and Central American beef ends up in luncheon meats, hamburgers, frankfurters, chili, soups, beef stew, hash, sausages, TV dinners, frozen pot pies, baby foods, and pet foods. In some countries, especially the United States, some of it is mixed with fatter, domestic beef to appear on supermarket shelves as ground beef for homemade hamburgers and meat loaf (Nations and Komer 1983a, Myers 1984).

Unfortunately, producing this beef on cleared rain forest land is a short-lived phenomenon. The effects of overgrazing and torrential rains soon turn rain forest pastures into weeded, eroded wastelands. As a result, although cattlemen may be able to raise one head of cattle per hectare during the first year of production, within five to 10 years they must dedicate five to seven hectares of land per head. After fewer than 10 years of production, the cattlemen, like the farm families before them, must move on in search of new forest lands. Throughout Mexico, Central America, and Panama, this system of extensive beef cattle production is destroying forest resources, wildlife, and rain forest peoples with equal disregard (Nations and Komer 1983b).

When confronted with this issue by environmental groups, the McDonald's Corporation rightfully denied involvement in the deforestation of Mexican and Central American forests. Other fast food operators, including Burger King, were forced to admit that they did use imported beef in their U.S. outlets. In a letter to one U.S.-based ecological research center, a Burger King spokesman wrote that, "Burger King has large imported quantities of beef from Costa Rica and has found it to be of consistently high quality."

The public pressure that resulted from this admission led Burger King to cease its imports of Central American beef during the mid-1980s. Caught in the cross fire of this public relations debate, McDonald's issued a rain forest pamphlet clearly stating that the company had never used imported beef. However, as recently as 1993, McDonald's United Kingdom was prompted to sue a radical London-based environmental group that persisted in printing materials wrongly accusing McDonald's of clearing rain forest for cheap beef production.

4.3. THE BIRTH OF THE AMISCONDE INITIATIVE

Even while confusion persisted among activist groups about which fast food companies had once purchased rain forest beef, McDonald's was planning to do more to demonstrate the corporation's commitment to the environment. Using the well-publicized partnership with the Environmental Defense Fund as a model, McDonald's joined with Conservation International to help meet the rain forest challenge.

Ray Cesca, the current head of McDonald's Global Purchasing and World Trade division, helped to create a new vision for this partnership. During a trip to Costa Rica, Mr. Cesca envisioned how McDonald's and other committed companies could help to bring a new way of life to the families who lived and worked on the edge of the rain forest. The idea, simple yet profound, argued that if families were taught new agricultural techniques and were encouraged to apply their new knowledge via subsidies, their need to clear away additional rain forest would be diminished.

The resulting project focuses on pilot programs of sustainable economic development in Costa Rica and Panama by working with rural families who live along the edges of the La Amistad International Park and Biosphere Reserve. La Amistad is a million hectare complex of national parks, forest reserves, and indigenous territories that straddle the border between the Central American countries of Costa Rica and Panama. The reserve protects a rare combination of cloud forest, lowland rain forests, and coastal plains that provide habitat to some of Latin America's rarest plant and animal species, as well as being the traditional territories of the Bri-Bri, Guaymi, and Teribe indigenous communities (Boza 1989).

During the 20th century, Spanish-speaking Ladino farm families had moved into the forests of this binational region to produce corn, coffee, and beef cattle. Their traditional slash-and-burn agriculture threatens to invade the reserve's boundaries. As their population expands and agricultural lands degrade ecologically, the families are faced with two choices. They can expand the amount of land under cultivation by invading the forested boundaries of the reserve, or they can increase crop yields on land they have already cleared. The AMISCONDE project was conceived to achieve the latter option. AMISCONDE stands for Amistad Conservation and Development Initiative, and it combines the efforts of the McDonald's Corporation with Clemson University, Conservation International, Costa Rica's Tropical Science Center, and Panama's Fundacion para el Desarrollo Sostenible (FUNDESPA). The project concentrates on three programs of action. The first component seeks to improve the agricultural yields of farmers by conserving soil and water through improved agronomic techniques. Project personnel teach farmers to plant crops along contour lines on hillside fields and to use vetiver grass to create live barriers on farming terraces, thus preventing soil erosion and resulting loss of agricultural potential. Similar activities concentrate on restoring unproductive pasture lands. Project staff are introducing farmers to forage-feeding techniques, in which improved grasses are harvested and taken to cattle held in small pastures, rather than having cattle graze extensively over large plots of seeded land.

A second component of the AMISCONDE project focuses on the reforestation of degraded land along the edges of the La Amistad reserve through the establishment of community tree nurseries and tree planting campaigns. Simultaneously, farmers are introduced to agroforestry techniques, in which tree crops are interspersed with food crops to hold the soil in place, serve as wind breaks, and enrich the soil with nitrogen.

Finally, the third component of the AMISCONDE project helps community members organize into farmer's cooperative groups to increase income by cutting out high-cost middle men who charge usurious fees to transport crops to market. Other community programs organize environmental education activities among community schoolchildren, women's groups, and the public at large through plays, clean-up campaigns, and radio spots.

Another effective activity of AMISCONDE is the provision of small-scale loans to farmers who have traditionally been cut out of credit programs for lack of collateral. A community-based bank, using funds provided by AMISCONDE, loans funds that allow farmers to invest in land improvements and reforestation programs without risking the only capital goods they have—their land.

Funding for the AMISCONDE project is provided largely by the McDonald's Corporation and its business partners—McDonald's-United Kingdom, the Coca Cola Foundation,

OSI Industries, Keystone Foods, L & O International, and Nestlé, for example. During 1994, other international corporations began to buy into the project. SONY Corporation of Japan has provided $90,000 of video equipment for environmental education programs.

Clemson University handles administrative duties for the project, routing funding and keeping accounts. The university also provides scientific and research expertise, linking the project to the broader scientific community. Conservation International, a field-based non-governmental organization, provides environmental oversight for the project based on eight years of experience in the La Amistad reserve. Staff from the Tropical Science Center, based in San Jose, Costa Rica and a local Panamanian organization, FUNDESPA, carry out direct actions with the involved farming communities.

The project has begun to generate a great deal of interest in the scientific community. Project scientists have been invited to give presentations on the initiative at a number of conferences and scientific meetings, and papers on technical aspects of the project have been published in conference proceedings and books (Lacher and Calvo-Alvarado 1994, Lacher et al. in press, this volume). ABC News covered the project in a five minute piece on the ABC Evening News on September 7, 1993 and the McDonald's Corporation featured the project in the company publication (McDonald's 1994). McDonald's Corporation also sees some potential long-range plans for AMISCONDE, including the possibility that one day McDonald's restaurants in Costa Rica and Panama might be customers for the lettuce, tomatoes, and onions grown by AMISCONDE farmers.

5. The Future of Corporation Environmentalism

To conservationists, the AMISCONDE project, and McDonald's support of it, are exceptionally positive actions for two reasons. First, the project serves as a solid demonstration of the concept of sustainable economic development. During the last two decades, conservation organizations have realized that providing logical arguments for the protection of biological diversity are not enough to prevent vital ecosystems from being degraded or destroyed by rural families desperate to produce food and income and keep their families alive. Instead, poverty-stricken families must see the benefits that accrue to them when they protect biological resources. By working to improve the lives of farm families who live on the borders of national parks and other protected areas, conservation groups benefit biological diversity by benefiting human communities. The AMISCONDE project serves as an example of how such projects can function.

The second reason conservationists support such projects is precisely because a major corporation is involved. Having seen the benefits brought to the corporation through its support of conservation efforts in the developing world, McDonald's has become the best possible advertisement for sustainable development efforts. In speeches and written materials, McDonald's representatives have promoted the AMISCONDE project before dozens of other U.S., European, and Japanese companies. Their support for the project and their direct involvement in its design and activities has convinced other business enterprises to approach sustainable development and conservation with a more objective perspective. Rather than seeing all environmentalists as anti-progress obstructionists, these companies are seeing the benefits of partnerships where experts from both sides met in the middle in a mutually-beneficial relationship. Although they might not see eye-to-eye on every issue, the environment can emerge the true winner, as McDonald's decisions to abandon polystyrene and to support AMISCONDE have demonstrated.

With the exception of a few hold-outs, most conservation groups now see corporate involvement in environmental endeavors as crucial to progress in the conservation of the earth's ecosystems. Conservationists can not provide the jobs nor generate the income that the rapidly expanding number of developing world families will require over coming decades if they are to survive and simultaneously keep alive the biological systems their very lives—in fact, all human lives—depend upon. Corporations can provide these jobs and income. If they are willing to follow the example of the McDonald's Corporation and the other companies supporting AMISCONDE, they can provide these jobs and income in a way that benefits human communities, biological diversity, and the corporation itself.

6. Acknowledgments

We thank the corporations that contribute their support to the AMISCONDE Initiative, including McDonald's-Latin America, McDonald's-United Kingdom, the Coca Cola Foundation, Keystone Foods, OSI Industries, L&O International, Nestlé, and SONY. Liliana Madrigal and Manuel Ramirez of Conservation International are responsible for the coordination of all in-country activities in Costa Rica and Panama. Luis Murillo and Federico Selles are the country coordinators for Costa Rica and Panama, respectively. We also thank the technical staff in Costa Rica and Panama for their tireless and difficult work at the sites. We are especially grateful to the people of San Jeronimo, Zapotal, Fatima, and San Rafael, Costa Rica; and Cerro Punta and surrounding communities, Panama. The success of the AMISCONDE Initiative has ultimately derived from their sincerity and a dedication to making a dream for a better life a reality.

7. References

Boza, M.A. 1989. *Costa Rica National Parks.* Incafo, S.A., Madrid.
Brandt Commission. 1989. *North-South: A Program for Survival.* Pan Books, London.
Dixon, C.J., D. Drakakis-Smith, H.D. Watts. 1986. *Multinational Corporations and the Third World.* Westview, Boulder, CO.
Donaldson, T. 1989. *The Ethics of International Business.* Oxford University Press, New York.
Fearnside, P.M., N. Leal, Jr., and F.M. Fernandes. 1993. Rainforest Burning and the Global Carbon Budget: Biomass, Combustion Efficiency, and Charcoal Formation in the Brazilian Amazon. *Journal of Geophysical Research* 98: 16,733-16,743.
Gillis, M. 1987. Multinational Enterprises and Environmental and Resource Management Issues in the Indonesian Tropical Forest Sector. In *Multinational Corporations, Environment, and the Third World: Business Matters.* C. S. Pearson, ed. Duke University Press, Durham, NC, pp. 64-89.
Gladwin, T.N. 1987a. A Case Study of the Bhopal Tragedy. In *Multinational Corporations, Environment, and the Third World: Business Matters.* C. S. Pearson, ed. Duke University Press, Durham, NC, pp. 223-239.
Gladwin, T.N. 1987b. Environment, Development, and Multinational Enterprise. In *Multinational Corporations, Environment, and the Third World: Business Matters.* C. S. Pearson, ed. Duke University Press, Durham, NC, pp. 3-31.
Gradwohl, J. and R. Greenberg. 1988. *Saving the Tropical Forests.* Earthscan Publications, Ltd., London.
Houghton, R. A. 1990. The Global Effects of Tropical Deforestation. *Environmental Science and Technology* 24: 414-422.
Jukofsky, D. and C. Wille. 1994. Trees in Pastures: Some Costa Rican Ranchers are Thinking Green. *Tropical Conservation Newsbureau* December 5.
Kefalas, A.G. 1992. The Global Corporation: Its Role in the New World Order. *National Forum* 72: 26-30.

Lacher, T.E., Jr. and J.C. Calvo-Alvarado. 1994. The AMISCONDE Initiative: Restoration, Conservation, and Development in the La Amistad Buffer Zone. In *Conservation Corridors in the Central American Region,* A. Vega, ed., Tropical Research and Development, Inc., Gainesville, FL, pp. 315-322.

Lacher, T.E., Jr., J.C. Calvo-Alvarado, M. Ramirez Umaña, and J.D. Maldonado Dammert. In press. Incentivos económicos y de conservación para el manejo de las zonas de amortiguamiento: la iniciativa AMISCONDE. In *Abordagens Interdisciplinares para a Conservação da Biodiversidade e Dinâmica do Uso da Terra no Novo Mundo.* Anais da Conferência Internacional "On Common Ground: Interdisciplinary Approaches to Biodiversity Conservation and Land Use Dynamics in the New World," Belo Horizonte, Brasil, Dezembro de 1993. G.A.B. da Fonseca, M. Schmink, L.P.S. Pinto, and F. Brito eds., Conservation International, Universidade Federal de Minas Gerais, e University of Florida, Belo Horizonte, Brazil.

Leonard, H.J. 1984. *Are Environmental Regulations Driving U.S. Industry Overseas?* The Conservation Foundation, Washington, DC.

Lugo, A.E. 1988. Estimating Reductions in the Diversity of Tropical Forest Species. In *Biodiversity,* E.O. Wilson, ed. National Academy Press, Washington, DC, pp. 58-70.

MacKerron, C.B. and D.G. Cogan, eds. 1993. *Business in the Rain Forests.* Investor Responsibility Research Center, Washington, DC.

McDonald's. 1994. McDonald's and the AMISCONDE Project: Reclaiming the Rain Forest. *Crew* April/May: 2-3.

Myers, N. 1980. *Conversion of Moist Tropical Forests.* National Research Council, Washington, DC.

Myers, N. 1984. *The Primary Source.* W.W. Norton & Company, New York.

Nations, J.D. and D.I. Komer. 1983a. Rainforests and the Hamburger Society. *Environment* 25: 12-20.

Nations, J.D. and D.I. Komer. 1983b. Central America's Tropical Rainforests: Positive Steps for Survival. *Ambio* 12: 232-238.

Pearson, C.S., ed. 1987. *Multinational Corporations, Environment, and the Third World: Business Matters.* Duke University Press, Durham, NC.

Pimenta, J.C.P. 1987. Multinational Corporations and Industrial Pollution Control in São Paulo, Brazil. In *Multinational Corporations, Environment, and the Third World: Business Matters.* C.S. Pearson, ed. Duke University Press, Durham, NC, pp. 198-220.

Redclift, M. 1984. *Development and the Environmental Crises.* Methuen & Company, London.

Reid, W.V., J.N. Barnes, and B. Blackwelder. 1988. *Bankrolling Successes: A Portfolio of Sustainable Development Projects.* Environmental Policy Institute and National Wildlife Federation, Washington, DC.

Sen, A. 1993. Does Business Ethics Make Economic Sense? *Business Ethics Quarterly* 3: 45-54.

Thrupp, L.A. 1990. Sterilization of Workers from Pesticide Exposure: Causes and Consequences of DBCP-Induced Damage in Costa Rica and Beyond. *World Resources Institute,* December.

Todaro, M.P. 1982. *Economics for a Developing World.* Longman, London.

United Nations Centre on Transnational Corporations. 1985. *Transnational Corporations in World Development: Third Survey.* Graham & Trotman, Ltd., London.

SUBJECT INDEX

affluence 116-117

agriculture 18-19, 165, 243-244

Alpach symposium 89

Amisconde Initiative 271-273

applied science 151-155

Aristotle 84, 131-132

autocatalysis 2, 78, 81-84

biocentric 221-223

biodiversity 5, 60-64, 67, 72, 127, 212, 218, 225, 240, 246, 257

biogeochemical cycles 39

biotic impoverishment 38, 44

Brookhaven symposium 90

Brundtland Commission 223, 254

burden of proof 8-10, 86, 179

Canadian forestry 230-232

Cartesian models 220-221

catastrophe theory 51

causality 79, 84-86, 93, 127

chaos 29, 51

chemical pollution 39, 43, 121

Clean Water Act 12, 42

Columbia River 34

community structure 2, 64, 69

complex systems 2, 5, 43, 130, 133, 177

conservation 62

core area 15, 209

corridors 127, 209

cost-benefit analysis 186-187

culture 19-23

cybernetics 90

disease 39-40

diversity-stability hypothesis 49-50, 134, 168, 170, 218

dynamical cohesion 80

Earth Summit Conference 12, 118

Easter Island 36, 43

ecoefficiency 259-262

Subject Index

ecofeminism 91

ecology 125-141

economic growth 44, 102, 104-105 255-257, 266

economic systems 90, 113, 266

ecoscience 92-97

ekistics 90, 95

energy 241, 246

environmental protection 37, 268

environmental trends 43

equilibrium 52, 66

ethics 4, 10, 20, 45, 126, 131, 134, 221-225, 269

evolution 40, 89, 172

extended peer communities 158-159

extinction 38

Florida Everglades 202-215

Gaia hypothesis 78

Glacier National Park 177

Grand Canyon National Park 177

Great Lakes 12, 97-98, 132, 223

Great Lakes Water Quality Agreement 64, 129, 223

Great Smoky Mountains National Park 177

gross national product 42

health 12-13, 15-16, 23, 39-43, 77, 136, 172, 254, 258

homeostasis 81, 166

humanism 221-223

hypothesis deduction 126-128

indicators 43, 66-71, 129, 183-185

instrumental 5, 14, 163, 165, 219-221

integrity 1-4, 6, 12-20, 23, 32-33, 40-43, 54-57, 60, 64, 66, 77, 97, 128-129, 131-141, 162-166, 172, 177, 180-181, 203, 208, 219, 224-231, 234-235, 239-240, 243, 254-257

intrinsic , 163, 173, 209, 219

island biogeography 127-128

landscapes 12, 14, 203-211

life support systems 165

Lifelands Project 18

living systems 39

McDonalds Corporation 269-273

Man and Biosphere Program 89, 177

military 92

Montreal Protocol 21

morality 1

multinational corporations 265-274

mutualism 80-81, 84

National Park Organic Act 189-192, 196

national parks 16, 177-198, 202

natural 4-5, 13-14, 181-182, 258

natural capital 105-109

natural selection 132-133

Netherlands 22, 24

nonequilibrium 2, 52, 65-66, 130, 177

nonliving systems 39

OECD countries 104, 106, 112, 117

OECD Eutrophication Study 89

open system 52, 78-82, 89-90

Paris Symposium 89

periphyton 2, 80

pesticides 19

Plato 20-21, 29

political ecology 44

population 115-116, 239-249, 256

positive feedback 2, 28, 80, 89

positivistic science 92, 126

post-normal science 146-160, 188

poverty 117,119, 257

precautionary principle 8, 11, 21

prediction 125-141

professional consultancy 153-155

racism 23

Redwood National Park 196

religion 37

resources 241-249

restoration 211-215

scientific method 49, 51, 147-148

self-organization 52-54, 57, 89, 97

soil 41, 244

species-area curve 209

Stockholm Conference on the Human Environment 61

stress 2, 130

succession 50, 65, 82, 168-175

sustainability 1, 12, 14-22, 60-64, 72, 102-122, 223, 229, 254, 257-258

Sustainable Biosphere Initiative 63

sustained yield 111

systems ecology 171

Talloires Declaration 37

technology 44, 90, 114-117, 163, 258-259, 267

thermodynamics 52, 130, 136

tropics 269-270

type I error 139, 185

type II error 139, 185, 188

Ultricularia 2, 80-82, 85

uncertainty 6, 8, 21, 147, 149-150, 170, 179-180, 185-186

United Nations Conference on Environment and Development 223

United Nations Environment Program 61

United States National Park Service 177

wading birds 202, 204-205, 207

water 36, 41, 178, 245

Wild and Scenic Rivers Act 194

wilderness 16-18, 24-27, 193-194, 234

Wilderness Act 193

Wildlands Project 18, 27

World Bank 105, 108-109, 118-120

Yellowstone National Park 177, 192

Yosemite National Park 177, 192-194

zoophytes 2, 80

Environmental Science and Technology Library

1. A. Caetano, M.N. De Pinho, E. Drioli and H. Muntau (eds.), *Membrane Technology: Applications to Industrial Wastewater Treatment.* 1995
 ISBN 0-7923-3209-1
2. Z. Zlatev: *Computer Treatment of Large Air Pollution Models.* 1995
 ISBN 0-7923-3328-4
3. J. Lemons and D.A. Brown (eds.): *Sustainable Development: Science, Ethics, and Public Policy.* 1995 ISBN 0-7923-3500-7
4. A.V. Gheorghe and M. Nicolet-Monnier: *Integrated Regional Risk Assessment.* Volume I: Continuous and Non-Point Source Emissions: Air, Water, Soil. 1995 ISBN 0-7923-3717-4
 Volume II: Consequence Assessment of Accidental Releases. 1995
 ISBN 0-7923-3718-2
 Set: ISBN 0-7923-3719-0

KLUWER ACADEMIC PUBLISHERS – DORDRECHT / BOSTON / LONDON